HUMAN-WILDLIFE CONFLICT MANAGEMENT

HUMAN-WILDLIFE
Conflict
Management

Prevention and Problem Solving

SECOND EDITION

RUSSELL F. REIDINGER, Jr.

JOHNS HOPKINS UNIVERSITY PRESS | BALTIMORE

© 2013, 2022 Johns Hopkins University Press
All rights reserved. Published 2022
Printed in the United States of America on acid-free paper
9 8 7 6 5 4 3 2 1

The first edition of this book was published as *Wildlife Damage Management: Prevention, Problem Solving, and Conflict Resolution*, by Russell F. Reidinger, Jr., and James E. Miller in 2013.

Johns Hopkins University Press
2715 North Charles Street
Baltimore, Maryland 21218
www.press.jhu.edu

Library of Congress Cataloging-in-Publication Data

Names: Reidinger, Russell F., Jr., 1945– author.
Title: Human-wildlife conflict management : prevention and
    problem solving / Russell F. Reidinger, Jr.
Other titles: Wildlife damage management
Description: Second edition. | Baltimore : Johns Hopkins
    University Press, 2022. | Includes bibliographical references
    and index.
Identifiers: LCCN 2022010111 | ISBN 9781421445250
    (hardcover) | ISBN 9781421445267 (ebook)
Subjects: LCSH: Wildlife management. | Wildlife pests—
    Control.
Classification: LCC SK355 .R45 2022 | DDC 333.95/4—dc23/
    eng/20220318
LC record available at https://lccn.loc.gov/2022010111

A catalog record for this book is available from the British
Library.

*Special discounts are available for bulk purchases of this
book. For more information, please contact Special Sales
at specialsales@jh.edu.*

For Carol, Stephanie, and Benjamin, and for James E. Miller, coauthor of the first edition of this book and one of "those exemplary heroes and mentors . . . in this honorable profession and avocation"

Contents

Preface

W as it only a decade since the last edition? It now seems widely accepted that humans are the underlying cause of all human-wildlife conflicts. The importance of human dimensions in resolving human-wildlife conflicts is also widely understood. We can now clearly see the impacts of human overpopulation on biological diversity, from impacts on climate to pollution of the oceans to fragmentation of wildlife habitat. COVID-19 pointed to the importance of science underwriting solutions to zoonotic pandemics, rather than conspiracy theories and disinformation.

We can see a plethora of human solutions, scientifically based and applied. CRISPR and Wolbachia technologies are examples. They are being used to help solve dengue fever and other human-wildlife conflicts. The first American chestnut tree (*Castanea dentata*) that can grow old enough to reproduce, transgenic-built Darling 58, might now be used to repopulate eastern deciduous forests. Drones report pollution on rivers and alert rangers to the whereabouts of poachers, facilitating their arrests. These unmanned aerial vehicles take samples from spouts of whales to assess their health and fly predatory patterns over orchards to deter birds from feeding on the fruits. De-extinction, including a woolly mammoth surrogate, is on the horizon.

The book is divided into six parts. First, I relate human-wildlife conflicts to wildlife damage management and related terms, summarize a conjectural evolutionary history of such conflicts, and point to resources available to resolve these conflicts. Second, I relate human-wildlife conflict (damage) management to its biological and ecological underpinnings. Third, I survey conflicts, including invasive species globally, proximate damaging species in North America, and wildlife diseases and zoonoses. Fourth, I review current practices, organizing them into physical, chemical, and biological methods. Fifth, I explore the human dimensions of wildlife management, including the economic, the social and cultural dimensions, the policy and legal dimensions, and the increased influence of public opinion on biologically and ecologically based practices. Sixth, to pull it all together, I look at overall management strategies and use current research directions to propose a vision of the future.

Each chapter or section within a chapter begins with a statement of concepts, principles, and terms. The statement is followed with an explanation. For each, the concepts are then illustrated with specific examples from human-wildlife conflict (damage) management experiences or case studies. Included, as appropriate, are examples from the entire spectrum of damaging species, from unicellular organisms to plants and animals, but vertebrate species are emphasized. A summary and questions are included at the end of each chapter to assist in reviewing and to further deeper discussion of selected issues.

PART I • AN OVERVIEW OF HUMAN-WILDLIFE CONFLICT (DAMAGE) MANAGEMENT

In this section, I introduce human-wildlife conflict (damage) management, explore its history, and point to resources that help practitioners choose proven solutions that are based on science and experience.

1

Introduction

This chapter relates wildlife damage management to human-wildlife conflict, human-wildlife interaction, coexistence, and other terms and describes the organization of the book.

Statement

Humans value wildlife, but humans also harm wildlife by illegally taking or killing them, by altering, polluting, or fragmenting their habitat, and by altering the planet's climate. In turn, wildlife are the proximate causes of activities that injure and kill people, make people sick, threaten their livelihoods, and damage their property, interests, and industry. Human-wildlife conflicts include all negative relations between wildlife and people whereas wildlife damage includes only those negative relations for which wildlife are the proximate cause (fig. 1.1 and table 1.1).

Explanation

What does the term *wildlife* mean to you? For some, it is a five-and-a-half-year-old buck, confident and cunning, that you have only glimpsed. Or a turkey (*Meleagris gallopavo*) you know would score a 66 with the National Wildlife Turkey Federation. For others, it is the American bald eagle (*Haliaeetus leucocephalus*) that you hope to videotape as it dives and plunges for its lunch. Or maybe it is the "ghost of the mountains," rarely seen and far away, the snow leopard (*Panthera uncia*) known only vicariously to you on its rocky cliff high in the mountains of Nepal. Or perhaps it is watching on television as a green sea turtle (*Chelonia mydas*) begins a 1,300-mile migration from Ascension Island to its feeding grounds off the coast of Brazil.

Would you have thought of **commensal** rodents such as the house mouse (*Mus musculus*) or the Norway rat (*Rattus norvegicus*), animals that live quietly behind storehouse walls and "come to the table" when you are not there? Or how about the fleas that carry a virus or bacterium on the backs of those rodents? How about the virus or the bacterium? And how about those dandelions (*Taraxacum*) growing outside your window among the cultivated grasses in your lawn? Game or nongame, abundant or rare, plant or animal, people-shy or not, these are all examples of wildlife as I use the term in this book. And although it would be too much of a stretch to formally label the virus and the bacterium as wildlife, I do include prions, viruses, bacteria, protista, and fleas in this book, but only as they relate to **wildlife diseases** and zoonoses.

As a working definition for this book, **wildlife** is any nondomesticated plant or animal. This includes formerly domesticated plants or animals that have successfully returned to the wild, such as **feral** pigs (*Sus scrofa*) and cats. It does not include companion animals or organisms bred for the laboratory. Fishes such as rainbow trout (*Oncorhynchus mykiss*), bluegills (*Lepomis macrochirus*), channel catfish (*Ictalurus punctatus*), and largemouth bass (*Micropterus salmoides*), which are bred and raised commercially in fish farms and bought and stocked by landowners for personal use, are not a focus of the book. Nor are deer and elk (*Cervus canadensis*) raised on "game farms," except as possible vectors for wildlife diseases.

Human-wildlife interactions refer to all types of interactions between people and wildlife, those perceived by people as both positive and negative ones (see fig. 1.1) (Nyhus 2016). Examples where people are seen

to interact positively with wildlife would include staff at a wildlife rescue center nursing an injured bobcat (*Lynx rufus*), first responders of the Disentanglement Network freeing a whale from a net, biologists seeking a cure for bat white-nose syndrome (*Pseudogymnoascus destructans*), electric companies seeking ways to reduce collisions of birds with wind turbines, and biologists seeking to list a tortoise as threatened or endangered so funding can be used to restore it. Examples where wildlife are seen to interact positively with people include the services provided by bees pollinating crops or gardens, bats pollinating desert flowers, a constructed wetland processing human waste, forests absorbing carbon dioxide and producing oxygen, and wildlife contributing to the culture and enjoyment of humans.

Human-wildlife conflicts are what people perceive as negative interactions between themselves and wildlife (Nyhus 2016; Messmer 2019) (see fig. 1.1 and table 1.1). These interactions can be caused by people—for example, a car driver hitting a deer or a boater hitting a manatee (*Trichechus*). These interactions can also be caused by wildlife—for example, a mountain lion (*Puma concolor*) attacking a jogger or a coyote (*Canis latrans*) preying on a pet dog. Interactions between people and wildlife that people see as negative with wildlife as the proximate causes are called wildlife damage (see fig. 1.1 and table 1.1). More formally, **wildlife damage** can be described as the perceived or actual harm caused by wildlife to humans, their property, or something they value.

Although it is conceptually clear that human-wildlife conflicts include all negative interactions be-

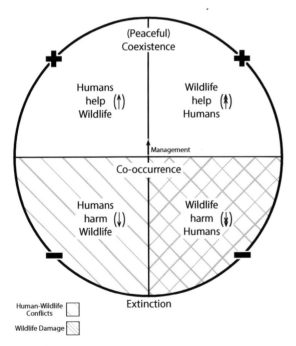

Figure 1.1 Relationships between terms commonly used in resolving human-wildlife conflicts. Note that human-wildlife conflicts (hatched) encompass all negative interactions between people and wildlife, including those negative ones proximately caused by wildlife (double hatched). Illustration by Lamar Henderson, Wildhaven Creative LLC.

TABLE 1.1   *Human-wildlife conflicts*

Type	Example
HUMANS	
Alter climate	Less ice for polar bears
Encroach on habitat	More mountain lion/people conflicts
Fragment habitat	More deer collisions with cars
Pollute habitat	More whale entanglements
Transport invasive species/ diseases	Fewer seabird species on islands
Illegally trade/harvest wildlife	Fewer rhinoceroses
WILDLIFE DAMAGE	
Property	Bird damaging airplane engine
Agriculture, including forestry and livestock	Rodents eating rice
Public health or safety	Ebola virus species jumping to humans
Other wildlife	Brown-headed cowbird parasitizing warbler nest

tween people and wildlife, including those caused by people as well as those caused by wildlife, be aware that in practice this is not always how the term is used. Rather, human-wildlife conflict is often used as a synonym for wildlife damage, emphasizing only the negative impacts of wildlife on people and mostly ignoring the broad range of impacts that people have on wildlife. This usage also tends to emphasize damage caused by the larger-sized carnivores, herbivores, and reptiles. For a visual illustration of this, google "human-wildlife conflict" and click on "images." Note that most of the images illustrate damage caused by wildlife, mostly by larger-sized charismatic animals. Now google "wildlife damage" and click on "images." Note that these key words call up images of damage caused by some charismatic wildlife, but that the words also result in more images of smaller, less charismatic wildlife such as rodents, birds, and opossums.

In this book, I use the terms "human-wildlife conflict" and "wildlife damage" interchangeably to refer to the negative impacts of wildlife on humans. I try to use human-wildlife conflict explicitly when I believe the concept or thought being described applies to both the negative impacts of humans on wildlife as well as negative impact of wildlife on humans (see fig. 1.1), but it is sometimes difficult to make that assessment. The book itself emphasizes wildlife damage and its management.

Practicing managers and scientists within the Wildlife Services program of the US Department of Agriculture (USDA) categorize wildlife damage generally as damage to property, agriculture (including forestry and livestock), public health or safety, or other wildlife species (see table 1.1). If one considers the impacts of wildlife on other species to include groups of species as occur in communities, ecosystems, and landscapes, the approach is comprehensive. But it is also overlapping. For example, ingestion of Canada geese (*Branta canadensis*) or other birds by engines of a commercial airplane, with consequent damage to the engines, is both damage to property (jet engines and the plane on crashing) and possibly public safety (harm to humans if the plane crashes) (fig. 1.2).

Another approach to classifying damage is taxonomic—i.e., damage caused by plants, birds, mammals, or other taxa. **Vertebrate pest control**, which for many years was considered the whole of

Figure 1.2 US Airways Flight 1549 crashed into the Hudson River shortly after ingestion of geese on takeoff, January 15, 2009. All 155 passengers and crew survived. Photo by US Army Corps of Engineers.

wildlife damage management, focuses on damage caused by birds, mammals, reptiles, and amphibians. In this book, I will organize chapters describing damaging species around their taxa.

Two types of wildlife damage have received particular attention: **invasive species** (nondomesticated species that invade new habitat, adversely affecting it) and **zoonoses** (diseases that are transmitted between wildlife and humans or their livestock or pets). Both invasive species and zoonoses have probably affected human enterprise and ecosystems since early humans first migrated and traded. But problems with invasive species—such as feral pig damage and zoonotic injuries and death as have been caused by **COVID-19** (a viral disease that causes severe acute respiratory syndrome believed to have jumped from bats to humans)—now occur with increasing frequency and intensity, prompting public concerns for better understanding of the causes and demands for solutions.

Management, in the simplest sense, is getting something or somebody to do what the manager wants. The term is nonjudgmental, accommodating both "good" and "bad" managers. When used in wildlife management, it means getting wildlife to do what people want, such as changing behavior, population size, or community structure. But just as often, it means getting people to do what the manager wants, such as changing attitudes about wildlife or a wildlife conflict. The manager ultimately decides what is to be done, but his or her decisions are almost always based on a specific **conservation** purpose or on mitigating a conflict, always after considering the views of people affected by the management actions. Although the concept of management is itself straightforward, getting people to agree on what they want can be difficult. Achieving consensus on acceptable strategies and methods for managing a wildlife problem is a further difficulty. The case of hen harriers in the heather moorlands, as described by Thirgood and Redpath (2005) and others, illustrates these difficulties (box 1.1).

Given that wildlife damage refers to the negative impacts of wildlife on people, **wildlife damage man-**

BOX 1.1 The Case of the Hen Harriers

The heather moorlands are dominated by the dwarf shrub heather and are largely restricted to the United Kingdom. This habitat is home to a mixture of species, but moorlands carry conservation significance primarily because they support golden eagles and peregrine falcons (*Falco peregrinus*) as well as golden plovers (*Pluvialis apricaria*) and curlews (*Numenius arquata*). Hen harriers (*Circus cyaneus*) are predators, as are the red foxes (*Vulpes vulpes*), stoats (*Mustela erminea*), and carrion crows (*Corvus corone*) inhabiting the moorlands.

About half of the upland moorlands are privately owned and managed for red grouse (*Lagopus lagopus*) production. The management is generally acknowledged to provide ecological, social, and economic benefits to the area. Usually performed by professional gamekeepers, this management is complex and includes burning to provide a mix of heather and grasses that are optimal for grouse production and management of predators and parasites (Thirgood and Redpath 2008).

Hen harriers prey on adult red grouse and chicks in the summer, and hunters believe the harriers limit grouse harvests, although limited scientific studies support this notion. Despite legal protection of the harrier in the United Kingdom since 1952, enforcement is limited, and the risk of being caught for illegal kills is low. Harriers are illegally taken, limiting their

populations, so there are essentially no successfully breeding harrier hens in game lands managed for grouse (Redpath and Thirgood 2009).

Conservationists want the existing laws protecting harriers to be enforced, but logistics make this unlikely. Biologists have suggested that low levels of harriers could be supported on managed game lands without impact on the grouse. The low levels would have to be maintained by lethal management, however, and conservationists oppose such management. Hunters therefore refuse to recognize that illegal killing of harriers is an unacceptable practice, and conservationists refuse to consider possible solutions involving lethal management that would limit harrier densities (Thirgood and Redpath 2008).

Several options have been considered to resolve the conflict. Compensating hunters for grouse losses to harriers would be prohibitively expensive, and there is uncertainty as to who would pay. Habitat management, use of golden eagles to manage harriers, and zoning concepts have all been considered, but none have been fully explored. And according to Thirgood and Redpath (2008), both sides favor the status quo. Hunters continue to kill harriers without serious legal risk. Conservation groups continue to demand increased enforcement; they know it is unlikely, but they gain publicity from illegal kills.

agement can be viewed as the science and art of diminishing the negative perceptions of wildlife while maintaining or enhancing their positive ones. Because interactions between humans and wildlife can also be viewed as forces moving humans into more or less positive relationships with wildlife, it might help to visualize the effects of such interactions on human perceptions of wildlife using vectors (fig. 1.3). Managing wildlife damage can be seen as a vector moving the wildlife toward a relationship perceived as more positive by humans. The net gain can be visualized conceptually as the increase in positive perceptions that result from the management action, minus any changes that might have occurred concomitantly with the negative perceptions. Whereas wildlife damage constitutes the lower right-hand quadrant of human-wildlife interactions in figure 1.1, the vector component representing the effects of management can occur anywhere along the vertical axis representing perceptions of positive to negative interactions, hopefully moving the relationship away from extinction and toward peaceful coexistence.

As with the term "management" in general, it is important to note that the term wildlife damage management is nonjudgmental, accommodating both "good" and "bad" management. What constitutes good or bad wildlife damage management has been the subject of much intense and often emotional discussion, often centering around lethal versus nonlethal management, which will be a central undertone throughout this book. At this point, suffice it to say that human perspectives on this topic are complex, depend heavily on factors such as the context of damage, and relate to human values, beliefs, and attitudes that themselves vary by culture, society, and time.

Today when we think of wildlife that causes damage, we think of a plant or animal that knowingly or not has acted against the will of humans but is still desirable. For example, we recognize that dandelions may be unwanted—a pest, a nuisance, and a negative value in our front lawn. Many old-timers, however, attest that leaves of dandelions make great salads and the flowers great wines; the root sap of the dandelion may someday provide inexpensive rubber (Stolze et al. 2017). African elephants (*Loxodonta*) damage timber and agricultural crops but also disperse the seeds of timber and nontimber forest species, all important for forest-dwelling humans (Mamboleo et al. 2017). The American black bear (*Ursus americanus*) causes serious damage to commercial timber production but also provides important **ecosystem** services to forests, increasing the habitat for cavity-nesting birds and woodpeckers (Mendia et al. 2019).

Pest is a term sometimes used in wildlife damage management. The term refers to the individual, population, or species of wildlife that is causing damage. Elephants (e.g., Löyttyniemi and Mikkola 1980) and American black bears (e.g., Mendia et al. 2019) are called pests when they strip the bark from commercial trees

Management

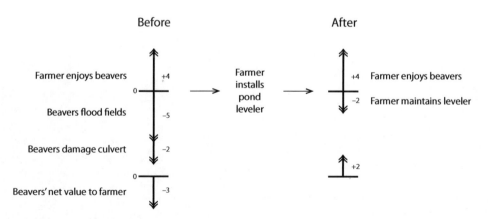

Figure 1.3 A hypothetical example of using vectors to visualize the impact of a management action on how humans value wildlife. Before management, the farmer sees beavers negatively as flooding fields (−5) and damaging culverts (−2) but enjoys watching them (+4): a net negative value to the farmer of −3. After the management action (installing a pond leveler), field and culvert damage ceases, but the farmer still enjoys the beaver (+4), and the leveler must be maintained (−2): the beaver now has a net positive value to the farmer of +2. On this hypothetical scale, the value of beavers moved from −3 to +2. Illustration by Lamar Henderson, Wildhaven Creative LLC.

or cause other agricultural damage (e.g., the Asiatic elephant, *Elephas maximus*; Tisdell and Zhu 1998). Common vampire bats (*Desmodus rotundus*), which lacerate skin and lap blood from donkeys (*Equus africanus*) and cattle (*Bos*) in Central America, transmit **rabies** and reduce cattle milk production or weight gain. European starlings (*Sturnus vulgaris*) are found at cattle feed lots and Eastern gray squirrels (*Sciurus carolinensis*) at some bird feeders. The term *pest* is a convenient one, and I will use it in this book but always with the understanding that all species have positive values to humans.

Technical solutions to human-wildlife conflicts fall within the applied aspects of biology, chemistry, and **ecology**, and their related human dimensions. In this book, I will review key biological and ecological concepts and principles that underwrite such solutions, and I provide examples in three chapters: one on the individual and species level of organization of life, one on the population level, and one on the community, ecosystem, and landscape levels. Human dimensional concepts are covered in an additional three chapters.

I close this chapter by reemphasizing the **anthropomorphic** (human-oriented) nature of human-wildlife conflicts, particularly as we are now living in the sixth great extinction of the Anthropocene (see, e.g., Lewis and Maslin 2015). When we use the term wildlife damage in its technical sense, I believe that most of us understand that wildlife is not to be blamed for the damage. Wildlife, at worst, is only the proximate cause of damage. Most of us understand that if one were to follow the famous "Five Whys" used first by the Toyota Motor Corporation to discover the true cause of a problem, *Homo sapiens* would inevitably show up as the root cause of any human-wildlife conflict long before we reached our last "Why." That is, humans or human activity underlie the whole of human-wildlife conflicts, including the damage attributed to wildlife as wildlife damage.

We understand that cars collide with deer or camels because people have cars and drive them in areas habited by these species. We know that elephants raid crops because the crops are grown in or near elephant habitat, or that elephants see the crops on their way to find water or some other resource and raid them opportunistically. We comprehend that the COVID-19 pandemic is as much a consequence of wet markets or research laboratories facilitating species jumps and of human transportation as it is a disease brought to us by bats or pangolins. To this end, then, I reiterate the need to understand not only the biological and ecological bases for managing human-wildlife conflicts but

also the need to understand their **human dimensions** (including economics, human perceptions and responses, public policies, and politics).

We are now only beginning to understand what cognition wildlife may have that makes them aware of their own existence (and therefore the certainty of their own death or that of other wildlife) or that might impart a concern for the extinction of their own or other species. It seems incumbent on humans—whom we know are capable of seeing and offering reasoned judgment on such issues—to make the best decisions about the needs of humans and of wildlife, and then to proceed with the wise management of wildlife resources. Perhaps in this manner we can yet truly earn our Linnaean name, *Homo sapiens*.

Examples of Human-Wildlife Conflict: Human Impacts on Wildlife

CLIMATE CHANGE. For years, the sow polar bear (*Ursus maritimus*) used the sea ice as a platform to stalk and hunt her favorite prey, ringed seals (fig. 1.4). From the vantage of the water, she could see the seals long before they saw her, often sneaking up from behind the ice where they lay resting. She even used the ice occasionally as a platform for her maternal den. The habitat had provided safe hunting with ample prey for both her and her young.

In the last few years, though, finding sea ice became more and more difficult. This year, for the first time, her efforts to catch prey without a suitable platform were futile. From a safe distance on the sea ice, she had seen whale carcasses at the outskirts of a town there. Unaware that grizzlies had already claimed the

Figure 1.4 Apparently healthy polar bear mother with cubs on Beaufort Sea ice in 2001. Climate change is reducing the ice used to have and rear young, forcing the bears to move to land and increasing the conflicts with humans and other bears. Photo by Steven Amstrup, USGS, Alaska Science Center.

carcasses as part of their food resource, and inexperienced in dealing with humans, the gravid bear was stressed and hungry. She began her swim to shore.

ILLEGAL TRADE OR HARVEST. The ranger had hiked trails for two days to find the gorilla troop. He now sat quietly, watching a gorilla watch him. He wondered what the gorilla was thinking. The ranger knew, strong as he was, that at best he might be able to press 350 pounds. The gorilla could press 4,000 pounds on a bad day. And that is why the ranger was watching the silverback so intensely. He was checking whether the gorilla was having a bad day. Any indications of a pain or fever? Body aches? Diarrhea or vomiting?

There had been rumors of poachers in the area. Odds are that some of the poachers had Ebola and, if they had gotten close enough, they might have given Ebola to the gorillas. Ebola was often fatal for a human—it was always fatal for a gorilla. If the silverback died, the rest of the troop would disperse, taking Ebola along with them to other troops. Trying to protect a fatally sick or dispersing troop from poachers would be an exercise in futility.

Satisfied that the gorilla was healthy, the ranger radioed back to his headquarters. It would take more rangers and supplies to protect the troop until the poachers left. The ranger looked at the silverback and smiled. It might be weeks before everyone was safe.

HABITAT FRAGMENTATION. The 8-inch black tubing was taller than the terrapin, effectively blocking its movement out of the brackish marsh and onto the runway. But that was before high tide. Now the water had risen just enough to assist the turtle in her effort to get over the tubing. In just a moment, she was on her way up the rise and onto runway 4L of John F. Kennedy Airport in New York.

On command from the air traffic controller, the pilot of the fully loaded passenger plane pulled to the head of the runway and stopped. Other planes taxied and stopped behind this one. The delay, though brief, would probably ripple throughout the nation's air traffic. Some passengers texted friends that their flight was being delayed by pregnant turtles. Most did not seem to mind.

Wildlife biologists scurried onto the runway. They picked up the diamondback terrapin (*Malaclemys terrapin*) and placed her in the back of a pickup truck where she could be marked and then released back into Jamaica Bay. She would have to find sandy soil above the tide and suitable for egg laying that was not adjacent to the runways or tarmacs of a busy airport.

Hundreds of feet of tubing had helped reduce the numbers of terrapins getting onto the airport—but it did not stop them all. In a few weeks, the biologists would be back, this time helping young hatchlings cross the runway and into the Jamaica Bay Refuge while planes again waited patiently.

NOISE POLLUTION. The bowhead whale (*Baleana mysticetus*) was making his way south through the Bering Strait, staying near the edge of the ice break on his way to the summer feeding grounds. A little over a century ago, when he had first achieved adulthood, the thickness of the ice had challenged him. Using his enormous bowhead, it took all his strength then to break and lift chunks of ice large enough so he could swim and breathe. That had changed. The ice had gotten thinner, easier to break.

Noise had now become the challenge. A hundred years ago, water in the Pacific Ocean was mostly free of human noises. And surface noises were attenuated for a whale living under a deep blanket of ice. The bowhead could efficiently use his own sonar to navigate, avoid obstacles, find prey, and communicate with other whales to have a social life. Now, the whale heard more sound, but it was mostly the repetitive noise of ship rotors, seismic air guns blasting for oil exploration, or military sonar. Sometimes the blasts were loud and unannounced, frightening the whale so that he surfaced too quickly, causing bends and risking organ damage. With his increasing confusion between the biological sounds and background noises, life in the dark waters of the Chukchi and Beaufort Seas had become much more difficult.

WAR. The impacts of war on wildlife are often collateral. Eagle River Flats is an 865-hectare estuarine salt marsh within the Joint Base Elmendorf-Richardson in Anchorage Borough, Alaska. It has been used for decades to train military personnel to handle munitions. In the 1980s, after dabbling duck die-offs were observed, a moratorium was placed on using the site for training. White phosphorus, a pelleted residue in the smoke rounds used for training, was eventually identified as the source of the kills. Thought to be unstable in air and short lived, white phosphorus retained its chemical stability and activity under the anaerobic conditions of the marshes. While foraging for seeds and insects in the water, the ducks, mostly mallards (*Anas platyrhynchos*), ingested small but deadly amounts of the substance.

The army reinstituted bombing but only during the winter season when the marshes were covered with ice, and it used munitions with no white phosphorus. The army also developed a 5-year program to reduce duck mortality by 50%. Sections of the marshes were drained so that white phosphorus pellets were exposed to air and oxidized. The sediments were capped. The army installed six pumping stations for use during the training season.

The short-term goal of 50% reduction in duck mortality was achieved by 2002. By 2006, duck mortality was less than 1%, lower than the natural mortality for the mallards. Pumping ceased in 2008.

Examples of Human-Wildlife Conflict: Wildlife Damage

DAMAGE TO AGRICULTURE. It is just past sunset as a Filipino farmer walks along a dike on his hectare rice farm, his steps slow and in unison with the rhythmic rattle of a stone within an old tin can, tied to a stick that he shakes. Resting momentarily, he hears the rustle of rice panicles as rats move from plant to plant, carefully inspecting each one and methodically removing and eating the rice. The farmer presses on, fearing that within a few days his rice will be gone, harvested not by him but by uninvited commensal rodents. The farmer will spend the next few nights walking his fields, rattling the tin can, guessing at the futility of his efforts. Perhaps in the morning he will seek the assistance of a local priest. Or perhaps he will negotiate with the rats. The rice is his family's main income and resource; without its availability and retail value, his family will suffer severely.

DAMAGE TO FORESTRY. The landowner looked on in dismay at the flooded woodland, thinking out loud that the flooded area must encumber at least 60 acres. When he was here two days ago, there had been no flooding. As he drove down a dirt road that quickly became impassable, he observed that water had begun eroding the rock surrounding one culvert and was already reaching for the substructure of the road overpass. The landowner hopped from his pickup and got his first glimpse of the source of the problem: "Beaver dam," he said. "Didn't know those little guys could cause me such trouble!"

A short time later, a local control specialist stood beside the landowner looking at the rising water. "You need to decide what you want to do," he said. "If you want to eliminate the flooding and prevent further damage to your timber, roads, culverts, bridges, and other crops, then we need to remove the dam, the beaver, and the debris. If you want to retain the beaver and the pond, then we can try a device called a pond leveler, although it will only work if you are willing to put considerable time and effort into maintaining it. If not maintained, damage will soon resume, and it may begin again anyway somewhere upstream."

The landowner thought it over. He considered the summer heat, the difficulty for him or anyone to work in ponds under those conditions, and the time and effort he would have to commit if he were to try the pond leveler. **Pyrotechnics** or mechanical/physical removal of the dam would be tough work, but he knew experienced trappers such as this agent were ready and able to do that work, and it offered a permanent solution to the problem. But he liked seeing the beaver, an animal that was a rare sight only 60 years ago. It was a tough choice.

DAMAGE TO PROPERTY AND HUMAN SAFETY. An American military pilot deftly banks his B-2 "Stealth" bomber into a holding pattern near the main runway at Whiteman Air Force Base in Missouri: the tower had instructed him to abort his landing route. Meanwhile, on the runway below a muffled sound is heard and a deer drops, downed expertly by a federal wildlife biologist sharpshooter. The deer was the unfortunate victim of being in the proverbial wrong place at the wrong time. Shortly after, a pickup truck emerges onto the runway, and the deer is removed. The plane lands safely.

DAMAGE TO OTHER WILDLIFE. A short distance from the edge of a Michigan pine-oak forest but beyond the easy or likely reach of humans, an endangered Kirtland's warbler (*Setophaga kirtlandii*) quietly builds its nest, spurred on by a growing urge to lay eggs. A few days later, the healthy eggs are laid, and the warbler incubates them. The incubation itself, aided by male feeding, is uneventful except for the visit of another bird—but its stay is brief.

In the ensuing days the warbler cares for the eggs, and when the young hatch, the male switches his energies to capturing and bringing insects for the voracious and rapidly growing young. At long last, the single surviving young is feathered and grown, and the "warbler" fledges—in the person of a brown-headed cowbird (*Molothrus ater*)!

This reproductive effort, commendable though it was, added nothing to the survival of the warbler species. It took the efforts of humans—first to assess the impact of the brown-headed cowbird on populations of Kirtland's warblers and then to aggressively trap and remove cowbirds from critical habitat—to reverse the warbler's slide toward extinction (DeCapita 2000).

DAMAGE TO PUBLIC HEALTH OR SAFETY. Within a tropical rain forest in the Democratic Republic of the Congo (formerly Zaire) is the small village of Yambuku. In August 1976, a 44-year-old schoolteacher returned to Yambuku from a trip to the northern part of the country. He felt ill, and he was treated in a local clinic for malaria. However, his symptoms worsened to include diarrhea, uncontrolled vomiting, headache, and dizziness. Eventually he began to bleed severely and died about 14 days later, on September 8.

Soon, nosocomial transmission resulted in others in the clinic dying. In all, 284 (89%) of the 318 reported cases of this illness were fatal (Pourrut et al. 2005). National and soon international assistance was sought to quarantine the village and prevent further spread of the disease. **Ebola virus disease** (formerly called Ebola hemorrhagic fever), a viral disease with malaria-like symptoms that is often fatal, was first recognized in the Democratic Republic of the Congo in 1976. Bats or other wildlife were its reservoirs.

Ebola virus disease has become an international concern. Sporadic outbreaks occur in Gabon, Africa, and the largest most recent outbreak occurred in West Africa (Guinea, Sierra Leone, and Liberia) from 2014 to 2016.

Summary

- Human-wildlife interactions include both positive and negative interactions between people and wildlife.
- Human-wildlife conflicts are the negative interactions between people and wildlife. Such conflicts include negative impacts of people on wildlife as well as negative impacts of wildlife on people (i.e., wildlife damage).
- Invasive species and wildlife diseases (including zoonoses) are two emerging concerns of human-wildlife conflicts.
- Wildlife comprises any nondomesticated living organism, including plants and animals.
- Damage is an affront, real or perceived, to humans. It is a global issue that can be categorized as damage to property; to agriculture, including forestry and livestock; to public health or safety; and to other wildlife species or their habitat.
- Management is getting both wildlife and people to follow the learned guidance of a manager.
- Wildlife damage management is the science and art of diminishing the negative consequences of wildlife while maintaining or enhancing their positive aspects.
- Today we see all wildlife as having positive and negative values to humans.

Review and Discussion Questions

1. Relate human-wildlife conflicts to human-wildlife interactions and wildlife damage management. Give examples of each.
2. Are wildlife species good or bad? How do descriptions of wildlife as pests or varmints relate to today's thinking about wildlife damage?
3. We describe human-wildlife conflicts, including wildlife damage, as human driven. Do you agree or disagree? Defend your view with specific examples.
4. List ten examples of damage where wildlife is the proximate cause. Now use the Japanese system of asking "Why?" up to five times, until you discover the primary cause of the problem. Was it humans? Explain your answer.
5. Try using vector analyses as described in this chapter to compare the theoretical net impacts of two management actions—say, building a wildlife crossing versus less effective signage to keep cars from colliding with Santa Monica mountain lions—on human perceptions of the mountain lion. Choose a human beneficiary, e.g., a car driver, and list positive and negative perceptions regarding the mountain lion. Estimate each of their values, before and after each management action. Was there a net difference between the actions? Did vector analysis help?

2

History

I consider the evolution of human-wildlife conflicts along with human evolutionary milestones, from early hominids to modern-day humans.

Statement

As humans evolved and reached milestones, such as using predatory weapons, transporting goods, and domesticating wildlife, corresponding conflicts with wildlife emerged, such as massive killings of predators to reduce **predation** on humans and **competition**, invasive species, and feral wildlife along with management such as shepherding and sheep dogs.

Explanation

The earliest efforts by humans to resolve conflicts with wildlife were probably to protect human life and limb. This speculation is based partly on the belief that humans inherited an instinct for survival, hardwired and ingrained, which also occurs in other animals. Perhaps a saber-toothed cat (*Smilodon* spp.), a crocodile (Brochu and Storrs 2012), or a giant short-faced bear (*Arctodus simus*) was killed preemptively by prehistoric humans. Although direct evidence for this view is limited, there is a growing body of circumstantial evidence.

Hart and Sussman (2005) summarized data suggesting that *Australopithecus afarensis*, an early hominid (the same taxonomic family as humans) that lived 5 million to 2.5 million years ago (YA), served ecologically more as prey than predator (table 2.1). The authors attributed this rather demeaning ecological function to three factors: small size (*A. afarensis* were mostly 3 to 5 feet tall in a world with much larger predators), ineptitude at making and using tools, and inability to use fire. As evidence, the scientists point to teeth marks on the bones of *A. afarensis*, talon scrapes on their skulls, and holes in their crania that just fit the size and shape of the fangs of a saber-toothed cat (fig. 2.1). The authors suggest that at least some of the marks were made in the act of predation rather than scavenging, and they estimate that about 6% to 10% of *A. afarensis* succumbed to predators, a percentage consistent with efficiencies observed for modern-day predators. It seems that, at least on a global scale, *A. afarensis* had little long-term negative impact on wildlife.

TABLE 2.1   *Evolution of human-wildlife conflicts (speculative)*

Period			Conflict/management	
Hominid/human	(YA)	Role	Activity	Impact
HOMINID		Prey	Defense from predators	Little impact on environment.
Australopithecus afarensis	5.0–2.9 M			
A. africanus	3.0–2.1 M	.		
Homo habilis	2.0–1.6 M			
HOMINID/ HUMAN		Predator	Massive killings of apex predators and megaherbivores; keystone species	Ecosystem disruptions.
H. erectus	2.0–0.5 M			
H. sapiens	0.5 M to now			
H. sapiens	100,000 to now	Predator, hunter/ gatherer	Trade, migration, war	Commensals move with humans, and some become invasive; disease vectors and reservoirs move with humans yielding zoonoses. Limited habitat fragmentation.
H. sapiens	15,000 to now	Hunter/gatherer	Domestication of wildlife (plants, fish, animals)	Domesticated species travel with humans, and some become invasive and/or feral. Some zoonoses emerge with localized distribution.
H. sapiens	10,000 to now	Agriculturist	Ecosystems simplified by slash and burn, plowing, and beginning of modern methods of crop protection	Insects, rodent, bird, and other crop pests emerge, with pest irruptions and feral plant releases, some invasive. Localized encroachment.
H. sapiens	10,000 to now	Pastoralist/ nomad	Herding and roaming, livestock protection such as shepherding, guard dogs	Livestock predation, zoonoses, and flock/ herd diseases emerge.
H. sapiens	4,000 to now	Urbanite	Cities spotty for most of history: 3% of humans in cities at start of Industrial Revolution versus 50% now	Commensal wildlife problems and zoonoses occur along with increased suburban/wildlife contact. Human/pet food needs intensify agriculture and livestock production, increasing energy and transportation needs that impact ecosystems. Encroachment, fragmentation, pollution of ecosystems become widespread as do human-wildlife conflicts.
H. sapiens	2,000 to now	Industrialist (Industrial Revolution)	Use of natural resources, including fossil fuels	Environmental degradation/simplification occurs along with global warming, which greatly reduces biodiversity together with increased human-wildlife conflicts.
H. sapiens	Recent	Conservationist	Beginnings of conservation efforts	Some island and other wildlife restored. Some endangered species protected. Some game species restored.

Abbreviations: M, million; YA, years ago.
Source: Hart and Sussman (2005).

Hunting skills began improving some 2.5 million YA as early humans, such as *Homo habilis*, learned to use stone tools. Fossil records indicate, however, that at about 4 feet tall, *H. habilis* remained a staple in the diet of predators, including leopards, lions, spotted hyenas, *Dinofelis* (a jaguar-sized saber-toothed-like cat), *Megantereon* (also a saber-toothed-type cat), and extinct hyenas (*Chasmaporthetes nitidula*). *H. habilis*, at least at first, probably still used tools mostly for scavenging. Self-preservation must also have dominated concerns surrounding human-wildlife conflicts of these early hominids and may even have been a driving selective force for larger size and brain capacity. Over time, *H. habilis* adapted, learning to create flint blades, stepping from scavenger to hunter, wearing hides, and understanding climate well enough to move from Africa into Asia and the borders of Europe.

As *H. habilis* was replaced by *H. ergaster* and *H. erectus*, early humanoids continued to move into a more predatory ecological niche. A dramatic shift in

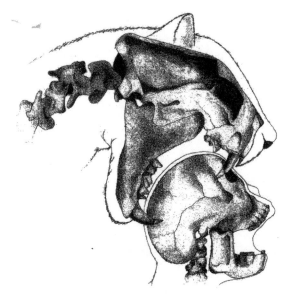

Figure 2.1 Early hominids were probably prey for predators such as saber-toothed tigers. Artwork by Ian Reid, based on earlier work by C. Rudloff, in Hart and Sussman 2005; Cavallo 1991.

their activity (and impact on the planet) occurred by 50,000 or so YA, a shift called by some the Great Leap Forward (Diamond 1997). Some humans began burying their dead and making more sophisticated clothing, indicators of spiritual and conceptual thinking. By this time also, hunting and protection from predatory wildlife included predator pits as well as organizational skills with sufficient hunting and communication techniques to allow mass killing of large herds of herbivores, such as mastodons (e.g., by forcing them off cliffs). *H. ergaster* and *erectus* also controlled fire. Negative impacts of humans on wildlife were underway.

By the time of early modern humans, *H. sapiens*, both they and their precedent species had made forays from Africa into Asia and Europe, possibly following large herbivores such as woolly mammoths (*Mammuthus primigenius*) into the plains of Asia and subsequently into North and South America as times and climate changes either demanded or allowed. A relatively short time after leaving Africa, some 60,000 or 40,000 YA, humans had migrated all the way to Australia. By this time also, modern humans began having major impacts on ecosystems across continents, starting in Australia, then in North and South America, then on the islands of Oceania, then Madagascar, and finally New Zealand at less than 500 YA (Roemer et al. 2007, Sandom et al. 2014). On each continent, major ecological impacts followed within 400 years of the advent of humans.

Paul Martin proposed that humans caused massive extinctions of Pleistocene megafauna, including its apex predators and larger herbivores—envisioning a direct predatory impact (Martin 1966). Today, proponents of human-caused mass extinctions believe there were too few humans to directly cause such massive die-offs. For example, Clovis people probably numbered only about 50,000 in Pleistocene North America. Rather, proponents now believe that the impact was indirect, a form of additive mortality. Larger herbivores were already under heavy predatory pressure and perhaps were stressed by climate changes as well when humans first arrived as a different (i.e., a weapon-toting, **reasoning**, socially organized) kind of predator.

Humans, with their unusual knowledge, skills, and abilities, likely depleted the supply of larger herbivores living in localized areas and then moved to easier hunting elsewhere as part of a nomadic lifestyle. The absence of keystone herbivores resulted in the subsequent demise of predators and abrupt and long-term changes in food chains and food webs. These changes may have set the ecological bases for some of today's problems with wildlife. In fact, some have suggested that "rewilding" or "trophic rewilding"—i.e., reintroducing large megafauna such as elephants, camels, and lions—might be an effective way to restore the complexity of ecosystems, bringing back more interspecies interactions, species diversity, and stability along with concomitant reduced conflicts between humans and wildlife (e.g., Alroy 2001; Donlan 2005; Martin 2005; Svenning et al. 2016).

As humans settled into a more pastoral existence, they domesticated wildlife that would serve them as food, pets, and beasts of burden. The dog was probably first domesticated between 11,000 and 35,000 YA (Morey 1994; Freedman and Wayne 2017). Dogs, both domestic and later feral, aided by humans who provided food and transportation, no doubt expanded their geographic ranges. Along with pastoral existence came shepherding, an occupation of wildlife damage management, as was breeding specialized varieties of dogs for guarding and working. Livestock-guarding dogs may have been used as long ago as 6,000 YA, probably in upland regions of Turkey, Iraq, and Syria where sheep (*Ovis aries*) and goats were first domesticated 9,000 to 10,000 YA. Although the first dogs probably matched the black, gray, or brown colors of sheep, white was preferred; and thus breeds such as the Kuvasz or Pyrenean Mountain dog were selected for their white color. The Romans also bred dogs and used them for different purposes, including for hunting and as guard dogs (e.g., MacKinnon and Erdkamp 2013;

Bennett and Timm 2016). Early Roman authors suggested that white allowed distinction of the guarding dogs from wolves and other predators (Rigg 2001, 7).

By 10,000 YA **domestication** of wildlife for both livestock and pets had begun. But along with this pastoral existence came the release of domesticated animals back into the wild and the birth of conflicts associated with feral animals.

With domestication of plants came a more sedentary agricultural lifestyle. An agrarian existence offered advantages to humans, freeing time for activities other than subsistence, but it also brought more conflicts with wildlife. The land was taken from habitat occupied by wildlife, so encroachment and habitat fragmentation had begun. Growing crops and keeping livestock in set locations increased the opportunities for crop damage by insects, diseases, and other pests. The house mouse (*Mus musculus*) likely became a commensal at this time (about 15,000 YA), attracted to farmland by cultivated fields and storage of grains (Cucchi and Vigne 2006). Movement of the mice into Europe was probably then facilitated by commercial and demographic expansion of Phoenicians and Greeks (Cucchi and Vigne 2006). Cats followed, perhaps as long ago as 10,000 YA; at first they fed on field mice in a commensal relationship with farm communities in Cyprus. Skeletal remains from Poland, dated at about 4,200 YA, serve as the oldest evidence of commensalism or domestication in Europe (see table 2.1) (Krajcarz et al. 2020).

Agricultural production provided the ecological **energy subsidies** that allowed the establishment of towns and cities: the beginnings of urbanization. Along with the benefits offered to people, living in towns and cities came with urban wildlife problems. The increased concentrations of people into small urban areas also meant further reaches into the surrounding ecosystems for food and other resources, for processing human debris, and for exchange of goods: e.g., for harvesting crops from farmland, harvesting timber, extracting minerals and fuels, processing human sewage in wetlands, dumping trash and other wastes at landfills, and moving products between urban areas. These reaches eventually extended globally. The reaches affected the quality of ecosystems for wildlife, often fragmenting or degrading them. Within cities where people and pets were densely populated, diseases spread more easily. These diseases transmitted to humans and pets from wildlife, called **zoonoses**, constituted another form of wildlife conflict.

Humans continued to disperse, establishing both land and sea routes that varied with time, changing patterns of land masses, and global climate changes (e.g., Veldhuis and Underdown 2017). Wherever they went, immigrating humans took with them domesticated livestock such as pigs, sheep, or goats for agriculture or perhaps pets or other animals for nostalgia. These animals sometimes escaped or were intentionally released and became feral (e.g., Hess et al. 2017). Humans also brought along agricultural crops as plants or seeds. Other organisms—unwanted but traveling nonetheless along with these domesticated plants and animals—included diseases and their vectors, insects, commensal rodents and weeds, and other wildlife that could damage crops and livestock, or impact endemic plants and animals.

Sometimes humans purposely released wildlife or domesticated plants and animals into new areas to survive on their own, perhaps in hopes of harvesting later generations of survivors. For example, seafarers sometimes released goats or foxes on islands along their sea routes in hopes of having meat resources available in the future (e.g., Campbell and Donlan 2005). Other organisms simply migrated onto the islands during landings or shipwrecks (e.g., Sherley 2000). In these ways, **exotic** species were introduced into new areas on both mainlands and on islands, sometimes becoming invasive and raising havoc in the ecosystems that they joined. Based on observations of today's impacts of roadways, it seems likely that the migratory paths themselves fragmented and otherwise altered habitat of wildlife, perhaps affecting their dispersal and movement patterns.

Trade may have been underway at 100,000 YA (e.g., Smith 2016). Routes included terrestrial and water, along riverways and across seas. Items traded might also have carried wildlife, such as weed seeds among crop seeds or commensal rodents and zoonoses among perishable goods. Direct evidence is wanting, but indirect evidence (e.g., as modern-day examples) of the movement of weed seeds or zoonoses along trading routes is plentiful. As with migratory routes, the trade routes likely altered and fragmented the habitat through which they passed, affecting the movement and distribution of wildlife.

As humans dispersed into new areas, they slashed and burned or otherwise cleared land for agriculture, fragmenting the landscape. The resulting agroecosystems simplified the checks and balances of natural ecosystems, turning them into **monocultural agroecosystems**. With species released from their natural checks, these simplified ecosystems responded with their own perturbations: "outbreaks" of wildlife that destroyed crops, brought pestilence, or caused other problems for the settlers.

At sometime within this conjectural narrative, humans began keeping written records. Written records sometimes allowed detailed accounts of human-wildlife conflicts occurring at the time. For example, the writings of Thucydides describe a **plague** that embraced the city-state of Athens in 430 BC and again in 429 BC and 427–426 BC. The human activity that caused the wildlife event was the Peloponnesian War. Athenians from surrounding areas were ordered to retreat behind the city walls, and the crowded and unsanitary conditions bred diseases, possibly plague, anthrax, or typhoid. Commensal rodents were likely carriers. Fear of getting the disease facilitated a retreat by the attacking Spartans but also encouraged the spread of the disease by Athenians who ignored the extant laws because they believed they were doomed to die anyway.

Interspersed among these other major human milestones are wars, ranging in scale from tribal to global. Warring people can have enormous, long-term impacts on human populations, ecosystems, and wildlife (see, e.g., Gaynor et al. 2016). Wildlife can be killed or maimed by the effects of bombs, pyrotechnics, or herbicides, and the habitat can be altered. Wildlife can be transported unintentionally along with the commodities of war. Beasts of war such as horses, elephants, or camels are sometimes taken into new areas, then released into the wild when no longer needed. Sometimes ecosystems respond with irruptions of wildlife that damage crops or spread diseases to animals or people.

Nelson (2009) has argued that because early human populations were relatively small and widely scattered in Europe, nonendemic diseases tended not to spread rapidly or easily, but Justinian's invading army brought a pandemic to Europe in 540–565. No further pandemics were recorded until the Black Death of 1347–1351; according to this account, the Black Death came to Europe after the disease broke out in the town of Kaffa, under siege by Mongols. The Mongols retreated, but only after the Mongol commander had "loaded a few of the plague victims onto his catapults and hurled them into the town." Merchants, along with fleas and rats, left Kaffa shortly after this, carrying the plague with them. The course of the plague can be followed along the established trade routes.

Sometimes warring protects wildlife from people. For example, Kay (2007) analyzed relationships between wildlife and contact with native North Americans described in the written records of the Lewis and Clark Expedition from 1804 to 1806. He found that abundance of wildlife was inversely related to contact with Native Americans. For example, wildlife was abundant in buffer zones between warring tribes in the areas avoided by tribal people. In fact, he argued that had it not been for warring tribes European settlers would have seen hardly any wildlife during the period of western expansion. Daskin and Pringle (2018), however, analyzed the impact of wars on large herbivores in Africa between 1946 and 2010 and found the effects were mostly negative.

The global industrial revolution began in England, spread to Europe and the United States, and later moved to Japan and other parts of the world (see, e.g., Mohajan 2019). This spread occurred quickly, and the global environmental impacts of the revolution in England were almost immediate. For example, to provide bread for British workers, industrialists financed massive wheat production in parts of Russia, South America, and California in the United States. Railways and ocean carriers were built to move the wheat to England. Global environmental impacts also occurred with the mining of coal and the mining and smelting of copper, key resources needed for the success of the British factories.

The revolution, which continues around the globe today, brought products at greatly reduced costs, including those associated with agricultural production and with the management of wildlife, and raised the standard of living for some people. The negative impacts on ecosystems and the environment, however, have been profound. For instance, the revolution facilitated the transport of wildlife, intentional or not, by air, land, and sea, accelerating impacts of invasive species globally. The industrial revolution has been driven by an unprecedented use of energy, mostly derived from fossil fuels. The consequence has been the most profound of all environmental impacts: human-facilitated global warming (e.g., Pachauri et al. 2014).

Wildlife, of course, respond to climatic changes. With global warming, we can anticipate range extensions by some wildlife and extinctions of others. We can expect greatly reduced biological diversity (Thomas et al. 2004) both in numbers of species within ecosystems and in genetic diversity within populations. Leigh et al. (2019), for example, has estimated a 5.4% to 6.5% decline in genetic diversity within populations of wildlife since the beginning of the industrial revolution. Given simplified ecosystems, we can expect more frequent wildlife irruptions (e.g., Kaeslin et al. 2012, 38–40) and more invasive species and more frequent wildlife diseases around the world.

Underlying the latter period of written human history has also been an impressive geometric expansion of the human population (see chapter 17). One consequence has been human encroachment on land that was marginal in its ability to support agriculture or provide stable human housing. Conflicts with wildlife often ensue. A burgeoning human population places further demands on natural resources; exploitation of these resources disrupts, simplifies, or destroys major natural ecosystems. Again, reduced biodiversity leads to oscillations of wildlife that damage crops, livestock, and property, facilitates disease outbreaks, and destroys the habitats of wildlife, crippling their ability to perform ecosystem services needed by people.

Although the time and order vary somewhat from country to country on a global scale, written records seem to show a common pattern for humans and wildlife. Partly because human populations start and sometimes remain low, and perhaps also because of local knowledge underwritten by an understanding of nature described by Wilson as biophilia (1984), the pattern begins with a period in which conflicts with wildlife are mostly localized and often short-lived. For some societies, this stage continues to the present day. For others, however, there follows a period in which human populations expand, urbanize, and sometimes industrialize, dramatically increasing impacts on wildlife and other natural resources. Some societies then enter a stage in which the limits of natural resources are recognized and concentrated efforts are made to conserve and sustain these resources. From this perspective, **conservation** of wildlife by societies with high populations of people seems a more recent effort of humans, often centered around public policy and law, as has the understanding that successes in conservation can themselves exacerbate conflicts. Examples in North America include restoration, with concomitant increases in damage, of game species such as deer, bear, or river otter. Global examples include protected areas for wildlife and consequent damage by protected species to surrounding local communities (e.g., Distefano 2005; Treves et al. 2009).

Beginning long before our advent as hunters in the Pleistocene (see table 2.1) and hardwired into our brains during the long eons when our early hominid forebearers were mostly prey, we inherited instinctive fears of creatures such as snakes and wolves (Öhman 1986). We can suffer occasional nightmares wherein we are chased by a tiger, a bear, or an eerie creature of the dark such as the chupacabra of Puerto Rico, the orang pendek of Sumatra, the shunka warakin of the Great Plains of the United States, or the waheela of the Northwest Territories of Canada. These, too, are part of our heritage, influencing our perceptions of wildlife

and therefore how we respond to human-wildlife conflicts (Castillo-Huitrón et al. 2020).

Examples

ANCIENT CIVILIZATIONS. Ancient Egyptians interacted with and had many conflicts with wildlife, from fleas and commensal rodents in their homes to crocodiles in their streams, birds in their fields, and lions in their pastures. The snake may have been introduced into homes and granaries to control rats (Talbot 1912). Brock (2005) summarized some of these ancient conflicts, as paraphrased in the Bible's book of Exodus, caused by irruptions of frogs, locusts, lice, and flies in Egypt. He provides the inscription in Menna's tomb (1400 BC): "The snake has seized half the grain, and the hippopotami have eaten the rest. Mice abound in the fields, the locusts descend, and the herds devour; the sparrows steal—woe to the farmers! The remains on the threshing floor are for the thieves" (Brock 2005, 8). Herodotus referred to children (and possibly priests) having their heads shaved to prevent lice and fishermen's use of casting nets as bed nets to exclude gnats (Thamis 2012). **Rinderpest**, a cattle plague that moves between livestock and wildlife and was recently extirpated from the planet, may have been present in Egypt since the third millennium BC (Spinage 2003).

Fleas (*Xenopsylla cheopis*) carrying bacilli that cause plague were found in the trash of an ancient workmen's village at Amarna, Egypt. According to Panagiotakopulu (2004), the fleas may have coevolved with the Nile rat (*Arvicanthis niloticus*) and then passed on to the black rats (*Rattus rattus*) in cities during Nile floods. If so, plague originated in ancient Egypt and spread from there throughout the ancient world. As additional evidence, Panagiotakopulu (2004) cited the squalid working conditions at the workmen's site and the descriptions of epidemics from Amarna letters, Hittite archives, and Ebers papyrus (including references to swelling buboes).

According to Dollinger (2000), sparrows and other songbirds attacked the ripening grain of Egyptian farmers. The hieroglyph for sparrow meant "common" and "small" but also "bad." Nets were sometimes thrown over trees or held up by poles to capture and destroy the birds. Boys threw stones or used slingshots to protect crops from birds. Rats and mice got into homes, which were often made of soft (unfired) clay that was easily gnawed. Archeological remains sometimes show rocks or stones placed in walls to block burrows. Petrie, a British Egyptologist working in Kahun between 1888 and 1890, found what may have been a pottery rat trap (Reeves 1992).

Rabies has been known from ancient times and written records of preventive measures date to as early as 1930 BC. In the Code of Eshnunna, owners were advised to confine dogs suspected of having rabies and to anticipate fines at different levels if aristocrats or slaves were bitten and infected. Although the viral nature of the disease was not understood in a modern-day sense, the saliva of the dog was thought to contain semen that transferred an organism carrying the disease (Tierkel 1975).

AUSTRALIA. The dingo (*Canis lupus dingo*), which likely originated as a feral dog, was brought from southern Asia about 5,000 YA and was followed by extinction from Australia's mainland of the Tasmanian hen (*Tribonyx mortierii*) and apex predators, including the thylacine (*Thylacinus cynocephalus*) and the Tasmanian devil (*Sarcophilus harrisii*). Dingoes and hunting by humans have been implicated in these extinctions (e.g., Johnson and Wroe 2003).

The list of invasive species released into Australian ecosystems is impressive and includes domestic cats (*Felis catus*), first released from Dutch shipwrecks in the 1600s with feral populations established by the 1850s; domestic dogs (*Canis lupus familiaris*), donkeys, goats (*Capra aegagrus hircus*), horses (*Equus ferus caballus*), and pigs (*Sus scrofa*), which came with the First Fleet and subsequent landings; the European red fox (*Vulpes vulpes*) in 1855 for hunting; the European wild rabbit (*Oryctolagus cuniculus*) also for hunting; and the camel (*Camelus* spp.) between 1840 and 1907 for riding and as a draft and pack animal for rail and telegraph construction as part of the effort to open up arid areas in central and western Australia.

Those species are just a start. Others include carp (*Cyprinus carpio*), first released in the 1850s though the "Boolara" strain released in the 1960s has had the greatest ecological impact (Koehn et al. 2000); the common myna (*Acridotheres tristis*), a member of the starling family, released in cane fields of Queensland in 1883 as a biological control for plague locusts, cane beetles, and other insects; the European starling during the late 1800s, with populations taking root in Victoria (1856–1871), New South Wales (1880), and South Australia (1881) so that most of Victoria was colonized by the 1950s; cane toads (*Bufo marinus*), native to South and Central America, in 1935 to control beetles that were infesting sugarcane crops; and over 220 plant species, over half of which were introduced as ornamentals.

Many of these species quickly became feral, establishing viable populations throughout Australia. Many, such as domestic dogs, the dingo, their hybrids, and the fox, became serious predators affecting the sheep, goat, and cattle industries. Others have become suburban and urban pests or pests of cropland. Many have impacted native Australian fauna and flora, raising concern for the long-term conservation of these resources.

Responses of the Australian people (see, e.g., West 2018) and their governments to these situations have also been impressive. One was the construction of fences, including rabbit fences beginning in the 1890s, to block the movement of wild rabbits from western Australia, and a dingo fence, starting in the 1880s, that extends about 5,320 kilometers (3,306 miles) from Jimbour, Darling Downs, to Eyre Peninsula, Great Australian Bight. The fences, which have been maintained, have affected both intended and some unintended species.

EURASIA. European fairy tales and other folklore abound with concerns about wildlife (e.g., the Big Bad Wolf), as do similar tales from other parts of the globe. Medieval Europeans developed poisons for rats (e.g., ratsbane) and wolves (wolfsbane) and built an array of wooden traps, sometimes elaborately designed. Rat traps were mentioned in the medieval romance *Yvain, the Knight of the Lion*, by Chrétien de Troyes, written in the 1170s.

Birds were hunted during medieval times in Europe using box and access traps, snares, nets, lures, techniques such as "dogging," and poisons such as birdlime. Dogging involved luring large numbers of birds onto water and into nets. Box and access traps usually involved propping a box on a stick with a pile of grain under it. A string attached to the stick was pulled to dislodge the stick and catch a bird. Traps with doors that opened in only one direction were also used. Netting was used with great variations in application; a common technique was to lure birds onto the ground where nets were set to be drawn over them as they fed (fig. 2.2). Birdlime, a glue made from boiled mistletoe berries and holly bark, was smeared on rods, in cones, or on layers of linen or grass; when birds fed on grain or other baits, they would stick to the birdlime. Sometimes bread soaked in beer or wine was used to stupefy birds so they could be captured. Alternatively, poisonous mixes, including henbane and foxglove, were used to stupefy birds (Porta [1658] 1959). Although putting food on the table was probably the main focus of these hunting efforts, the same methods were undoubtedly applied to reduce damage to crops or property.

NORTH AMERICA. The Inuit made "spring-baits" of coiled baleen held by sinew and covered with blubber. Upon ingestion and digestion, the baleen would spring straight, cutting the stomach of a scavenger, which

Figure 2.2 Bird netting technique illustrated by a fourteenth-century medieval artist. From Taccuino Sanitatis, fourteenth century. http://commons.wikimedia.org/wiki. File: 34 caccia tortore, Taccuino Sanitatis, Casanatense 4182.

would then die of internal bleeding. Reference to these spring-bait traps was made in an account by English explorers; on visiting a central Inuit village, the English had been troubled by "13 ravenous wolves that proved troublesome to the English and Eskimos alike. The wolves did not attack people but were adept at seizing anything edible, including unguarded dogs" (Oswalt 1999, 173).

Two species, the mountain lion (*Puma concolor*) and the timber wolf (*Canis lupus*), were a ubiquitous part of colonial life in the United States. These species, accustomed to feeding on deer and other wildlife, found the fat and instinct-challenged livestock of the colonists to be easy prey and took them readily. Many colonists feared the predators would also attack them or their children. Many saw it as a God-given responsibility to tame the wildness of colonial America. William Wood ([1634] 1865) said of the timber wolves "they be the greatest inconveniency the Countrey hath, both for matter of dammage to private men in particular, and the whole Countrey in generall" (27). Nineteen years after the landing of the pilgrims in 1624, William Bradford (1856) still saw the timber wolf as one of the greatest threats to successful colonization, stating his hope that "pyson, traps and other such means will help to destroy them" (163). In this context,

therefore, the mountain lion and the wolf were killed both to eliminate predation on livestock and for public safety.

Wolf pits surrounded many colonial towns. The pits had steep sides and often sharp stakes on the bottom. The pits were covered with light poles and leaves that gave way under the weight of a wolf or occasionally under that of an unwary traveler. Josselyn ([1674] 1986) described how four hooks were tied with thread, wrapped in wool, and dipped in melted fat to form a ball. The balls were placed near recent wolf kills. Sometimes a gun or pistol was tied to a tree to be fired by the pull of a trip cord or a baited line leading to the trigger. In addition, wolves were caught with ropes from horseback and brought to town tied to the backs of horses. After wolves were relegated to swamplands or rocky areas as denning sites, wolf drives were used to further reduce their attacks on livestock. Townsmen carrying firearms or pitchforks would surround the denning area and advance toward it, killing the wolves that failed to escape.

Summary

- Human evolutionary milestones included predatory toolmaking, hunting, migrating, warring, trading, transporting, domesticating, herding, farming, urbanizing, industrializing, facilitating global warming, and conserving wildlife and other natural resources.
- Each milestone affected human relationships with wildlife and human-wildlife conflicts. For example, trading and migrating affected wildlife habitat and led to exotic invasive species. Hunting led to impacts on wildlife populations. Domestication led to wildlife diseases and zoonoses as well as feral wildlife. A pastoral lifestyle led to zoonoses, shepherding, and guard animals. Agriculture led to habitat loss and fragmentation, and insect, disease, and wildlife damage, plus pest problems and irruptions from simplified agroecosystems. Establishment of cities and towns led to disease transfer within human populations, pressure on natural resources and ecosystems, reduced biotic diversity, and pest and damage irruptions and diseases.
- Responses to human-wildlife conflicts also evolved with humans. From a prey species acting in self-defense, humans became intelligent, socially organized, and weapon-bearing predators. Human societies evolved to shepherding and breeding and using guard animals; to developing sophisticated systems for the management of pests and disease

in managing crops, forests, livestock, and human diseases; and to eventually leading an effort for broad conservation of wildlife species.

- Having begun long before humans were hunters and hardwired into our brains, our heritage as prey manifests itself today as instinctive fears of creatures such as snakes and wolves or occasional nightmares in which we are chased by a tiger, a bear, or eerie creatures of the dark.

Review and Discussion Questions

1. What do you think were hominids' first human-wildlife conflicts? Altering wildlife habitat? Saving human life and limb? Protecting crops, caves, or other wildlife? Defend your view by including appropriate references.

2. Relate the notions of the Pleistocene extinction of mammals to the ecology of apex predators, keystone species, and foundational species. Relate "rewilding," proposed by some conservationists, to these notions.

3. It is said that "history teaches prudence." How might history teach us prudence in resolving human-wildlife conflicts? Give a specific example.

4. How might past activities of humans have influenced the development and expression of today's conflicts with wildlife? Give some specific examples.

5. Do humans have an instinctive fear of some wildlife? How might such a fear influence how people perceive and work to resolve human-wildlife conflicts? Support your view with a specific example.

3

Resources

I review the resources available to those resolving human-wildlife conflicts.

Statement

A vast scientific and secondary literature as well as practical experience and local knowledge are available globally to assist those resolving human-wildlife conflicts. This knowledge is accessible through private and public experts, local citizens, citizen science, libraries, databases, and the internet.

Explanation

The individuals with modest to great interest in resolving human-wildlife conflicts include students, practitioners, and others. Some are students enrolled in undergraduate or advanced natural resources programs, perhaps studying forestry, wildlife management, or human-wildlife conflict specifically at colleges and universities around the world. Some are professional natural resource managers and members of public or nongovernmental natural resources and conservation organizations. Some work for zoos or wildlife rehabilitation centers. Others are farmers, ranchers, landowners, or homeowners trying to resolve their own human-wildlife conflicts, or members of the public who just want to participate in resolving such conflicts.

The available resources include the primary scientific literature and secondary literature. **Primary scientific literature** refers to original reports of studies following the **scientific method** and related observational or experimental designs (fig. 3.1). By using the scientific method, researchers follow procedures that attempt to minimize the opportunities for personal or other biases that might lead to misinterpretation of results and to maximize the odds that the results observed are due to the factors being controlled and studied. The scientific method is a series of steps followed sequentially (in theory, though sometimes not in practice): gather background information and current literature surrounding an issue; form a hypothesis and its negative form, the "null hypothesis"; develop an experimental or observational design to test the hypotheses; conduct experiments and/or observations; gather data; analyze the data

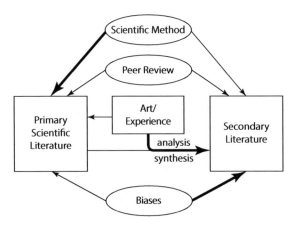

Figure 3.1   Relationships between primary and secondary literature in managing human-wildlife conflicts. Illustration by Lamar Henderson, Wildhaven Creative LLC.

following appropriate statistical methods; interpret the results; and communicate the findings to the rest of the world. Although the scientific method does not eliminate bias, misanalysis of data, or misinterpretation of results, the method is one of humankind's better attempts at objectivity. Primary scientific literature therefore tends to be a useful initial source of information for both the student and the practitioner.

An important aspect of the scientific method, not revealed by simply listing and describing the procedural steps, is critical review by peers who have knowledge and experience with the subject. For example, findings may be presented at scientific meetings where methods or results may be challenged by peers. If the findings are to be published in a scientific journal, the journal editor sends the manuscript to at least one and often two or more knowledgeable peers for critical review. The peers are invited to challenge every aspect of the study, from inclusivity of appropriate background information and current literature to formation of appropriate hypotheses and the correctness of the experimental design, statistical analyses, and interpretation of results. Language is criticized, especially the use of technical words, "scientific jargon," and lack of clarity. The reviewing peers recommend to the editor acceptance, rejection, or revision of the manuscript. If revision is required, the editor may subject the revised manuscript to a second **peer review**. The final decision on publication or rejection lies with the editor, who is also trained and experienced in scientific methodology.

In addition to journals, original scientific reports may be published elsewhere—in proceedings of professional meetings, in books based on topical symposia, or in governmental technical reports. The results might be made into extension educational publications or placed on websites for access by the public. Because there are few scientific journals devoted specifically to human-wildlife conflicts and because much research has been done throughout the world, the primary literature tends to be scattered globally.

Accessibility to scientific literature is greatly facilitated today through databases on wildlife, wildlife management, and related subjects, including human-wildlife conflicts specifically, and on specialized areas within it such as wildlife damage management. Such databases may be provided by private vendors, professional associations, educational institutions, governmental agencies, and international organizations. The databases often provide sufficient information for the student or practitioner to obtain an article from its published source, from the author, or from an agency or institutional website. Increasingly, the databases are electronic, connected through university or college information networks, and include electronic copies of the articles.

As an example of databases available through an international organization, the International Union for Conservation of Nature (IUCN), Species Survival Commission (SSC), Human-Wildlife Conflict Task Force, in partnership with People and Wildlife, maintains databases on human conflicts with "key species"—including elephants, bears, cheetahs, crocodiles, jaguars, leopards, lions, lynxes, otters, primates, raptors, sharks, snakes, snow leopards, tigers, and wolves. Included are references to scientific articles. This resource is available through the internet at https://www.hwctf.org/document-library. To find these references, look under "Resources," then the wildlife of interest. References are also sorted by "key topics" such as electric fences, compensation, or other financial instruments.

The Berryman Institute, Utah State University, publishes the international journal *Human–Wildlife Interactions*. The journal serves the needs of professionals in human-wildlife interactions, including human-wildlife conflicts and wildlife damage management. The journal encourages conversation in these areas and serves as a repository for information. It is accessible via the internet at https://digitalcommons.usu.edu/hwi/.

The National Wildlife Research Center (NWRC) is the only US federal research center whose mission is focused on human-wildlife conflicts. It maintains a database of "research articles, reports, factsheets, technical notes, data, and other materials authored or co-authored by NWRC scientists and colleagues." The database is available online for the public and can be accessed as the "NWRC Research Gateway" (https://www.aphis.usda.gov/aphis/ourfocus/wildlifedamage/programs/nwrc/SA_Publications/CT_Research_gateway).

The Digital Commons at the University of Nebraska in Lincoln, the Internet Center for Wildlife Damage Management (https://digitalcommons.unl.edu/icwdm/) has historic databases on human-wildlife conflicts that include some proceedings of symposia no longer commonly available. For example, the *Bird Control Seminar Proceedings* (1964–1983) can be found there. Individual articles in the database, identified as part of the Nebraska collection, can be located by using Google or other search engines on the internet.

Scientific literature may not contain the specific information needed for a management situation. Further, scientific literature may lack information and findings related to the "art" or the more experiential aspects of observing or managing human-wildlife conflicts (see fig. 3.1). Such information may include the practical experience and knowledge of professionals as well as "**local knowledge**" (e.g., Guerreiro, 2019, Sillitoe et al. 2019, Ivașcu and Biro 2020). It may be based on years, even generations, of experience and firsthand knowledge of both the issue and its resolution. Such information may lead to, indeed may be requisite to, particularly effective management actions.

Fortunately, human-wildlife conflict has abundant **secondary literature**: accumulations and summaries of information that often provide unique insights into conflicts and their resolution. Secondary literature is a place where one can find experience and knowledge relating to the art of understanding and resolving human-wildlife conflicts (see fig. 3.1), often synthesized with and underwritten by the scientific literature. The literature may be general or focused on specific conflicts. It may be packaged as professional books, textbooks, control manuals or pamphlets, videos, webcasts or documentaries, databases (as described earlier), popular writings in newspapers or magazines, classroom lectures and other materials, or research-based extension educational activities that might include experiential/hands-on demonstrations, personal instruction, publications, records of dialogues with local residents, and websites or podcasts.

Secondary literature tends to be focused on the practical aspects of conflict management. Management manuals are widely available through many sources. For example, the IUCN, SSC, Human-Wildlife Conflict Task Force, in partnership with People and Wildlife, referenced previously for its databases of primary literature, also maintains databases of secondary literature, including useful management and training manuals on conflicts with the key species. Governmental agencies, such as Wildlife Services in the United States, also offer literature such as manuals, brochures, and pamphlets, as do the state or provincial wildlife agencies and federal and state extension programs. International or national nongovernmental organizations such as the World Wide Fund for Nature (known as the World Wildlife Fund in the United States and Canada), the National Audubon Society, or Defenders of Wildlife also serve as sources of secondary information on human-wildlife conflicts. Again, the internet serves as a conduit for much of this information.

Secondary literature is usually subject to careful review and editing, often by peers experienced with the specific topic. However, these reviews can lack vigor regarding application of the scientific method. In fact, the review and editing may include systematic inclusion or exclusion of information based on a particular view held by the author, the publisher, or the sponsoring organization, whether private or governmental. With secondary literature, the reader should be particularly wary of information based only on anecdote, or that follows exclusively the view or policy of an organization, or that is provided to support a particular cultural, social, religious, economic, or political perspective. The student and practitioner alike are cautioned to uncover such biases and consider them carefully before taking a management action that could be misguided.

Operational programs, in which experts do the management, offer the advantages of both scientific base and artful experience because the management is practiced directly by professionals on behalf of the landowner or natural resources agency. Professional managers include those employed by for-profit entities as well as those working for governmental or nongovernmental private agencies or nonprofit associations. The entity itself may be any of a broad range of institutions, e.g., a university, a natural resource management agency, a zoo, or a wildlife rehabilitation center. Within the constraints of the policies, politics, religion, and economic situation of the individual or organization, the manager provides the most appropriate resolution of a human-wildlife conflict that his/her experience and knowledge allow.

Citizen science, the public participation of nonscientists in scientific research, has taken an increasingly important role in understanding and managing human-wildlife conflicts. Because citizen scientists often volunteer their time and there may be many volunteers, citizen science provides an opportunity to gather data on issues at a scale often otherwise unaffordable to the research scientist or a granting organization. Trained carefully and working within understood limitations, the participants can gather data used for testing hypotheses and scientific publication.

Further, citizen scientists may sometimes participate in management actions, gaining experience and local knowledge regarding conflicts that may affect their attitudes and willingness to engage with the broader community. (See, e.g., Shirk et al. 2012, for a general framework; Toomey and Domroese 2013, for the Earthwatch Coyote Project in New York City; Larson et al 2016, for a project to mitigate human-wildlife conflict in Sierra Leone; and Drake et al 2021, for human-coyote conflicts in Madison, Wisconsin.) Bonnell and Breck (2017) used resident-based hazing to evaluate the effectiveness of hazing coyotes in the Denver Metro Area in Colorado. The program produced 207 trained citizen scientists who generated 96 documented hazing events. Further, the citizen scientists reported "improved understanding and acceptance of coyote management tools as well as increased confidence and capacity to deal with human–coyote conflict in their communities" (148).

As another example, citizen scientists have played a key role in our understanding of the distribution of the invasive lionfish (*Pterois*) along the southeastern coast of North and Central America. Citizens already trained in scuba diving have contributed vacation time to participate in transect and timed surveys along coastal waters that have documented the exploding population of this invasive species, saving funding while contributing valuable information (Scyphers et al. 2015) (fig. 3.2). Some citizen scientists, e.g., spearfishers, are also participating in the removal of lionfish.

How does one filter out incorrect information while keeping the useful? How does one avoid the waste of time, energy, and expense associated with misguided or ineffective methods of control? Peer-reviewed scientific information available through primary literature seems an essential starting point. Reliance on experienced and knowledgeable expertise available through the secondary literature and other media is also important to ensure that both the art of resolving human-wildlife conflicts is incorporated and the information used is reliable. Beyond this, the student and the practitioner alike are left to their own prowess.

Examples

ELEPHANTS AND BEE FENCES. I will illustrate a search for relevant information on a human-wildlife conflict and a solution by searching for information on elephants and bee fences. I began by looking for primary scientific and secondary written literature.

Although Google opens a broad range of article types, Google Scholar tends to narrow the field to mostly primary literature. For example, a Google search for "human-elephant conflicts" yielded 5.4 million results on January 15, 2021. The same search using Google Scholar yielded 5,760 results, about one-third (1,970) of which were posted in 2017 or later. The more refined search for "elephants beehive fences" provided 758 references on Google scholar. Abstracts or full texts of many of these articles were available for free (often through university subscriptions for students or faculty) or for a charge.

I then tried some other sources, looking for focused information. For example, for the African elephant the website for the IUCN SSC Task Force on Human-Wildlife Conflicts listed nine "Guidance" documents, including a field guide on safety around elephants, a field guide on protecting a field from elephant damage, a decision support system for managing human-elephant conflicts, and an assortment of manuals related to training programs. The site included a link to one downloadable guiding document on constructing a beehive fence. The site referenced 56 "Key Papers" and four chapters in books, two of which specifically mentioned beehive fences in their titles, both authored by Dr. Lucy King.

Information gained in this way then served to launch a more probing exploration. In this instance, I joined an online forum for People & Wildlife (with permission of the group), which I expect will be a source of additional detailed information on elephants and bee fences as well as related information. One might also gather more detailed information at websites such as Save the Elephants: Elephants and Bees Project (https://elephantsandbees.com/) or by contacting directly some of the scientists or staff listed in the published literature. For instance, one might want more information on Dr. King's statement that local knowledge played an important role in the development of bee fences.

Secondary literature in the form of popular articles, audio recordings, and videos also abound on human-wildlife conflicts in sources such as newspapers, magazines, television shows, podcasts, websites, and other

Figure 3.2 Invasion of lionfish (*Pterois*), from their discovery along the Florida coast from 1995 (top) to 2015 (bottom). Citizen scientists helped gather the data, saving funding while contributing valuable information. Data provided by the National Oceanic and Atmospheric Administration and the US Geological Survey (nas.er.usgs.gov).

popular media. In searching Google using the keywords "elephants and bee fences videos," I found nine videos listed as of January 15, 2021, including one presented by Dr. Lucy King (2019): "How Bees Can Keep the Peace between Elephants and Humans."

COMMERCIAL COMPANIES. Rollins, Inc., is an example of an international enterprise that offers worldwide pest control and wildlife control services. Its subsidiaries include entities such as Orkin, Orkin Canada, Critter Control, and Aardwolf, which offer services to over 2 million customers in the United States, Canada, Australia, Europe, and Asia. Additional services are provided by Orkin and Critter Control franchises in the United States, Canada, Mexico, Central and South America, the Caribbean, Europe, the Middle East, Asia, Africa, and Australia, totaling over 700 locations.

Other enterprises offer similar services throughout the world. Leo Technova India is an example of an entity offering bird and other wildlife management services in India.

GOVERNMENT ORGANIZATIONS. The major federal agency of the United States focused entirely on research and management expertise in human-wildlife conflicts, particularly wildlife damage management, is Wildlife Services. However, many other agencies also offer expertise in some aspects of human-wildlife conflicts and their resolution, including the US Fish and Wildlife Service (USFWS), the US National Park Service (USNPS), the US Forest Service (USFS), the US Geological Survey (USGS), and the US Natural Resources Conservation Service (USNRCS).

Resolving human-wildlife conflicts is often included as part of the programs offered by state agricultural cooperative extension services directly linked to land-grant universities. Most state fish and game agencies offer such services, either as technical programs or as direct assistance programs.

Other US federal agencies, such as the Federal Aviation Administration (FAA) and the National Aeronautics and Space Administration (NASA), have their own expertise or contract for conflict management. Other countries have analogous agencies, as appropriate for their own political structures.

NONPROFIT ORGANIZATIONS. In addition to for-profit entities and businesses, many not-for-profit, nongovernmental organizations also offer expertise in understanding and managing human-wildlife conflicts. These include entities ranging from conservation organizations, zoos, and wildlife rehabilitation centers to funding organizations established by for-profit businesses. Examples include Defenders of Wildlife, the Sierra Club, and the Audubon Society.

Many zoos have expertise in human-wildlife conflicts and are involved in programs globally. For example, the Oakland Zoo has provided expertise and supported fund raising for a human-carnivore conflict management program in Uganda that is community based.

The Disney Conservation Fund has supported the resolution of a wide range of human-wildlife conflicts, including using bee fences to discourage conflicts with elephants (Baxter 2016; Baxter and Hancock 2020). Their support includes expertise from Disneyworld scientists. For example, Soltis, who leads the bioacoustical unit at Disneyworld's Animal Kingdom, is also on the staff of Save the Elephants: Elephants and Bees Project, and has contributed to their scientific understanding of the meaning of elephant "grumblings" and body language (e.g., Soltis et al. 2016).

Wildlife rehabilitation centers sometimes offer expertise in managing human wildlife conflicts as well. For example, the Sonoma County Wildlife Rescue, a not-for-profit entity in California, offers barn owl houses as well as an installation and maintenance program in addition to its central focus on rehabilitation of wildlife at the center.

PRODUCTS. Products are also available to the public, including traps, pesticides, chemical and electronic repellents, scare devices, pyrotechnics, and exclosure nettings for birds and mammals. Again, the student and the practitioner need to be wary of questionable claims. For example, some ultrasonic or electromagnetic devices are sold to the public with claims that they repel only unwanted insects or rodents, not desirable animals or pets. Such claims are mostly devoid of any scientific basis and are suspect.

Summary

- Students include those in colleges and universities who study human-wildlife conflicts in designated courses or as part of related courses.
- Practitioners include professional managers associated with public and private organizations with responsibility for or interest in natural resources management and human-wildlife conflicts specifically, as well as farmers, ranchers, landowners, and homeowners with a conflict and interested members of the public, including citizen scientists.
- Primary literature includes mostly publications that result directly from peer-reviewed studies following the scientific method. With human-wildlife conflicts and their resolution, the literature tends to be scattered globally.

- Secondary literature includes written, audio, or video forms and synthesizes scientific information with practical experience of experts and sometimes local knowledge. It is packaged as professional books, textbooks, control manuals or pamphlets, videos or documentaries, databases, popular writings in newspapers or magazines, classroom lectures and other materials, and research-based extension educational activities that might include experiential/hands-on demonstrations, personal instruction, publications, records of dialogues with local residents, websites, and podcasts.
- Operational programs, in which experts do the management, offer the advantages of both scientific base and artful experience, in that the management is practiced directly by professionals on behalf of the landowner or natural resources agency.
- Access to resources is facilitated through local, regional, and international organizations, public or private, for-profit or nonprofit. Such entities include colleges and universities, governmental service and extension education programs, public entities such as zoos and rehabilitation centers, and private-sector businesses.
- The internet is often a good place to begin accessing information on human-wildlife conflicts. Use of scientifically based information is an essential starting point. Reliance on experience and expertise found mostly in secondary literature is also important.
- Cleverness and scrutiny in separating useful information from biases based on culture, society, economy, religion, and political views are essential. To this end, the student and the practitioner are left largely to their own judgment.

Review and Discussion Questions

1. Explain why you are unlikely to find anecdotal recommendations for solving human-wildlife conflicts in primary literature.
2. Explain why you are unlikely to find tips, based on personal experience, on removing netting from entangled whales in primary literature.
3. You have been told that bees effectively repel African elephants from fields of crops and gardens. Do bees do that? Base your opinion on both primary and secondary literature.
4. You have been told that Wrigley's Spearmint gum effectively repels the eastern common mole (*Scalopus aquaticus*). Does it? Base your opinion on both primary and secondary literature.
5. You have been told that mice effectively repel African elephants from fields of crops and gardens. Do mice do that? Base your opinion on both primary and secondary literature.
6. You have been told that ultrasonic whistles attached to automobiles effectively prevent collisions with deer. Do these whistles work? Base your opinion on both primary and secondary literature.
7. Find a website that contains useful information on the management of damage by larger mammalian predators but carries biases. What is the citation? What are the biases? How might you use this information?
8. Using the internet, locate literature from somewhere you would consider remote from where you live. For instance, an American student might check publications of the Bangladesh Agricultural Research Council or Institute. Does the source offer publications on human-wildlife conflict?

PART II • BIOLOGICAL AND ECOLOGICAL CONCEPTS

In this section, I probe the concepts and principles of individual, population, and community ecology that underpin human-wildlife conflict (damage), its assessment, and its management.

4

Organismic and Species Systems

I describe habitat and ecological niche, sympatry and ecological equivalents, domestication and feral wildlife, biological clocks, and behavior as attributes of organismic and species systems that help to understand human-wildlife conflicts and underwrite their resolutions.

HABITAT AND ECOLOGICAL NICHE

Statement
Habitat and ecological niche describe places where organisms can be found and the roles of organisms in the environment and help practitioners understand, anticipate, and manage human-wildlife conflicts.

Explanation
Habitat is where an organism, species, or population lives. The broadest ecological term it encompasses (excepting the biosphere) is a **biome**, an assemblage of organisms distributed over an extensive area, such as a tropical monsoon forest or temperate grassland. One can also describe the spring habitat of the American robin (*Turdus migratorius*) as open meadows, often adjacent to deciduous forests. Depending on context, "habitat" here refers to the robin as a species, a population, or an individual organism. Human-wildlife conflict practitioners use habitat in all of these senses.

Ecological niche is a larger concept than habitat. **Ecological niche** is described as the multidimensional array of environmental factors (i.e., the address) within which an organism must live and how it interacts with its environment (i.e., its role or function). For example, some plants live in a constructed wetland and act to remove minerals and excessive nutrients as part of a water treatment program; the address and role combined describe the ecological niche. Muskrats (*Ondatra zibethicus*) may also live in the wetland and remove plants. Here, the address and role of the muskrat constitute its ecological niche, although the role may disfavor people and be viewed as damage.

The breeding habitats of weaver finches (*Lonchura* in Asia and *Quelea* in Africa) and red-winged blackbirds (*Agelaius phoeniceus*) are marshes and swamps near ponds, lakes, estuaries, or river deltas. However, the ecological niche of weaver finches includes their role as grain eaters, foraging extensively on rice during its milky-dough stage of development in Southeast Asia and Africa. The niche of the blackbird includes its roles as a grain eater, feeding on rice, corn, or sunflower broadly in North and South America during emergent and maturing stages, and as an insectivore that sometimes benefits humans.

An organism's **fundamental** or **theoretical ecological niche** includes all possible roles in the broadest possible habitats. Although most organisms never achieve the full potential of their theoretical ecological niche (i.e., some interactions between the individual organism and other biota reduce the size of the fundamental niche, resulting in the organism's **realized ecological niche**), the concept is useful. For example, theoretical ecological niche may describe the broad damage caused by newly invasive species, unrestrained by interactions of other organisms in a new ecosystem.

Organisms interact with both the **abiotic** (nonliving) and **biotic** (living) components of their environment. Interactions with abiota are described by Liebig's law of the minimum and Shelford's law of tolerance. **Liebig's law of the minimum** states that the material available in lowest quantity in relation to need will limit the growth of an organism. Liebig was a botanist concerned mostly with plants and nutrients, but the concept applies as well to other organisms. **Shelford's law of tolerance**, often termed **range of tolerance**, states that excessive amounts as well as deficiencies can limit the habitat within which an organism can live. Again, the concept works well with plants but can be applied to other organisms. An application to human-wildlife conflicts is the use of climatic niche models to assess the potential range expansions of exotic species invading new habitats. Another is the **exclusion** of wildlife from habitats such as airports and landfills by reducing one or more factors needed for their presence (such as food or water).

An organism may be an ecological generalist and live within a niche having broad ranges, or it may be an ecological specialist and live within a niche that is defined by a narrow range of tolerance for one or more factors. The prefix "**eury**" describes a broad range or function and "**steno**" a narrow one. Thus, **eurythermal** organisms live within a broad range of temperatures, whereas **stenothermal** organisms are restricted to a narrow range of temperatures. Likewise, **euryhalic** organisms tolerate a broad range of salt concentrations while **stenohalic** organisms live within a narrow range, perhaps in brackish marshes. Range of

diet is described by adding "**phagic**," meaning "to eat." **Euryphagic** organisms have broad diets whereas **stenophagic** organisms have specialized ones.

Some stenophagic wildlife affront people—for example, the great economic and health costs due to the **sanguinivorous** habits and transmission of rabies by vampire bats in Central America. As a rule, however, eury wildlife species are most likely to conflict with people. Commensal rodents have some of the broadest ecological niches among mammals. Coupled with a high reproductive capacity, commensal rodents are among the most damaging, both historically and globally, of wildlife species today. Other generalist mammals that sometimes affront people include bears (*Ursidae*), coyotes, and jackals. Although classified as carnivores, these mammals have broadly defined niches, are widely distributed, and are euryphagic, eating a broad range of plants and animals as opportunities allow. The feral pig, another euryphagic mammal, is often in conflict with people.

Among birds, blackbirds, starlings, and weaver finches tend to feed preferentially on grains but take insects and other plant materials given the opportunity. These adaptable birds live within broad niches and are distributed globally. One weaver finch, the red-billed quelea (*Quelea quelea*), is one of the world's most costly vertebrates because of the damage it causes to cereal crops in Africa (e.g., Bruggers and Elliott 1989; Cheke and Sidatt 2019).

Fishes such as carp and suckers, which can tolerate extreme ranges of chemicals and nutrients, particularly low levels of dissolved oxygen, tend to be more broadly distributed in warm, freshwater ponds, lakes, and streams, sometimes replacing more desirable game or ecologically important species.

Niches differ not only among individuals and species but also with factors such as stage of metamorphosis, sex, stage of reproductive cycle, season, and geographic location. The ecological niche of a mosquito (*Culicidae*) larva is in stagnant water, feeding on nutrients and microorganisms within its size range, a euryphagic phase lasting from four days to a month or more for some species. During the pupal stage, mosquitoes cease feeding. The emerged female is stenophagic, seeking blood that provides protein for her first eggs, whereas the emergent adult male is short lived and **nectivorous** (eating nectar).

A practical constraint on using the niche concept to resolve human-wildlife conflicts is the unlimited number of physical and biotic factors that can describe niches. It is difficult to know if the most important factors have been discovered for a given organism. In practice, one focuses on factors that seem limiting, recognizing that a critical aspect may have been overlooked.

Examples

Kikillus et al. (2010) used data on climate from the present distribution and known breeding sites (163 locations) of the red-eared slider turtle (*Trachemys scripta elegans*) plus 12 alternative models to assess its "climatic envelope." The turtle is invasive, distributed globally as a pet. From the analyses, the researchers deduced that Southeast Asia still had large areas of unoccupied habitat suitable for the turtles, a prediction supported by at least one later study (Zhang et al. 2020). Sliders are farmed in China and are commonly found in Chinese markets, potential sources of feral sliders. Further, Kikillus et al. (2010) noted that religious ceremonies in Singapore resulted in vast numbers of sliders being released into the wild. The authors suggested other "hotspots" where effort might be made to detect the presence of sliders, thereby helping practitioners make strategic use of limited management resources.

SYMPATRY AND ECOLOGICAL EQUIVALENTS

Statement

Concepts of sympatry and ecological equivalents help to predict possible outcomes of efforts to suppress populations, efforts to introduce species for biological control, and unintended introductions of exotic species.

Explanation

When two species occupy the same niche at the same time and place, they are called **sympatric**. When they occupy the same niche but are located in different places, they are termed **ecological equivalents**.

For sympatric species, **Gause's law** applies: if two organisms (or species) occupy the same niche at the same time, one of the organisms (or species) will move, change its role, or die. If the organism (or species) changes its role, morphological or behavioral changes may also occur over time within the population, known as **character displacement**. Because direct competition is not occurring with geographically separated ecological equivalents, selective forces may move the species closer together in morphological, physiological, and behavioral characteristics, a process known as **convergent evolution**.

The mammalian family *Canidae* includes wolves, coyotes, foxes, dingoes, dholes (*Cuon alpinus*), various jackals, and the dog. Canids have ecological niches that overlap with each other and serve as ecological equivalents in geographic regions where they occur separately (fig. 4.1). In locations where members of this family are sympatric, the ecological functions of one or both of the species may change.

Ecological equivalence has obvious applications in understanding and resolving human-wildlife conflicts, e.g., in **restoration** (or rewilding) attempts (see, e.g., Pereira and Navarro 2015 for European examples). The concept is somewhat analogous to a habitat equivalency analysis sometimes conducted to mitigate habitat injury at the ecosystem level. Ecological equivalence can also help to predict whether an introduced species will become invasive. The absence of an equivalent species and the presence of an unoccupied niche would favor success of the introduced species—for example, mammalian predators on islands previously without predators. The presence of an equivalent spe-

Figure 4.1 The coyote (*Canis latrans*, above) and black-backed jackal (*C. mesomelas*, below) as ecological equivalents. Photos by USFWS/R. H. Barrett and ©Hans Hillewaert/CC-BY-SA-3.0, respectively.

cies would not necessarily ensure failure of the exotic; instead, the outcome would follow Gause's law, as with the introduction of dingoes onto Australia's mainland (see chapter 2).

Sympatry also has obvious applications. For example, wolves probably preyed on about half of the ducks produced in the prairie pothole region of North America prior to western expansion and settlement in the 1800s. After settlement, however, most wolves were removed, and one might therefore predict an increased duck population for the region. The duck populations actually remained steady, however, because sympatric red foxes, coyotes, and other, smaller predators assumed the ecological functions of the wolves. As surrounding habitat became increasingly agricultural, access to nesting waterfowl became even easier for smaller predators. With predator management, nesting success has reached 70%–90%; without such management, nesting success remains about 50% (Sargeant et al. 1993).

Example

Gosselink et al. (2003) radiomarked 28 coyotes, 16 rural red foxes, and 19 urban foxes in east-central Illinois. They utilized over 10,500 locations to gather information on habitat use, analyzing home range (the area within which the animal conducts its daily activities) from animal locations and resting sites within the study area. The researchers found that red foxes avoided coyotes by moving closer to humans and using less favorable habitats like tilled farmland, active farmsteads, and culverts. The authors saw urban areas as refugia where foxes could avoid coyotes, and urban landscapes provided more stable habitats for the foxes than did the rural landscapes. In a different study that looked at sympatry between gray fox (*Urocyon cinereoargenteus*) and coyote in the eastern United States, Egan et al. (2020) found that presence of the coyote itself had a greater impact on presence of the gray fox than did influences of urbanization. Data for the study came from camera traps, and citizen scientists allowed sufficient data to be collected to allow an analysis of the eastern United States. The situations illustrate Gause's law.

DOMESTICATION AND FERAL WILDLIFE

Statement

Domestication provided both guard dogs that help protect livestock from predators and feral wildlife that can injure people and other wildlife, transmit diseases, and damage property.

Explanation

Domesticated plants and animals are often bred for characteristics without regard for the ability to protect themselves. Domesticated plants tend to be susceptible to herbivory, and domesticated animals can be "instinctively challenged," prone to lose battles with wild predators. Domesticated species can therefore exacerbate human-wildlife conflicts and make management difficult. Even guard dogs and other guard animals, bred to protect themselves and livestock, are sometimes challenged by the capabilities of wild predators.

Feral wildlife result from release or escape of organisms into the wild after domestication and subsequent successful breeding and include feral pigs, goats, horses, camels, cats, and dogs. These formerly domesticated plants and animals, now wild again, can conflict with human interests. For example, a recent concern with feral plants is that they may allow the escape of genetically modified crops, making it difficult to contain the genetic modifications (see, e.g., Bagavathiannan and Van Acker 2008).

Examples

Examples of conflicts with feral wildlife appear throughout this book.

BIOLOGICAL CLOCKS

Statement

Understanding the biological clocks and rhythms of offending species can help practitioners' time management actions; disruption of such rhythms can be a management tool.

Explanation

Those of us who dislike the sound of an alarm clock often find that we can wake up just before it triggers and turn the alarm off. Possessing such internal **biological clocks** is common among organisms and underwrites circadian, lunar, and circannual rhythms.

Circadian (*circa* meaning "about," *diem* or *dies* meaning "day") **rhythms** are internal biological clocks based on 24-hour cycles in synchrony with the earth's daily rotation. These may be the most fundamental of rhythms. **Lunar rhythms** are based on 28-day cycles tied to the rotation of the moon around the earth. **Circannual rhythms** arise from yearly cycles tied to the movement of the earth around the sun, including both seasonal and annual changes.

Although they are internal, these rhythms are adjusted by external cues called **zeitgebers** (German for "time giver"). **Photoperiod**—i.e., day length—is detected by organisms, and changes in the photoperiod are the primary cues for daily to seasonal changes. Because the day length varies less near the equator, tropical organisms are particularly sensitive to small changes in day length. Organisms are classified by their responses to the photoperiod. Those active by day are called **diurnal**; by night, **nocturnal**; and by dusk or dawn, **crepuscular**.

Whereas photoperiod might stimulate physiological and behavioral preparations for circannual events such as hibernation, reproduction, molting, or migration, or budburst in plants, their actual onset may be initiated by secondary cues such as rainfall. Belmain et al. (2008), for example, related outbreaks of rodents in parts of Bangladesh to the flowering of bamboo, an event that may occur in cycles that span decades.

Cues and rhythms have been intensely studied (**chronobiology** or **ecophysiology**) over the years, but details of their relationships remain unclear. In experiments where animals were isolated from daylight, circadian rhythms continue but drift a bit longer or shorter each day. In such studies, subjects may still sense external cues that are difficult or impossible to eliminate, such as oscillations in magnetic fields or electromagnetic forces. Regardless of the source of cues, rhythms remain critical for the proper timing of many organismic activities, such as timing birth to the availability of food for young or timing daily activity to the availability of prey or lack of predators.

It would seem essential to ensure that circadian rhythms overlap before introducing a new predator to manage a prey animal. Yet the literature on human-wildlife conflicts attests that this fundamental consideration was sometimes overlooked or ignored. On the Marshall Islands monitor lizards (*Varanus indicus*) were introduced before World War II, probably for their skins and food but also to manage rats. Because the lizards were diurnal and the rats mostly crepuscular or nocturnal, the intended predators rarely met the intended prey. After the lizards began raiding chicken houses, giant toads (*Bufo marinus* or *Rhinella marina*) were introduced to control the lizards. As the rat populations continued to rise, both the monitor lizards and the giant toads became pests (Bennett 1995). Although photoperiod (diurnal lizards and toads, crepuscular and nocturnal rats) was not the only issue facilitating these consequences, it is one that could have been assessed a priori, and it would have weighed heavily against such introductions. The introduction of the small Asian mongoose (*Herpestes javanicus*) to control rats in Hawaii and other tropical islands in the Pacific had the same flaws and similar consequences; the mongoose turned to endangered birds and their eggs (Stone and Anderson 1988).

Although disruption of biological rhythms offers potential as a means of managing damaging species, and some laboratory studies have supported the concept, I know of few practical uses. Haim et al. (2007) found that altering the day length during winter impeded the thermoregulation of the social vole (*Microtus socialis*) in confined populations under natural conditions, a lethal effect for the voles. The researchers suggested that precision agriculture (e.g., the use of global positioning satellites to precisely locate burrows and assess microenvironmental conditions) might be used to pinpoint photoperiod alterations at active burrows. The social vole can cause extensive damage to alfalfa crops in Israel.

Examples

Sicard et al. (1999) studied the biological rhythms of four rodent pests in African habitats ranging from humid to arid. For each species and habitat, the researchers looked particularly for nonphotic zeitgebers, such as temperature, relative humidity, water, and chemical signals (e.g., signals that emanate from germinating plants), that stimulated endogenous circadian clocks. Their methods included both long-term field monitoring and laboratory studies. From these studies, Sicard et al. were able to predict when critical biological events would occur, in turn suggesting times when specific rodent management actions might be most effective. For example, the researchers were able to help the Sahelian-Sudanese better time the use of sound and physical or chemical barriers to coincide with phases when rodents were dispersing or regrouping in new areas (i.e., when rodents were most mobile).

BEHAVIOR

Statement

Modification of undesirable behaviors, both wildlife and human, underlies successful resolution of many human-wildlife conflicts.

Explanation

Behaviors have been classified by increasing complexity: **tropism**, general attraction to or avoidance of an environmental stimulus such as temperature or light intensity; **taxis**, attraction to or avoidance of environmental stimuli but in a more directed manner; **reflex**, response of an organism or part of an organism to an environmental stimulus, similar to but more sophisticated than a taxis and both modifiable with learning; **instinct**, unlearned response to an environmental stimulus consisting of a sequence of encoded, **stereotyped behaviors**, often observed with insects, amphibians, reptiles, and birds; **learning**, complex behavioral responses that are modified by experience; and reasoning, in which behavioral responses are based on rational thought and strategy (e.g., Dethier and Stellar 1964).

Attenuation, reduced response to a stimulus after repetition (some believe because of pairing with safety rather than danger), is an attribute of certain behaviors important for managing human-wildlife conflicts. For example, if an explosive sound is used continuously at a bird roost, birds will soon return and ignore the sound. Attenuation is often a problem with behaviorally based devices and approaches (e.g., Khorozyan and Waltert 2019). As a consequence, their effectiveness may often be measured in days rather than months or years. Effectiveness can sometimes be extended by changing timing or the locations of the devices, or by modifying the devices themselves. Sometimes the effectiveness of behaviorally based methods can be **enhanced**—that is, the intensity of responses can be increased or their duration prolonged—by occasionally pairing the stimulus with a different, perhaps more stressful one. For example, studies (Lance et al. 2011; Bruns et al. 2020) have found that electrifying **fladry** fences greatly extended their effectiveness in keeping wolves from food sources and prey (fig. 4.2). Even with improved effectiveness, given the limited

life expectancy of such methods, the practitioner should be planning the next management actions.

One way to alter the behaviors of a plant or animal is to kill it. This approach is straightforward and direct and, for the offending individual, permanent. The approach becomes complicated, however, because it involves potential social, cultural, and legal issues that are best considered a priori, particularly if the wildlife is a charismatic species that has garnered public favor rather than one that has not, such as a commensal mouse or a rat at a local landfill. Complex ecological responses may also occur at the population and the community levels, such as increased fecundity or undesired adjustments in the food web, also best considered beforehand.

Some methods, such as exclusion, physically prevent an animal's undesired behavior. Fencing out coyotes can stop predation on livestock (e.g., Bruns et al. 2020), just as netting can prevent bird damage to structures or crops. For instance, Anderson et al. (2013) found netting was cost-effective in preventing bird damage for high-cash-value crops like wine grapes, blueberries, and Honeycrisp apples grown on small acreage.

Other methods may be based on responses (tropism, taxis, reflex, or learned) to stimuli such as visual, auditory, olfactory, or gustatory (taste) cues. For example, ultrasonic sound has been tested (with limited success) as a possible repellent to prevent bats from flying into wind turbines (see, e.g., Arnett et al. 2013). Laser lights can frighten birds from roosts or discourage them from roosting in a grove of trees (see,

Figure 4.2 Fladry, strips of fabric or flags hung at regular intervals along a fence, have been used traditionally in Eastern Europe for hunting, and have been rediscovered as a physical method for keeping wolves from livestock. If the fence is electrified (termed turbo-fladry), the effectiveness of fladry and duration of the effect can be enhanced. Photo courtesy of the Oregon Department of Fish and Wildlife.

e.g., Seamans and Gosser 2016) or discourage bold coyotes (Darrow and Shivik 2009). The effectiveness of **Mylar tape** (see, e.g., Seamans and Gosser 2016), a red and silver plastic tape used to repel birds, might be because the tape gives the appearance and sounds like fire from the air (Bruggers et al. 1986). Sound systems designed to broadcast alarm or distress calls of birds have at least short-term repellent effects on **conspecifics** (other individuals of the same species) and closely related species, and sound has been used to discourage seals from preying on fish at fish farms (see, e.g., Götz and Janik 2015).

Waterfowl hunters have long relied on decoys, using positions of their wings and heads to signal safety for landing, loafing, and feeding. A repellent technique is to put decoys or dead animals in postures that signal danger to overhead conspecifics. Tillman et al. (2002), for example, found that hanging dead vultures or their effigies at roosts or sites of damage was effective in repelling black (*Coragyps atratus*) and turkey (*Cathartes aura*) vultures, protecting both property and agriculture. In one situation, a roost of about 800 vultures was dispersed using effigies. Effigies have also been used to reduce egg and chick predation by corvids (*Corvus*) of the federally listed snowy plover (*Charadrius nivosus*; Peterson and Colwell 2014).

Linhart et al. (1992) reported results from testing an electronic device designed to be hung near a herd of sheep to repel predators. The battery-operated device, called the Electronic Guard randomly emitted light from a strobe light and a siren, and it could be set to operate from dusk to dawn. In tests, the device was effective from 8 to 103 days. A similar device, called the Foxlight, is now available with LED lights and sirens. In recent studies, devices placed directly on sheep or lambs can turn on during an attack, emitting a sound that startles the predator. Another device, a Radio-Activated Guard Box, has been designed to be placed near a herd of sheep. A wolf equipped with a radio collar triggers the device to emit sound and light. The devices can be effective for smaller areas (60 acres or less) for relatively short periods of time, e.g., during lambing season (Breck et al. 2002).

Methyl anthranilate and dimethyl anthranilate, grape flavorants, stimulate pain receptors in the trigeminal systems of many birds and therefore serve as bird repellents (Mason et al. 1992). The effect is unlearned and does not readily attenuate, so the birds continue to avoid treated areas or items. These chemical compounds are active ingredients in commercial repellents such as Bird-X GC-PT Goose Chase Goose Repellent and Liquid Fence Goose Repellent.

Odors can attract or repel offending animals. Odors may mimic those of a predator (i.e., a **semiochemical**), prey, or a conspecific (see information on pheromones shortly) or simply be attractive to the animal—e.g., the rotten odor of synthetic fermented egg is attractive to scavengers. Semiochemicals can repel herbivores directly or attract additional predators, further discouraging the presence of herbivores. Lindgren et al. (1995) reported that odors from feces, urine, and anal glands, particularly those from the weasel family, repelled voles, pocket gophers (*Geomyidae*), and snowshoe hares in both laboratory and limited field trials. Damage to apple trees was reduced in the trials.

Management may also be based on reproductive, foraging, feeding, or social behaviors, such as grooming. Systems of **hormones**, nerves, and behaviors regulate reproduction in higher animals. The systems also regulate secondary sexual characteristics, including how animals look and behave. The behaviors can sometimes cause problems for people or their interests. For example, deer and elk bucks rub trees to remove velvet from their antlers, causing damage reported by forestry and Christmas tree industries. People have been injured when stepping between parenting wildlife and their young. Deer, elk, and moose collide with vehicles more frequently during rut, the fall mating season.

In birds, reproductive behavior is often strongly stereotyped, following a carefully orchestrated sequence of behavioral and physiological events that leads from mating to fledging. As Lehrman (1964) showed with ringneck doves (*Streptopelia risoria*), one specific behavioral event, courtship, produces physiological changes that induce the next behavioral event, nest building. Nest building then engenders physiological changes that induce egg laying. Doves are hardwired in this sense and cannot avoid any steps. For example, doves could not accept a premade nest; instead, the pair would break down the nest and rebuild it, thereby stimulating the physiological changes needed for egg laying. Disruption of any event or alteration of a hormonal response or level could therefore disrupt the entire cycle, a weakness sometimes exploited by wildlife management specialists.

Although less stereotypic than birds, mammals also have sequential physiological and behavioral events that lead them through the reproductive process. The behaviors can be affected by environmental factors. For example, Christian and Davis (1964) and Davis (1966) showed that Norway rats build strongly concave nests, effective in retaining pups, when population densities are low. With high densities, the nests are shallow and flatter, allowing pups to escape; in these situations, rats are more likely to share nests and less likely to retrieve wandering young that are emitting distress ultrasounds. Under high densities, mothers tend to snip into the gut of a pup when cutting the umbilical cord. Survival of young pups under high densities is greatly reduced when coupled with the mothers' reduced ability to lactate. Such adjustments to crowding have received only limited applications managing wildlife, but they have great potential.

Synthetic hormones, such as artificial steroids, can directly induce sterility and reduce growth of over-abundant bird or mammal populations. For example, baits containing the hormone-like compound **diethylstilbestrol** have been used to block egg production in birds. Others like nicarbazin act by disrupting an embryonic membrane. Immunological and other techniques can also induce temporary or longer-term sterility in wildlife. In fact, there are effective antifertility compounds available for managing increasing numbers of wildlife species (see chapter 11). Feeding young can strongly motivate food gathering and predation (e.g., Blejwas et al. 2006). Bromley and Gese (2001b) found that surgical sterilization of coyote packs would reduce predation on sheep, a cost-effective method for small sheep operations.

Pheromones are substances, usually highly volatile, that are secreted into the environment and have behavioral or physiological effects on conspecifics. Broadly used for insect management, pheromones occur in other animals as well, including mammals. In the 1950s, pheromones in rodents were described that blocked pregnancy (Bruce 1959), synchronized estrus among females (Whitten 1956), and induced pseudopregnancy (Van der Lee and Boot 1955). Pheromones influence management methods such as trapping. Stoddart and Smith (1986), for example, found that woodmice (*Apodemus sylvaticus*) were more likely to enter traps treated with conspecific odors than untreated traps set beside them. Scientists have explored uses of pheromones in managing human-wildlife conflicts—e.g., to keep beavers from building dams in New York State (Welsh and Muller-Schwarze 1989)—but only recently has their potential in managing human-wildlife conflicts begun to be fully realized (e.g., Schulte 2016). For example, McCann et al. (2019) found that use of a suppression pheromone could improve culling of the invasive cane toad; the pheromone suppresses a "backfire" effect of trapping by preventing increased viability of tadpoles that remain untrapped.

Social behavioral concepts include dominance, subordination, home range, and territories. These behaviors are sometimes part of sexuality and reproduction. **Dominance** is a social "pecking order" wherein one or more members of a population claim more resources than are made available to others. The resources might be food, water, nesting or denning sites, or a mate. The arrangement can be linear, in which A is dominant over B, B is dominant over C, and so on. Dominance, both conspecific and **interspecific**, has long been a consideration in managing human-wildlife conflicts, e.g., at bait stations and traps. More recently, it has been suggested that dominance, or more accurately the suppression of intraspecific dominance, may also be a factor contributing to the success of some invading species. The invasive crazy ant (*Anoplolepis gracilipes*), for example, has caused extensive damage to island ecosystems such as Christmas Island (Abbott 2005) (see chapter 7). The ants are able to form "supercolonies" that include as many as 300 queens. The invasive success is probably due to a number of factors, one of which may be a reduced intraspecific competition wherein worker ants see ants from other queens as part of the same colony. Warren et al. (2019) found this to be the case for some invasive forest ants, allowing them to outcompete native species.

Other hierarchies include **despot**, in which A dominates equally subordinates B, C, and D; circular, in which A dominates B, B dominates C, and C dominates A; and **coalitions**, in which B and C join to subordinate former despot A (Feldhamer et al. 2007).

Territorial behavior occurs when one member, usually a male, establishes a geographical area and defends it against other male conspecifics. Territories may help regulate overall population size by spreading members out according to available resources. Birds often use calls and visual displays to establish and defend territorial boundaries. Mammals too may use calls, but more commonly they mark edges with scents such as pheromones found in scat or urine, using specialized scent glands, and sometimes rubbing points such as branches or rocks.

Territories are defended, whereas **home ranges** (where an organism carries out its normal daily activities) are not. Both have core areas that provide food, water, and shelter. Blejwas et al. (2006), for example, found that most sheep kills occurred within home ranges of coyote pairs.

Some management applications using territory and home range have been straightforward. For instance, practitioners sometimes choose trapping locations according to territorial boundaries and markings of wildlife. Other applications are more subtle. Conner et al. (2008; see also chapter 5) found that including territorial behavior and social structure as part of models of coyote populations led to predictions that sterilization can be a more appropriate long-term strategy than several lethal methods of management.

Foraging (searching for food) and feeding (eating) include herbivory and predation. Strategies for herbivory can be important to managing human-wildlife conflicts (box 4.1), as can strategies for foraging. One foraging theory is called optimal foraging. Models based on this theory predict how animals maximize energy intake in relation to foraging behavior. Models usually include three factors: type of food or prey; currency, such as energy or time; and constraints or limits, including risk of predation. Another theory type is marginal value. These models are based on energy intake versus energy used. Marginal value models are designed to predict, for example, when an animal decides to abandon a patch where food resources are depleted and invest time and energy in locating a new patch with more food. Food hoarding (storing food) is another strategy used by some birds and mammals, including some rodents, shrews, and carnivores.

Foraging theories and strategies can predict movement and distribution of wildlife. For example, Feldhamer et al. (2007) suggested that optimal foraging theory might predict when feral pigs give up a depleted patch of acorns and move to the next. Amano et al. (2004) used optimal foraging theory to predict damage to rice grains or wheat by white-fronted geese (*Anser albifrons*) around the lake Miyajima-numa in Japan, and they used the predictions to reduce damage to wheat. Wilkinson et al. (2020) maintained that optimal foraging theory can help to explain the ecological bases for carnivore-human conflicts as well as solutions. They argued that livestock are a "low-cost, high-reward prey item for large predators, at least where human involvement is low" (5). Because most livestock are in good condition, they offer a high-calorie reward for predators. Human interventions such as fencing or electronic repellents increase the real or perceived risks for predators, leading to suboptimal foraging conditions. Sterner (1994) used foraging theory to adjust baiting rates of zinc phosphide for vole control. Cached rice is removed from burrow systems in dikes in countries such as Bangladesh to reduce the survivability of pest rodents during catastrophic events such as floods. Alternatively, hoarded foods are sometimes treated with toxicants to manage rodents.

BOX 4.1 Herbivory and Human-Wildlife Conflicts

Two basic strategies underlie herbivory. One uses a **digastric** (two stomach) digestive system and is called **rumination** or **foregut fermentation**. This is used by artiodactyls (even-toed mammals), including camels, giraffes (*Giraffa camelopardalis*), hippopotamuses (*Hippopotamus amphibius*), antelope, cervids, and bovids such as cattle and bison (*Bison bison*). Kangaroos (some *Macropus*), sloths (*Folivora*), and colobus monkeys (*Colobus*) also use foregut fermentation. Digestion of food through foregut fermentation is elaborate and time consuming, but plants are efficiently ingested by grazing or cropping. Important for wildlife damage management, microorganisms in the rumen detoxify poisons, including alkaloidal compounds used in plants that protect them from herbivory and active ingredients for some pesticides.

The other strategy is **hindgut fermentation** and uses one stomach—**monogastric**. Hindgut fermentation is used by perissodactyl (odd-toed) mammals, including horses, zebras (*Equus zebra, E. quagga, E. grevyi*), asses (such as donkeys), tapirs (*Tapirus*), and rhinoceroses (*Rhinocerotidae*). Elephants (*Elephantidae*), lagomorphs (i.e., rabbits and pikas, *Lagomorpha*), and rodents also use the system. Almost all herbivorous birds use hind gut fermentation. Hindgut fermentation is faster and less elaborate than foregut fermentation but also less efficient. For some hindgut fermenters, the inefficiency is compensated for by **coprophagy**: ingesting partially digested feces for redigestion. Hindgut fermenters absorb toxins into the bloodstream and count on the liver to detoxify them.

Foregut fermentation allows the animal to forage and feed quickly then find a place safe from predators to digest the food. Hindgut fermenters need to forage and feed more slowly, carefully selecting foods that allow a more efficient digestion but at greater risk of predation. Wildlife damage practitioners sometimes take advantage of this by visually "opening areas"—making foraging wildlife, such as the Canada goose (*Branta canadensis*), feel unsafe.

Birds are baited or trapped at cattle feedlots, around feeding areas in farms, and near crops. Knowledge of feeding behavior helps the practitioner place baits where they are likely to be used selectively by targeted species. Treated grains may be placed at elevated bait stations in cornfields. Effectiveness of traps can sometimes be enhanced by placing live birds or silhouettes of feeding birds in the cages.

Feeding behaviors include preferences and avoidance and can be learned or unlearned. Feeding preferences underlie many wildlife damage management recommendations, from developing varieties of corn and sunflower seeds that resist bird damage to lists of "deer-resistant" ornamentals for landscapes. Hockings et al. (2012) studied diet preferences of wild chimpanzees (*Pan troglodytes*) as a means of reducing crop raids. As a first rule, when wildlife species are given a feeding choice, they will select one food; use of **decoy crops**, where a crop of lesser value is planted to attract pest species and protect a crop of greater value, is an application of this rule. Refuge managers might plant crops preferred by wildlife to reduce damage to crops of neighboring farms. The approach is effective because it offers wildlife a better choice. There is an important second rule, however: given no choice, wildlife will take whatever is available. Hence, the tops of young trees or bark and cambium of trees may be undesirable to deer until alternative foods are covered with snow. This situation is well known to horticulturists, foresters, and Christmas tree growers.

Feeding preferences may be expressed as **specific hungers** or drives, including for wildlife **geophagia**, the tendency to eat soil, and sodium or **salt drive**, the urge to eat salt. Salty baits containing a toxicant are sometimes used to control North American porcupines (*Erethizon dorsatum*; Anthony et al. 1986) and have been evaluated to control Indian crested porcupines (*Hystrix indica*; Mushtaq et al. 2013), a species exhibiting strong salt drives. Deer, moose (*Alces alces*), and other animals lick deicing salts off roadways, increasing chances for collisions with vehicles (e.g., Grosman et al. 2009). Nonsodium-based deicers are sometimes used in areas with high deer densities to reduce the potential for collisions (e.g., Bruinderink and Hazebroek 1996).

Feeding preferences and avoidances can be influenced by **social facilitation**—when the presence of one animal improves the performance of another. Some young learn food preferences by associating the flavors in mothers' milk with a positive feeding experience; weaned young may seek and eat foods having similar flavors (Galef and Henderson 1972). With some animals, the young also learn to avoid food avoided by adults. For example, Avery (1996) showed that seeds avoided by adult house finches (*Carpodocus mexicanus*) were also avoided by their offspring. Social facilitation

is a promising behavior that has seen limited applications to resolving human-wildlife conflicts.

Behavioral defenses against dietary poisoning must be overcome if wildlife are to ingest the amounts of chemicals needed for pharmacological activity, whether to induce toxicity, contraception, or avoidance of foods or crops. Garcia et al. (1974) summarized such defenses, subsequently describing them in the form of a prototypic gastronome. With this gastronome, Garcia argued that most organisms were consummate dietary gourmets within their own needs and environments. When encountering a new food, the prototypic gastronome first exhibits **neophobia**—fear of something novel—and avoids it. After neophobia is extinguished, the gastronome samples food for taste (**gustation**) or flavor (gustatory and olfactory components) but does not necessarily swallow the food. If the flavor is one instinctively associated with illness, usually bitter, or sufficiently stimulates pain receptors in the trigeminal system (cranial nerve V, the **common chemical sense**), the food is rejected without swallowing, called **primary flavor aversion**. If the food contains a poison and is swallowed, a post-ingestional gastric illness (stomach illness causing nausea or vomiting) will be associated with the food; if the gastronome survives, this will result in subsequent avoidance of that flavor, called **flavor aversion learning** (fig. 4.3).

Flavor aversion learning has attributes important to application, including one-trial learning, attenuation, and association of the flavor with illness up to six hours after ingestion. Further, animals learn to avoid the sight (with visual learners such as birds), taste, or flavor but do not necessarily learn to identify

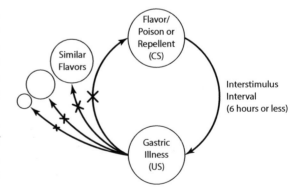

Figure 4.3 If an animal develops a gastric illness within six hours of ingesting a substance, the animal will learn to avoid the flavor of the substance as well as similar flavors, particularly if the flavor is novel. Adapted from Reidinger 1995, with permission. Illustration by Lamar Henderson, Wildhaven Creative LLC.

the agent causing illness. Thus, a novel flavor followed by radiation (causing nausea) results in subsequent avoidance of the flavor, not radiation. Some flavors are more **salient**—i.e., more likely to be associated with the illness—than others. When several flavors are presented, the most salient **overshadows** the others, and the gastronome learns to avoid only the salient one.

The behaviors described by the prototypic gastronome apply directly to managing human-wildlife conflicts. Flavor aversion learning, for example, describes **bait-shyness**, avoidance of a bait after consumption of a sublethal amount of a toxic bait. Neophobia explains the tendency to avoid novel baits, and primary flavor aversion explains avoidance of bitter-tasting baits and compounds. **Prebaiting** (providing bait without poison, allowing a positive post-ingestional consequence) is sometimes used to overcome these defenses. Flavor aversion learning has been explored extensively for applications to human-wildlife conflicts. Gustavson et al. (1976) explored its applications to predation on sheep and other livestock, and others have used flavor aversion learning to induce avoidance of bird or turtle eggs by raptors. The concept has been used to help determine flavors of baits and to assist in the formulation of prebaits. Although application of flavor aversion learning is an appealing concept, success with this approach has varied. The search continues for effective illness-inducing agents (e.g., Tobajas et al. 2019a, 2019b).

At Camp Pendleton, a US Marine base in coastal California, ravens take endangered least tern eggs. Surrogate eggs have been injected with methylcarbamate, a compound that induces nausea and learned aversions in ravens. The aversions generalize from surrogate eggs to other, nontoxic eggs, including those of the least tern, thereby affording them protection (Avery et al. 1995). Similar approaches have been used previously and more recently by other scientists to protect eggs from predation (e.g., Nicolaus et al. 1983; Forys et al. 2020). Predation is a centrally biased stereotypic behavior that is difficult to stop. Even if consumption of a prey is prevented through aversion learning methods, predatory behavior may continue. Under those circumstances, a predator will attack and kill a prey, then become nauseated and avoid eating the quarry.

Grooming is another behavior that can be strongly stereotypic and sufficiently strong to overcome learned flavor aversions—during grooming, animals will ingest flavors that they have learned to avoid by flavor aversion learning. The Felixer (Moseby et al. 2020), a "grooming trap" designed for control of feral cats in southern Australian, is designed to take advantage of this behavior. Nonetheless, grooming remains an underexploited means of getting management substances (toxins, repellents, contraceptives) onto the fur of wildlife and ingested (Reidinger et al. 1982).

Example

Bear damage was observed on forestry plantings in the western United States in the 1940s and became a concern as early as the 1950s. Bears peel the bark of Douglas fir, western hemlock, and western red cedar, causing economic losses valued at millions of dollars annually (fig. 4.4) (Nolte and Dykzeul 2002). Douglas fir develop buds first and are usually the first damaged.

Figure 4.4 Bears peel the bark and scratch trees, sometimes causing economic losses valued in millions of dollars. Supplementing with a fabricated food until a more desirable one becomes available naturally can reduce or eliminate the behavior. This bear is just scratching his back. Photo by Neal Herbert, National Park Service.

The bears prefer berries when they become available during the early summer, and then they switch to cedars as berry abundance declines (Ziegltrum and Nolte 1995).

A supplemental feeding program was initiated as an alternative to lethal control in 1985. Bears were provided with a pelleted food supplement designed to be more attractive than the phloem of trees but less attractive than berries, and damage ceased (Flowers 1986). The program was expanded to other sites until it protected over 400,000 hectares of Douglas fir (Ziegltrum 2004). Although this supplemental feeding program has been effective and other programs have also been successful, some attempts have yielded mixed or negative results (Garshelis et al. 2017). Because of this, Garshelis et al. (2017) have recommended that such programs be viewed as experimental, carefully organized with collection of relevant data, conducted by professionals, and include adaptive management practices.

Summary

- Ecological niche is used to assess potential range expansion of an exotic invasive animal and underlies exclusion of wildlife from airports and landfills.
- Ecological equivalents, sympatry, and Gause's law are used to predict outcomes of removals or introductions of wildlife, such as predators, in ecosystems.
- Some organisms with narrowly defined niches, such as the sanguinivorous vampire bat, can harm humans or their interests, but most offending wildlife, such as rodents, coyotes, feral pigs, and weaver finches, are ecological generalists.
- Domestication makes plants and animals susceptible to wildlife diseases and predation, exacerbating human-wildlife conflicts. Feral wildlife, including feral pigs and dogs, are among the most damaging species found worldwide today.
- Biological rhythms—circadian, lunar, and circadian—of prey should be compared with predators' behavior rhythms before the predators are introduced.
- Altering biological rhythms can disrupt behavioral events of wildlife. Understanding rhythms can help time management activities, such as baiting rodents during their dispersal, for optimal effect.
- Modifying behaviors, wildlife or human, underpins many successful resolutions of human-wildlife conflicts.
- Physical exclusion, such as fencing, can prevent unwanted behaviors such as herbivory or predation.
- Some lights, sounds, odors, and flavors offer temporary repellency by taking advantage of the responses of wildlife; some, such as effigies and semiochemicals, are biologically based.
- Reproduction underlies some conflicts with wildlife, such as deer colliding with vehicles during rut. Reproductive behaviors can be disrupted by altering cues that stimulate reproduction or by using chemicals that interrupt reproduction, such as antifertility compounds or pheromones.
- Foraging and feeding behaviors and theories help explain or predict some conflicts with wildlife, such as excessive predation on livestock during the reproductive season and movement of feral pigs from one patch to another.
- Flavor preferences and aversions are used to manage damage, as when decoy crops attract wildlife away from high-cash-value crops; when preparing baits, repellents, or lists of ornamental or crop varieties avoided by wildlife; and when understanding and managing bait-shyness or inducing repellency by flavor aversions.

Review and Discussion Questions

1. What might be the ecological equivalent of the passenger pigeon (*Ectopistes migratorius*) in North America today? Explain, with references.
2. How does the domestication of wildlife species relate to their causing damage around the globe?
3. It is commonly believed that attenuation to auditory or visual repellents, such as "Scarey Man" and propane cannons to scare birds, can be reduced by occasional reinforcement with a negative experience, usually shooting some birds. Is there scientific evidence supporting this belief? Provide specific references and supporting evidence.
4. Has flavor aversion learning been used to induce coyote aversions to livestock? To reduce livestock depredation? Summarize scientific studies supporting and not supporting the application.
5. How might you use an animal's behavioral defenses against dietary poisoning to your advantage in the management of wildlife conflicts? Give a specific example.

5

Populations and Their Applications

In this chapter, I look at population size and its critical minimum, population density, and crowding. I review the fundamental equation for population growth, types of growth curves, life tables, and survivorship curves, age pyramids and population models, and their underlying equations. I show how these aspects of populations are used in assessing or managing wildlife conflicts. I examine relationships between population size and extent of wildlife damage.

POPULATION SIZE, DENSITY, OVERABUNDANCE, CROWDING, AND THE FUNDAMENTAL EQUATION FOR GROWTH

Statement
Attributes of populations—including size, density, and growth, and overabundance, and crowding—sometime underlie causes of human-wildlife conflicts and are the bases for assessing and managing wildlife populations.

Explanation
Populations are actually or potentially interbreeding groups of organisms—i.e., species—living in the same place at the same time. The **population size** at any moment in time is usually designated as N. For example, if a total population of geese in a city park is counted at 145, then $N = 145$. If the intent is to reduce wildlife damage by reducing the size of the goose population, then N must be brought below 145. This might be done by rounding up the geese during the summer molt and relocating or harvesting them, or perhaps by addling their eggs.

Sometimes populations are assessed by density. **Population density** is the number of individuals per unit area. **Ecological density**, the number of individuals per unit habitat, is sometimes more useful. For example, the density of the geese in that park is 145 total geese per total 10 acres of park, or 14.5 geese per acre. However, if half of the park is wooded and another quarter of it is used for parking lots and buildings, the actual habitat available to the geese may be only 2.5 acres, making an ecological density of 145 geese per 2.5 acres or 58 geese per acre of habitat. Such a high density raises questions of overabundance or crowding.

Caughley (1981) said populations are **overabundant** when they threaten human life or livelihood, impact densities of (other) favored wildlife species, are too numerous for their own good, or impact ecosystems and cause their dysfunction. **Crowding**, related to overabundance, occurs when population density exceeds a level at which the habitat can maintain healthy population members, and they become stressed. Selye (1936) postulated that animals placed in an untenable environment, with no hope of escape, would become stressed, going through stages including alarm, resistance, and exhaustion. Selye later suggested that low levels of stress might be beneficial, expressed as **eustress**, but if continued without hope of relief would lead to **distress**. Selye tied stress to enlarged adrenal cortices and the hypothalamic-pituitary-adrenal axis, a concept picked up and expanded by Christian and Davis (1964). They found symptoms of crowding among many mammals, including rodents and red deer (*Cervus elaphus*).

Factors other than density affect crowding, so it cannot be assigned an exact threshold density above which crowding occurs. For example, animals may tolerate higher ecological densities in environments with high plant densities. I observed densities of over 10,000 rats per hectare in marshes in the Philippines, and the rats seemed unstressed. Possibly the thick, tall vegetation in the marshes minimized contact among conspecifics and reduced stress. Regardless, crowding has been observed in situations where populations of mammals, e.g., deer or elephants grow unchecked, often leading to sick animals, damage to the environment, and problems for the wildlife and for wildlife managers.

Whereas N provides population size at a moment in time, it may be important to know whether a population is growing, stable, or diminishing over time. Such information can predict whether damage will likely abate or worsen, or whether actions taken to reduce a population have worked. Growth rate (r) measures such trends.

Growth rate (r) is the change in numbers of population members over time. It is also the sum of four other rates: **natality** (birth, b), **mortality** (death, d), **immigration** (i, movement into), and **emigration** (e, movement out of), described by the following **fundamental equation for growth**:

$$r = (b+i) - (d+e),$$

where $(b+i)$ is sometimes called **recruitment**.

I illustrate this concept by hypothetically gathering more information on the geese in the city park. I now decide to legally place numbered neck rings around the geese. Further, I mark each new member and record the loss of each old member and, having unlimited funds (this is truly hypothetical), I gather information every day for a year. I find, summed for exactly one year, that 180 goslings were hatched and survived, 25 geese died, 20 new adult geese moved into the park, and 15 marked geese left the area.

There were 180 hatched and 20 immigrants, or 200 new members. Added to the original 145 members, this makes 345. But 40 members died or emigrated, so the actual size is reduced to 305. That is 160 new members per 145 total original members in one year, or a growth rate of 110.3% per annum! It looks like a burgeoning population of geese at the park! Maybe neighbors are feeding them, despite posted signs asking the public not to feed wildlife. Perhaps there are few local predators like cats, fox, dogs, raccoons (*Procyon lotor*), and urban coyotes.

As with natural changes in population size, those induced by management actions must also follow the fundamental equation for growth. Therefore, the actions of wildlife managers must impact natality, immigration, mortality, or emigration rates or some combination of these. If the intent is to eradicate, the action must push population size below its **critical minimum** (the size below which the population fails). If the intent is to reduce damage, and population size is directly related to damage, the overall impact must reduce growth rate (r) even if one or more individual factors, such as natality or immigration rate, increases. For example, natality rate may increase because of reducing coyote populations; the consequence is nevertheless reduced damage if net growth is reduced and if population size relates to damage.

Each type of rate can underlie human-wildlife conflicts or might be used to help manage them. Immigration rate is often important with exotic and invasive species (and may be seen as part of propagule pressure, see chapter 7). It can also underlie solutions, such as public policies that restrict movement of undesired wildlife between governmental units (e.g., state or provinces) or countries. Emigration and mortality rates contribute to reduced population size and can sometimes help to reduce conflicts. Increasing mortality is a common way to reduce populations. Wildlife managers sometimes use translocation as a form of emigration (see Germano et al. 2015 for uses and issues surrounding animal translocations). Natality rates contribute to the growth of populations and can exacerbate conflicts. Reducing natality rate can sometimes reduce conflicts as with addling eggs, sterilizing wildlife, or denning.

Examples

OVERABUNDANCE. In a 10 year study of impacts of deer density on forests of northwestern Pennsylvania, DeCalesta (1992) found that deer densities exceeding 7.9/km² resulted in a significant decline in forest species richness, abundance, and diameter of saplings at breast height. Six woody tree species were missing among the saplings (DeCalesta 1992). Thus, deer at higher densities negatively impacted the ability of the forests to regenerate. Cascading effects were found for songbirds (e.g., reduced woody vegetation meant reduced foraging, escape from predators, and nesting) and the richness of other herbs and shrubs. Further, DeCalesta (1997) suggested that higher densities of deer might eliminate eastern hemlock regeneration and prevent American beech (*Fagus grandifolia*) and sugar maple regeneration by furthering an insect and disease complex.

Deer health can also be impacted by overabundance. Davidson and Doster (1997) reviewed the two major health problems of deer in the southeastern United States: hemorrhagic disease and a syndrome caused by malnutrition and parasitism. The researchers found that one of the diseases, the syndrome due to malnutrition and parasitism, was density dependent. As an example of a different form of impact, Nussey et al. (2007) found high rates of senescence in female red deer that experienced high levels of competition for resources early in life. And Kilpatrick et al. (2001) reported improved health of white-tailed deer (*Odocoileus virginianus*) herds, and reduced deer browsing, following a two-year program of deer population reduction by hunting at Bluff Point State Park, Connecticut.

CULLING. Wildlife damage practitioners sometimes **cull** (kill individuals) in overabundant populations, usually by hunting or shooting. For example, to minimize environmental damage and stress among densely populated herds of elephants in Kruger National Park in South Africa, culling was begun in 1967. Based on the collective opinions of experts from different parts of Africa when culling was started, a limit of about 7,000 elephants was established for the park, roughly a density of one elephant per square mile. Most experts felt that the environment and biodiversity of the park could be sustained with that density, as could the health of the herd. Culling continued until a moratorium was put on it in 1994, based partly on concerns from the international community that questioned the humaneness of culling and that applied political pressure to the South African government. Alternative control approaches such as elephant contraception were considered (Whyte 2004).

Today's view, expressed by Whyte in 2004, is that expanding and diminishing elephant populations add an important dynamic component to the park. The dynamic nature of elephant populations is now seen to add ecological value in diversity and richness, provided that the populations are not extremely low or high for too long a time. Kruger Park was home to about 12,000 elephants in 2009, 5,000 more than the estimated carrying capacity for the park in 1967. Ferreira et al. (2017) estimated at least 17,086 elephants in the park based on aerial surveys conducted in 2015. The scientists calculated a growth rate of 4.25%, down from 6.5% in 1994. Midgley et al. (2020) noted that persistent elephant damage had resulted in sterility of at least one widespread African palm (*Hyphaene petersiana*), because the trees are unable to achieve the size needed to reproduce. The park continues to monitor and to manage populations and to reassess their critical minimum and maximum densities as scientists gather more information on herd conditions and impacts on habitat.

POPULATION GROWTH CURVES

Statement
Growth curves provide theoretical and practical information on patterns and limits of growth of populations; the curves are used by wildlife conflict practitioners to predict when action is needed, to time management actions, and to monitor their effectiveness.

Explanation
Growth of wildlife populations is often a logarithmic (geometric) function of time rather than linear. Suppose, nonetheless, a linear relationship exists between time and growth of the hypothetical goose population in the park. A linear relationship follows the equation for a straight line, as in figure 5.1a, and is more readily understood than a logarithmic one. Here, the ordinate is N and the abscissa is t, and the relationship between the two is described as

$$N = m\,t + N_0,$$

where m is the slope, or change in N per change in unit t, and N_0 is the t intercept, or, the value of N when t is 0.

Reviewing the goose population when we last left it, there was an increase of 160 population members when one unit of time (a year) passed, or a slope of 160/1 or 160. At time 0 (i.e., time N_0, the t-intercept),

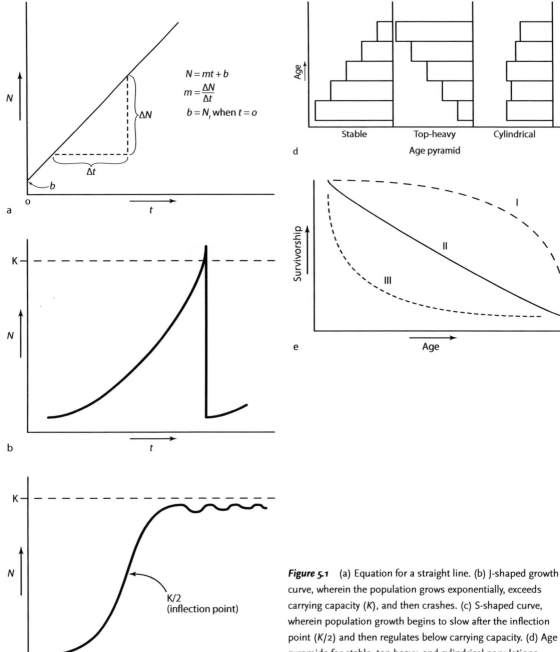

Figure 5.1 (a) Equation for a straight line. (b) J-shaped growth curve, wherein the population grows exponentially, exceeds carrying capacity (K), and then crashes. (c) S-shaped curve, wherein population growth begins to slow after the inflection point (K/2) and then regulates below carrying capacity. (d) Age pyramids for stable, top-heavy, and cylindrical populations. (e) Three types of survivorship curves. Illustrations by Lamar Henderson, Wildhaven Creative LLC.

the population had 145 members. Using the equation for a straight line, then, I would predict that in 1 year, the population would be 305, as follows:

$$N = (\Delta N/\Delta t)t + N_0,$$

or

$$N = (160/1)1 + 145 = 160 + 145 = 305$$

members (as deduced in the first section of this chapter).

Further, if population size continued in the same linear fashion (although this is unlikely in reality)— that is, if the population increased at a steady rate of 160 geese per year—we would predict that in six years, the park would have 1,105 geese, as follows:

$$N = (160/1)6 + 145 = 960 + 145 = 1,105.$$

Now, suppose one made smaller and smaller the interval of time for analyzing the measurements of population size, from once a year to every half year, every three months, every month, every week, every day, every minute, every second, every millisecond, and so on. If this trend continued, the intervals would eventually become so tiny that they would be dimensionless and instantaneous (represented by d). One would then be measuring **instantaneous growth rate**, which gets me (and hopefully you) to dN/dt, the instantaneous change in size of the population for an instantaneous unit of time. (You could use calculus to calculate the same rate.)

An advantage of instantaneous growth rate is that it can be used to calculate growth and related population parameters for relationships between N and time that are nonlinear, such as dimensionless points along **J-shaped** or **sigmoidal growth curves**. In such cases, the instantaneous growth can be visualized as the slope of the tangent to that point on the curve.

Two equations that describe population growth are unrestricted (**Malthusian**) growth,

$$dN/dt = r_{max} N,$$

and logistic or sigmoidal growth,

$$dN/dt = r_{max} N(K - N)/K.$$

In both equations, dN refers to the change in population size at any moment in time, dt, and dN/dt to its instantaneous growth rate. K is **carrying capacity**, and r_{max} (see the first section of this chapter) is defined here as the **maximum growth rate**: the growth rate under ideal conditions. Note that, unlike instantaneous growth, which varies with time and which has been the basis for the discussions so far, maximum growth rate is a constant that remains unchanged. Growth rate is usually given on a per member basis (or some multiple, such as per 100 or 1,000 members). For example, if r_{max} is 0.25, each member contributes on average one-quarter of a new member to the population under ideal conditions during the unit of time (e.g., a year or a reproductive season).

When unrestricted, the population size (N) passes right through carrying capacity (K) and exceeds it (see fig. 5.1b). Because the size of the population now exceeds the number that the habitat can sustain (K), the population crashes. This was the point of Malthus' original description of the human population in which he predicted "the power of population is indefinitely greater than the power of the earth to produce sustenance for man" (Malthus [1798] 1993, 13). The pattern resembles a "J" and hence is called a J-shaped curve

(see fig. 5.1b). Many wildlife populations—e.g., microtine rodents such as voles—follow a J-shaped growth curve, also known as an **exponential growth curve**. So do some populations of "released" invasive species that have moved into new habitats with essentially unlimited resources and without naturally occurring diseases, parasites, and predators.

With populations having logistic or sigmoidal growth, the population responds to carrying capacity in a density-dependent fashion—i.e., environmental conditions that reduce growth, such as disease and starvation, collectively called **environmental resistance**, apply more and more pressure against ideal growth (and achievement of the biotic potential for the population) as the population density increases and the population size approaches K (Verhulst 1838). At one specific point, the inflection point (see fig. 5.1c), $K/2$, density-dependent factors impact growth sufficiently that growth rate (r) slows. This point is where maximum growth occurs, sometimes considered by wildlife biologists as **maximum sustainable yield**, the population size at which maximum harvest can be sustained for the longest time.

Maximum sustainable yield is also based on **compensatory mortality**—that is, many game animals removed by hunting or trapping do not reduce net population size because these animals, had they not been hunted or trapped, would have been removed anyway due to other factors such as disease or starvation. To take full advantage of compensatory mortality, timing of hunting or trapping is important. In temperate countries, hunting and trapping are usually permitted in the fall before winter stresses further affect populations through starvation or disease.

For wildlife conflict managers, who sometimes want to reduce a population as much and as inexpensively as possible, additive mortality may be more important than compensatory mortality. With **additive mortality**, hunting, trapping, and poisoning are conducted so as to add to, rather than compensate for, other mortality factors. To accomplish this, management actions might be taken in late winter after disease and starvation have already taken their toll and compensatory mortality is no longer possible. Or, if practicable, the management actions might be taken before the expanding population reaches its $K/2$ inflection point, again to maximize impacts on N.

Density dependence is accounted for in the equation for sigmoidal growth by the expression $(K - N)/K$. The expression decreases geometrically as N approaches the carrying capacity, as illustrated below. If the carrying capacity of a habitat is 500 members for a given population, then as N increases, $(K-N)/K$ decreases as follows:

N	K	$(K-N)/K$
10	500	$(500-10)/500 = 0.98$
50	500	$(500-50)/500 = 0.90$
100	500	$(500-100)/500 = 0.8$
200	500	$(500-200)/500 = 0.6$
300	500	$(500-300)/500 = 0.4$
400	500	$(500-400)/500 = 0.2$
450	500	$(500-450)/500 = 0.1$
490	500	$(500-490)/500 = 0.02$

In fact, as N approaches K, the growth rate approaches 0. Now, putting the expression back into the whole equation for sigmoidal population growth, and, assuming a maximum growth potential of 0.5 (i.e., r, the number of new population members contributed per member per unit time), the impact on instantaneous growth (dN/dt) can be seen as the following:

$dN/dt =$	r	N	$((K-N)/K)$
4.9	0.5	10	0.98
22.5	0.5	50	0.9
40	0.5	100	0.8
60	0.5	200	0.6
62.5	0.5	250	0.5
60	0.5	300	0.4
40	0.5	400	0.2
22.5	0.5	450	0.1
4.9	0.5	490	0.02

Note that, although r_{max} remains constant, instantaneous growth is slow in the beginning, when there are few population members to find each other and mate, and it slows again as N approaches K.

In another form the equation can be used to calculate population size (N) for any time interval (t), if the population size of the previous interval (i.e., $t-1$) is known, as follows. For unlimited growth,

$$Nt = N_{t-1} + r_{max}N_{t-1}$$

For sigmoidal growth,

$$Nt = N_{t-1} + r_{max}N_{t-1}(K-N)/K$$

So what can an understanding of population growth curves add to the understanding of human-wildlife conflicts or their management? First, growth curves serve as the bases for most computer models of population estimation. Such models are used extensively in managing human-wildlife conflicts to assess the status of populations, e.g., before and after management actions have been taken.

Second, knowing the type of growth curve a particular species follows—e.g., J-shaped versus **S-shaped**—can help predict likelihood of the population becoming a problem and provide insights into how to manage the population. For example, populations such as microtine rodents and hares, which follow J-shaped curves, also tend to exhibit irruptive cycles, damaging their environments and crops when their population numbers exceed carrying capacities. Such irruptive cycles have also been observed with ungulates such as deer, as summarized by McCullough (1997).

Third, both inflection point and biological carrying capacity are useful concepts that have special application in wildlife management. Using carrying capacity, for example, one can envision how wildlife populations might be reduced indirectly, without directly managing natality, mortality, emigration, or immigration rates. Recall that carrying capacity is determined by a density-dependent limiting factor (the factor needed for existence that is least available) that might potentially be food, water, or shelter. By severely restricting its presence, the manager can potentially make that factor the limiting one. In doing this, the manager would have also indirectly forced the population to self-regulate to a lowered N that now corresponds to its reduced carrying capacity. Self-regulation of the elephant population below carrying capacity is what managers are hoping will occur at Krugers National Park in South Africa.

David Lack, a prominent biologist who particularly studied populations, argued that food was the single most important factor limiting growth of many bird populations (Lack 1954). By reducing the presence of food in areas where a damaging bird population is present, the population would either self-regulate to a lowered number (populations with sigmoidal growth) or crash when the carrying capacity for the population was exceeded (for populations with J-shaped growth). In practice, food might be limited by different methods, such as use of netting to exclude orchard grapes or other fruit from the birds, use of chemicals that make the flavor of potential foods distasteful, or use of soil to cover landfills and physically separate birds from edible wastes.

Regarding rodents, successful control of rodents in large cities such as Baltimore, Philadelphia, and Chicago has often hinged on public campaigns to sanitize areas by containing trash and garbage, thereby reducing their availability as food for commensal rodents. Repair of houses to prevent access by rodents further reduces habitat. Airports are examples of places where surrounding habitats are made as undesirable as possible to reduce carrying capacity for wildlife.

Examples

Conner et al. (2008) developed a simulation model to evaluate the effectiveness of differing coyote management strategies. They began with a population model already designed by Pitt et al. (2003) that was nonspatial and stochastic (random based). The model also incorporated some behavioral features, such as dominance and sociality. Subordinate male coyotes are often nonterritorial and nonbreeding. They live along the interstices of other male territories. The subordinate males also, however, serve as a population reserve, in that they can replace dominant males that have been removed, becoming both territorial and reproductive. Thus, the behavioral responses, a social feature incorporated into the model, add resilience to coyote populations. The model was designed to simulate management actions on a pack of 100 coyotes; in the model, parameters were adjusted to match those known from already existing field data.

Conner et al. (2008) added a spatial component, also important for coyotes and coyote management, and refined the social components to make them even more realistic. The researchers then used the model to assess different management actions. They found that spatially focused and intense lethal removal was more effective and lasted longer than less intense, random removal of the same numbers of coyotes. Sterilization appeared to offer the longest lasting and largest impact on the coyote population. They suggested that the model serve as a tool for developing more effective and socially acceptable predator management strategies and recommended some changes to further improve the model.

LIFE TABLES, SURVIVORSHIP CURVES, AND AGE PYRAMIDS

Statement

To facilitate interpretation, age-dependent natality, immigration, mortality, and emigration rates are often organized and displayed as life tables, survivorship curves, or age pyramids. The displays are useful to portray the dynamics of wildlife populations and assess management impacts on populations.

Explanation

Growth rate (r) and its four component rates are influenced by ages of population members. Consider, for example, a hypothetical population having three age groups: young, middle, and old. Young and old members are often unable to reproduce and have natality rates near zero, leaving higher natality rates to the age group in the middle. Young members take more risks and therefore tend to have higher mortality rates than more experienced population members. Members of older age groups also tend to die, so the older group also has a high mortality rate. And emigration from a population is often a characteristic of its younger members.

Life tables help us understand the impacts of age on populations (box 5.1). The overall shape of a life table, portrayed as a figure with age groups as the y coordinates and numbers of members for each group as the x coordinates, can indicate the health of the population. For example, a pyramid-shaped life table would indicate a stable or even expanding population; a top-heavy or inverted pyramid would indicate a declining population; a cylindrical-shaped life table would indicate a stable population or one that is about to decline (see fig. 5.1d).

Data from life tables can also be used to construct survivorship curves and age pyramids (see fig. 5.1e). The curves are derived by plotting survivorship (l_s in table 5.1) against age at death. Wildlife populations are often described as fitting one of three types of survivorship curves. In Type I, often after an initial die-off of young, the odds of surviving are high until an individual becomes old. Localized human populations often fit a Type I curve. With Type III curves, mortality rates are highest in earlier age groups. Type III curves characterize many plant, invertebrate, and fish species, but a Type III curve is rarely seen with populations of mammals and birds. With Type II, survivorship rates remain fairly constant over the age groups within the population; many mammals and birds have populations fitting Type II curves.

Age pyramids are derived by stacking age groups on top of one another, from youngest on the bottom to oldest on the top. Sometimes the pyramid is split down the middle, with age groups of one sex on the left, the other sex on the right. A triangular pyramid usually indicates a stable population, whereas a top-heavy one indicates a diminishing population.

In managing human-wildlife conflicts, understanding the age-related dynamics of populations can be important. Life tables, survivorship curves, and age pyramids can help assess the likelihood that a species will be invasive, the most appropriate management strategy, and the costs for management in relation to anticipated benefits. See, for example, the studies done

BOX 5.1 Life Tables

Life tables have been used to estimate human population demographics since Roman times and were further developed by modern insurance companies to account for age in the costs of life insurance. The age-dependent approach has been incorporated into many models for wildlife populations. One of the first wildlife studies using life tables was conducted by Murie (1944). He collected horns of Dall mountain sheep (*Ovis dalli*) from Mount McKinley, Alaska, and used the horns to estimate the age at death of sheep. Over time, he collected horns from 608 sheep, a sufficient sample to establish age-dependent mortality rates for the sheep, as reported by Deevey (1947).

The whole of Murie's life table was based on horns collected opportunistically over years—that is, from all individuals, regardless of age, found dead over one period of time, often called a **static** or **vertical life table**. Sometimes all members of a population born within a period, say a year, are followed until they have all died. A different type of life table, called a **cohort** or **horizontal life table**, can be constructed from these data. Static tables require an assumption that the population is stationary, with constant birth and death rates. This is unlikely. Cohort tables require an assumption that the age group selected represents accurately the whole population, also unlikely. Attempts to fix these and other problems result in more complex mathematical models than are considered here.

When interpreting life tables, be sure to know the bases on which the rates are calculated. For example, a growth rate of 3.2 might mean growth per 1,000 population members, per 1,000 females, or per 1,000 reproductively active females.

TABLE 5.1   *Life table for the raccoon dog in a primal forest in Poland*

Age (years)[a]	n_x	l_x	d_x	q_x	p_x	e_x
0	250[b]	1.000	205	0.820	0.180	0.796
1	45	0.180	26	0.578	0.422	1.144
2	19	0.076	13	0.684	0.316	1.026
3	6	0.024	4	0.667	0.333	1.167
4	2	0.008	1	0.500	0.500	1.500
5	1	0.004	0	0	1.000	1.500
6	1	0.004	1	1.000	0	0.500
7	0	0	—	—	—	—

Source: Redrawn from Kowalcyk et al. 2009, with permission.

[a]Across row, all age-dependent, respectively, age, birth rate (n_x), survival rate (l_x), death rate (d_x), mortality rate per 1,000 (q_x), survival per 1,000 (p_x), and expectation of life (i.e., lifetime remaining, e_x).

[b]Zero frequency calculated from fecundity ratio.

population dynamics supported the raccoon dog as a species well suited for invading new areas, including Western Europe.

Example

Dolbeer (1998) used four population models (PM1 to PM4) to predict relative responses of vertebrate species to reproductive and lethal control methods. He contrasted fruit bat populations with rat populations, both damaging species on the Maldives (fig. 5.2). He showed that fruit bat populations, with low reproductive rates, could be reduced more efficiently with lethal compared to reproductive control. Rat populations, with high reproductive rates, could be controlled more efficiently with reproductive rather than lethal methods. Dolbeer (1998) also compared brown-headed cowbirds to laughing gulls (*Leucophaeus aticilla*), again demonstrating that the birds with relatively low reproductive rates (i.e., the laughing gulls) could be more effectively managed with lethal rather than reproductive control. He suggested that red-billed quelea would respond like cowbirds.

This study (Dolbeer 1998), based partly on analyses of age-dependent population dynamics of offending populations, provides important insights in both theoretical and practical approaches and limitations of differing management actions. For example, Dolbeer (1998) proposed that the relative efficacies of lethal and reproductive methods for vertebrate species could be generalized based on adult survival rate and age of onset of reproduction. Specifically, he pointed to the long period of time for effectiveness of contraceptive methods when used for species, such as deer, that reproduce relatively slowly.

in the mid-1990s in Finland on the invasive raccoon dog (*Nyctereutes procyonoides*; e.g., Helle and Kauhala 1993). Kowalczyk et al. (2009) constructed a life table for the raccoon dog in the Białowieża Primeval Forest, Poland, based on age of death recorded between 1996 and 2006 (see table 5.1). These animals showed a high mortality within the first year, longevity up to seven years, and among the highest reproductive rate seen with raccoon dogs in Europe (the authors felt it might have been compensatory). Despite a high mortality rate, Kowalczyk et al. (2009) postulated that overall

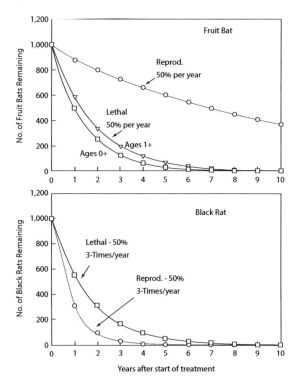

Figure 5.2 Relationships between reproductive potential and success of lethal or reproductive-based management strategies. Species with low reproductive potential, such as fruit bats, tend to respond to lethal methods, whereas species with high reproductive potential, such as black rats, tend to respond to reproductive methods. Redrawn from Dolbeer 1998, with permission. Illustrations by Lamar Henderson, Wildhaven Creative LLC. With permission of the Vertebrate Pest Council.

MEASURING POPULATION SIZE AND AGING

Statement

Estimating actual population size depends on direct counting or indirect sampling methods, often expensive, whereas population indices can inexpensively provide information on changes in population sizes. Adding estimates of ages of members into population analyses requires some age-dependent factor such as reproductive status, tree rings, eye lens growth, tooth development or wear, or annular rings of scales or horns.

Explanation

Most ways to measure dynamics of wildlife populations are variations of a few basic methods. One approach is to use distance sampling methods such as line or point **transect surveys**. The technology can be as simple as a walk or drive along a transect during

which one counts population members. The count might be of members themselves or of their signs, such as sightings of coyote tracks or scat or numbers of bird calls. Sometimes aerial photographs and digital analyses are used. The photography might be accomplished with fixed-wing aircraft or a helicopter or might come from satellite images. The photographs can be analyzed with methods used in geographic information systems to provide estimates of population density and size.

For example, the North American Breeding Bird Survey uses point counts to assess overall trends of bird populations in the continental United States, Alaska, southern Canada, and more recently parts of Mexico (e.g., Sauer et al. 2017). The Audubon Christmas Bird Count also uses point counts, in this case to look at long-term trends in populations of birds in the United States and Canada and in parts of Mexico, Central America, and the Caribbean Islands (e.g., Meehan et al. 2019). The surveys provide trends useful for assessing wildlife conflicts. For example, long-term declines in songbird populations in North and South America have led to investigations of causes, including feral and domestic cats and brown-headed cowbird parasitism.

Gese (2004) summarized methods for estimating sizes of canid populations. Methods included counts of scat or tracks along transects, den and burrow surveys, response to vocalizations, frequency of complaints, harvest data, road mortality, drive counts (where the canids are driven past a point where an observer counts them), and spotlight surveys. Most of these methods could be applied to either estimate population size or for indexing populations (see below), but their precision and accuracy vary.

Blackwell et al. (2006) summarized applications of infrared technologies, including the use of forward-looking infrared (FLIR) devices in wildlife surveys. Infrared technologies sometimes allow better detection of targeted wildlife, improving the sensitivity of surveys (see, e.g., white-tailed deer surveys for management of a confined population at Plum Brook NASA Station in Sandusky, Ohio; Blackwell et al. 2006). FLIR devices have been used successfully to estimate seasonal abundance of raccoons in north-central Ohio (Blackwell et al. 2006) and to detect polar bear (*Ursus maritimus*) dens in Alaska (Amstrup et al. 2004) and have been recommended for detecting feral goats and removing them from islands (Campbell and Donlan 2005).

The Howard Hughes Medical Institute describes the roles and importance of citizen scientist divers in assessing the expanding population of invasive lion-fish in the Atlantic Ocean off the coast of Florida. Volunteer divers follow established transects to count numbers of lionfish or count the numbers of lionfish for a defined length of time. The data garnered from the volunteers have been used in models to establish a density-dependent growth pattern for the lionfish. The Institute has used the information to develop an interactive program for teaching population dynamics (see Question 4 in the Review and Discussions sections of this chapter).

As a second approach to estimating population numbers, animals can be temporarily removed, marked, released, and then recaptured (box 5.2). Gese (2004) encouraged the use of mark and recapture methods when accurate estimates of population sizes are needed. Ear tags, radio collars, dyes, and physiological markers have been used; alternatively, individuals might be recognized by facial or other features. Capture and recapture might involve the physical capture of the wildlife, or it might be done unobtrusively with such devices as **camera traps** (i.e., individuals can be recognized from photographs). Karanth and Nichols (1998), for example, used camera traps to estimate populations of tigers (*Panthera tigris*) in India. Martorello et al. (2001) used bait-triggered stations and cameras to gather capture and recapture data on black bears in coastal and mountain habitats. Goswami et al. (2007) used similar methods to assess population sizes and other demographic factors for the Asian elephant (*Elephas maximus*). Camera traps can have problems. Séquin et al. (2003), for example, found that alpha males were wary of cameras and were under-represented in their studies on coyotes, a factor that may require consideration when interpreting results for other territorial and social species. Identification of individual DNA from fecal or hair samples have also been used to estimate population sizes (fig. 5.3). Ebert et al. (2012) used DNA in fecal samples to identify enough individuals to allow capture recapture methods for accurate estimates of population size of wild boar in the Palatinate Forest in southwestern Germany.

With a third approach, sometimes called **depletion** or **removal trapping**, members of the population are intensely trapped and permanently removed. Trapping must occur over a brief interval, say three to five days, to minimize effects of immigration, emigration, or nontrapping mortality. The daily take is plotted against cumulative take totals, and regression analysis is performed to describe the relationship. Extrapolation of the resulting equation to the point where no more takes occur (i.e., one of the intercepts) will yield an esti-

---

BOX 5.2 Lincoln-Petersen Index

One capture-mark-recapture method is called the **Lincoln-Petersen Index**. Assuming that the marked individuals stay within the population and mix randomly, and that the marking does not differentially affect mortality of the members (e.g., by making them more visible to predators), the proportion of marked to unmarked members that are taken during subsequent collections can be used to estimate total population size, as below:

$N = M * C/R$, where

N = population estimate in numbers of individuals,

M = total numbers of animals captured, marked, and released,

C = total numbers of animals captured on the second visit, and

R = number of animals captured on the first visit that were marked and recaptured on the second visit.

For example, if 100 green sunfish (*Lepomis cyanellus*) were captured and tagged, then released (M), and if 200 of the sunfish were captured on the second visit (C), of which 50 had tags (R), then

$N = (100) * (200)/(50) = 400$ sunfish.

The population estimate is 400 sunfish. If the area of collection is known, an estimate of density might be given, such as 400 sunfish per surface acre of pond. Repeated samples might be used to give an estimate of variance. Wildlife may be physically captured and marked, or identified unobtrusively using techniques such as identification of unique individual markings with wildlife cameras or DNA sequences with hair samples. Although the concept underlying the Lincoln-Petersen Index is straightforward, models intended to accurately represent real populations can become complicated quickly.

---

mate of population size (N). Again, repeated sampling gives an estimate of variance that can be used to calculate confidence levels around the curve. Related removal methods, including those based on change-in-ratio or catch per unit effort, have also been useful in assessing wildlife populations, particularly those that are trapped or hunted (e.g., Wilson et al, 1996).

Depletion trapping has limited use for species such as canids because it can affect social structure of the

Figure 5.3 Hair samples are one of an increasing number of ways to unobtrusively identify wildlife. This sample is from a grizzly bear at Grand Teton National Park. DNA analyses can be used to identify the individual bear. Credit: Mike Ebinger, USGS. Public domain.

population (Gese 2004). Engeman and Linnell (1998), however, showed how trapping data for removal of brown tree snakes (*Boiga irregularis*) from Guam could be modeled as exponential decay and how the model could be used to plan control strategies. Specifically, the analyses showed that perimeter trapping around areas to be protected was relatively more effective in relation to trapping costs. And the analyses indicated that once snakes were removed from fragmented habitat their recovery was slow. Hence, depletion trapping these areas could be both economic and effective.

Population indices are a fourth, and often most practical, approach to assessing wildlife populations. When it is too difficult or expensive to get estimates of absolute population size or densities, estimates of relative population size might still be obtained. A **population index** (Engeman and Witmer 2000) assesses a change in activity, such as movement to and from a station having an attractant, active burrows, nests, or feeding sites. One might measure presence of signs or tracks around stations, for example, or use tracking tiles to document footprints or tail movements. Although the approach does not yield absolute information, such as population size (N) or ecological density, it allows indirect measurement of changes in population by measuring related changes in activity. For example, by counting the number of active rodent burrows in an orchard before and after use of a rodenticide, one could assess the treatment's effectiveness. Population indices have been used widely to monitor damage and effectiveness of management actions (Engeman and Allen 2000; Engeman and Witmer 2000; Gese 2004, for carnivores).

Engeman and Campbell (1999) used reoccupation of burrows after application of rodent baits to assess the success of the treatment in reforestation areas of Oregon where pocket gophers damage seeds or seedlings. In North America scent stations, used over a 30-year period, have offered insight in the relative changes of coyote populations (Linhart and Knowlton 1975). Engeman et al. (2000) used a passive tracking index to monitor population changes due to trapping coyote and bobcat on two ranches in Webb County in southern Texas. They established tracking plots along dirt roads at about 5-mile (0.8-km) intervals and examined them for spoor (a track, a trail, a scent, or droppings) for two to four consecutive days before and after periods of trapping. Allen et al. (1996) found that a passive activity index would reflect changes in dingo populations while it also monitored other species, including macropods, fat-tailed dunnarts (*Sminthopsis crassicaudata*), feral cats, brushtail possums (*Trichosurus vulpecula*), and rabbits.

Determining ages of wildlife is necessary when relating age to other population dynamics. To do this, one must find a factor that changes in a manner that can be related to age. Often, hard body parts such as bones, teeth, scales, or spines are used. Sometimes the relationship is developed by measuring the trait in captive wildlife of known age. For some wildlife, such as white-tailed deer, irruption, loss, and wear of teeth is commonly used. Bone growth is sometimes used to age rabbits and rodents, particularly the epiphyseal plates at the ends of long bones where growth occurs or the baculum (*os penis*) in some marine mammals. Weight of eye lenses increases over the lifetime of some animals and can be related to age. Sometimes annular rings can be used, as seen in trees, fish scales, and the cementum layers of teeth. Recession of gum lines from the incisors has been used to age mountain lions.

Example

Blackwell et al. (2006) used FLIR cameras to assess and manage white-tailed deer at the Plum Brook Station of the National Aeronautics and Space Administration in Erie County, Ohio. The station is 22 hectares in area and fully enclosed. Beginning in 1998, deer population estimates were standardized at the station using spotlight counts. Beginning in 2005, Wildlife Services began comparing the spotlight method with one using a FLIR camera, finding 11% more animals with the camera. In the winter 2006 estimate, 271 deer were counted by spotlight, whereas 378 were counted with the FLIR cameras. The difference may be due to increasing vegetation within the station, which brought into question the reliability of the spotlight surveys. Public hunts are permitted to manage the herd, with harvest goals based on the most recent estimate of herd size, so it is important to have accurate population counts.

POPULATION MODELS

Statement

Management of human-wildlife conflicts increasingly rely on population models to predict irruptions, to anticipate conflicts, to simulate impacts of control methods, and to evaluate effectiveness after management actions have been taken.

Explanation

Populations can be modeled by modifying the fundamental equation for population growth to include an iterative component, as follows:

$$N_{t+1} = N_t + (b+i) - (d+e),$$

often called the BIDE equation.

While the BIDE equation is itself straightforward, bases for calculating component rates can be more difficult, complicating the model. For example, an estimate of N_t might be based on a ratio coming from one of the capture-recapture methods just described. Compensatory or additive mortality or breeding season and age-group related natality also adds complexity, sometime best simulated using matrix algebra.

Examples

Chapron et al. (2003) used stochastic computer models of wolf population dynamics to examine two approaches to management of wolves in Europe. One approach simulated **zoning**, wherein wolves were left untouched within predesignated areas called wolf zones. They were removed from areas near livestock, called nonwolf zones. The second approach simulated removal of a portion of wolves once a preset **population growth rate** was achieved. With this approach, there were no zones.

This research team used life information for wolves to create their computer simulations, refined with specific information on European wolves when it was available. They divided populations into four age groups, with pack leaders being at least 18 months old. In the model, winter mortality affected the whole population and accounted for annual mortality; subordinates dispersed only if the breeding pair survived; dispersing young sought vacant territory and partners; reproduction occurred in the spring, with first reproduction at 22 months; one litter was produced per year, with pup mortality occurring in the summer to simulate deadly disease; and a census of distribution and status of wolves was conducted each fall. Analyses and simulations were conducted using a computer program called Unified Life Models.

Chapron et al. (2003) drew two important conclusions. First, under zoning scenarios, the long-term viability of the population of wolves is extremely sensitive to the number of packs. Second, an adaptive strategy in which a moderate percentage of wolves is removed when the population reaches a threshold growth rate maximizes the effects of control to reduce livestock losses, while minimizing the risk of extinction of the population.

Sea otter (*Enhydra lutris*) populations in southeast Alaska have increased from about 400 in the 1960s to over 8,000 by 2003. Through their cascading effects as apex predators, the increased numbers of otters have helped restore kelp ecosystem function. The increased numbers of otters have also improved indigenous hunting. The high population of otters competes with people for shell fishing and become entangled in nets. Effective management depends on accurate population estimates, historically difficult to get because of logistical problems. Past attempts at data gathering have been erratic and included a mix of skiff-boat and aerial-based surveys. Strip-based aerial surveys began in 1999.

Tinker et al. (2019) used existing survey data and a Bayesian state-space model to estimate past and current trends in the otter populations. By partitioning sources of error, they were able to provide refined estimates of populations sizes and carrying capacities at the regional and subregional levels. At the regional level, the scientists found the sea otter population had increased to over 25,000 by 2011. Based on patterns generated by the models, the scientists were able to describe a pattern of temporary population declines at subregional levels as members moved into available habitat, followed by population surges in the subregion. They were able to calculate carrying capacities at the subregional levels and at the regional level for the first time for sea otters. Given available unoccupied habitat and current population dynamics, the model predicted regional carrying capacity of three times the present population size. The scientists suggested localized surveys as the most effective way to get data needed to make population estimates for future management actions.

RELATIONSHIPS BETWEEN POPULATIONS AND DAMAGE

Statement
A clear relationship cannot be assumed between the size of a population and the level of damage; such a relationship should be established before population reduction is undertaken.

Explanation
Population reduction is often viewed as the approach of choice for reducing wildlife damage. There are many examples of clear relationships between population increases and wildlife damage. Taylor and Dorr (2003) summarized general relationships between a burgeoning double-crested cormorant (*Phalacrocorax auritus*) population and increased economic losses to catfish farmers and recreational fisheries in North America. Dolbeer et al. (1993) reported an increase in numbers of laughing gulls at Jamaica Bay Wildlife Refuge from 15 nesting pairs in 1979 to 7,629 in 1990, along with concomitant increases in birds striking airplanes at nearby John F. Kennedy Airport, New York. These researchers also reported a decline in bird strikes that correlated directly ($r^2 = 0.97$) with a reduction of the birds by shooting ones attempting to enter airspace of the airport.

Beasley and Rhodes (2008) assessed population densities of raccoons using mark-recapture methods, and Program MARK, at 14 locations in Indiana. They showed a clear correlation between population estimates and damage to field corn; in addition, these scientists demonstrated that the presence of edges between forests and crops were also important in predicting damage. In a study conducted in China, Giefer and An (2020) found a relationship between wild boar damage reported by farmers in household surveys and population size as measured by a model using data from camera traps, remote sensors, and aerial photographs. The model was more effective in predicting damage to households in coniferous and bamboo forests than in broadleaf forests.

Although such relationships between population size and damage often occur, they cannot be assumed.

A direct relationship may not exist, or it may be sufficiently obtuse to prevent its effective use in limiting damage. The presence and relative abundance of a species does not prove it is causing damage. Dolbeer et al. (1979) observed large flocks of grackles in sprouting winter wheat whose stomachs were filled with leftover corn residue, not wheat. Starlings, less abundant than grackles, were removing the germinating wheat seeds.

Further, unless the population is considered in the broader context of the community (see chapter 6), reduction of one population might result in the "release" of another. Zavaleta et al. (2001) provided a useful summary of possible ecosystem responses to consider before eradication of an invasive plant or animal, including food-chain and food-web effects, predator–prey interactions, and herbivore–plant interactions. A wise manager will consider such relationships before embarking on a major campaign of population reduction.

Example
On Stewart Island, New Zealand, exotic cats prey on the endangered flightless parrot *Strigops habroptilus* (Zavaleta et al. 2001). The cats, however, prefer exotic rats. Although it seems desirable to remove the cats, their absence might allow "release" of the rats and create the potential for an even greater impact on the parrot. That concern has prevented management action to eradicate the cats from the island. Zavaleta et al. argue that concerns for mesopredator release is not unique to Stewart Island; similar mixes of cats, rats, and mice occur on at least 22 other islands.

Summary
- Overabundance occurs when a population damages itself, humans or their interests, or the environment. Crowding is related and occurs when population density exceeds a level at which the habitat can maintain healthy population members and the population becomes distressed.
- Populations may need to be reduced when they become overabundant or crowded.
- If eradication is the goal, the population must be reduced below its critical minimum. More often, reducing damage is the goal, and there is an assumed relationship between extent of damage and size of the population.
- Natality and immigration add to populations, whereas mortality and emigration reduce populations. Each can contribute to conflicts, and any action to regulate populations must fall

within the fundamental equation for growth and impact one or more of these rates.

- In resolving human-wildlife conflicts, when the intent is to reduce a population to the lowest size practicable, additive mortality rather than compensatory mortality might be sought.

- Some wildlife, such as microtine rodents or some invasive species, follow a J-shaped growth curve wherein they exceed carrying capacity and crash, often causing damage. Other wildlife self-regulate just below the carrying capacity and can sometimes be managed by lowering the carrying capacity.

- Population growth and each of its component rates are affected by age of population members, often visualized using life tables, survivorship curves, or age pyramids; such displays help researchers to understand dynamics of populations and to assess effectiveness of management.

- Direct and indirect sampling methods are used to measure population sizes and to determine whether management is needed or whether management was effective. Population indices, which provide information on change without measuring population sizes, are economical and often used in managing human-wildlife conflicts.

- Aging a population requires knowledge of some age-dependent factor such as lens curvature or annual rings or scales.

- Computer-based population models can perform complex iterative calculations quickly and accommodate important ancillary factors such as compensatory or additive mortality; they are often used in managing human-wildlife conflicts.

- A relationship between size of a population and level of damage should not be assumed.

Review and Discussion Questions

1. If a population had a natality rate of 3.4 per thousand per year, an immigration rate of 1.2 per thousand per year, an emigration rate of 0.1 per thousand per year, and a mortality rate of 2.0 per thousand per year, what is its growth rate? Suppose a control action increased the mortality rate to 6.4 per thousand per year, but the population responded with an increased natality rate of 4.2 per thousand per year. What was the overall impact of the control action on growth rate? Does this impact on growth rate necessarily have the same impact on damage?

2. Contrast the responses to carrying capacity of a population following a J-shaped curve with that following an S-shaped curve. Explain how this information might be helpful in applying a control method that is based on reducing carrying capacity of a population and indirectly controlling a population.

3. If you were reducing damage by muskrats to a marsh used for treatment of wastewater, and you trapped the muskrats intensively for five consecutive days each month, might the data on trap success be used to assess changes in size of the muskrat population? How? If you gathered data as well on the ages of the muskrats trapped, what shape of life table would you hope to see to confirm a diminishing muskrat population?

4. Howard Hughes Medical Institute (HHMI) Bio-Interactive offers an excellent overview of population modeling, using the invasive lionfish as an example. It is entitled "Lionfish Invasion: Density-Dependent Population Dynamics" (https://media.hhmi.org/biointeractive/click/lionfish-invasion/index.html), and I encourage you to "dive in."

6

Communities, Ecosystems, and Landscapes

More and more, practitioners focus on whole ecosystems or landscapes to both understand and manage wildlife conflicts. With this **synecological** approach, the practitioner effects change within the ecosystem or the landscape often without necessarily paying regard to a specific species. Checks and balances within the system then correct the problem.

Here, I look at attributes of communities, ecosystems, and landscapes that help to understand and manage human-wildlife conflicts. I consider community and landscape composition, diversity and stability, energy flow, food chains and food webs, and succession. I show with examples how each characteristic might be used in resolving such conflicts.

COMMUNITIES, ECOSYSTEMS, AND LANDSCAPES AND THEIR APPLICATIONS TO RESOLVING HUMAN-WILDLIFE CONFLICTS

Statement
Long-term solutions to wildlife conflicts often rely on understanding and manipulating ecosystems or landscapes, viewed by some practitioners as functional units of conflict resolution.

Explanation
Ecosystems are communities interacting with their abiotic components. A **community** is all organisms living in an area at a given time. Ecosystem is a concept that may represent a system that is huge or tiny, one that is either largely independent (i.e., gets most of its energy from the sun) or one (e.g., a city) that is largely dependent on other ecosystems for its energy. The greatest ecological community on earth, constituting all living organisms on the planet, is the **biosphere**. The biosphere interacting with its abiotic components, powered by energy flowing from the sun, is called the **ecosphere**. Biomes, described in chapter 4 as types of habitat, are also communities dominated by particular types of vegetation and associated flora and fauna.

Ecotones are boundaries between ecosystems. The boundaries may be sharp or fuzzy, straight or irregular. Because they are edges, ecotones usually have species from all contiguous systems. In addition, ecotones may provide habitat for species not found solely in any of the systems. Some species, such as white-tailed deer and coyotes, thrive in ecotones. The increasing presence of coyotes in the midwestern and eastern United States during the past few decades might be partly explained by ecotones created by humans. Other species, such as pileated woodpeckers (*Dryocopus pileatus*), martens, and flying squirrels, require space within the deeper recesses of the ecosystem, away from ecotones. The brown-headed cowbird (see chapter 1) is an ecotonal species whose presence in North America has increased along with the proliferation of fragmented landscapes.

Cities are examples of ecosystems that require enormous **energy subsidies** to maintain overall stability. For example, human and pet foods are energy subsidies moved from agroecosystems, sometimes located in distant parts of the globe, into the city. Transportation of the goods is also an energy subsidy, often a considerable one. City dwellers—humans and their pets—are therefore major energetic and economic forces underlying efforts to make modern agroecosystems more productive.

Within cities, the diversity (number of species) of wildlife species is often reduced. Remaining species are often "eury"-type generalists that sometimes conflict with human interests (e.g., McKinney 2002). Thus starlings, the house sparrow (*Passer domesticus*), house finch, striped skunk (*Mephitis mephitis*), eastern gray squirrel, and commensal rodents—as well as feral animals such as cats and dogs—typify urban wildlife in

many cities worldwide. Nonnative species are often more prevalent while species that do poorly with humans, such as the larger carnivores, tend to disappear (e.g., McKinney 2002; see Braczkowski et al. 2018 for an important exception). Overall, species diversity declines from suburbs to city centers. Urban wildlife can damage property, transmit diseases, cause fires, and sometimes attack people or take pets as prey. These wildlife also offer city dwellers glimpses of nature and remind them that humans are irrevocably tied to ecosystems. Cities can be viewed as novel ecosystems with wildlife adapting (e.g., gleaning food from bird feeders, streets, and exposed garbage, Clergeau et al. 1998) to their sometimes-unique environmental circumstances (fig. 6.1, urban foxes). By understanding this, human residents can manage urban environments to increase the diversity of species as well as the number of positive experiences with them.

As cities sprawl into countrysides, suburbs overlap wildlife habitat. Thus, the mountain lion, once removed from urban areas, is again becoming familiar with *Homo sapiens* and their ways (and their pets). The increased presence of mountain lions along with an increased inclination to protect them, as exemplified in California's Proposition 117 (which protects mountain lions and was passed in 1990), and the stage is set for increased human-wildlife conflicts. It is unsurprising that mountain lion attacks on people and pets are on the rise in California and along the eastern foothills of the Rocky Mountains.

Figure 6.1 Fox investigating a garden pet in Birmingham, United Kingdom. Competition between coyotes and foxes is forcing foxes into suburban and urban habitats in the eastern United States. If other countries are an indication, urban foxes will do well in the United States. In the United Kingdom, for example, urban foxes increased from about 30,000 in 1995 to about 150,000 in 2017. Photographed by Oosoom.

Climate is a dynamic part of ecosystems. Planetary climate changes have occurred long before human presence and force corresponding changes in ecosystems. The changes can be dramatic, resulting for instance in perturbations and irruptions of populations that increase human-wildlife conflicts. As one specific example, climatic shifts resulting from movement of the El Niño-Southern Oscillate increased rainfall and vegetation and subsequently caused a population irruption of the white-footed mouse (*Peromyscus leucopus*) in the Four Corners region of the southwestern United States in 1993. The mouse carried hantavirus, which subsequently sickened and killed people in the area. Hantavirus has since spread to other parts of North America.

Global climate change, as defined by the Intergovernmental Panel on Climate Change (Pachauri et al. 2014), is any change in climate over time due to either natural variability or human activity. Measured as radiative forcing, changes in abundance of greenhouse gases (carbon dioxide, methane, nitrous oxide) and aerosols (primarily sulfate, organic carbon, black carbon, nitrate, and dust) in both solar radiation and land surface properties drive warming or cooling of the global climate. Other radiative forces include tropospheric ozone and halocarbons. **Global warming** is the increase of the earth's air and oceanic temperatures. The IPCC (Pachauri et al. 2014) summarized a broad base of evidence for a warming planet, facilitated by human activity.

Climate change can impact wildlife in at least five ways. (1) Wildlife species change their geographic distributions, keeping within geographically shifting niches, as climatic factors change. (2) Species alter timing of seasonal events, including migration, molting, and reproduction, in response to shifting climatic cues such as temperature and rainfall. (3) Biotic relations within communities and ecosystems change in nature and complexity as new species, such as exotic invasive species and diseases, enter the mix. (4) More frequent catastrophic weather events, such as droughts, floods, and precipitation, influence the structure and function (e.g., the diversity) of populations and communities. (5) Humans respond to climatic changes with adjustments in land use forcing further wildlife adjustments.

Increased conflicts between humans and wildlife can be anticipated as global warming proceeds. For example, extinctions of species and catastrophic climatic events will reduce diversity of ecosystems, simplifying checks and balances and facilitating surges in numbers of some species. Changing climate will also favor introducing more invasive species and the

spread of zoonotic diseases such as avian influenza, malaria, and dengue fever (e.g., Shope 1992).

Landscape ecology (Troll 1939) is a form of **biogeography** (study of geographic distribution of organisms in space and time). **Landscape ecology** is the geographic description of ecosystems, their interrelations in time and space, and the interrelations of their functions. The term emphasizes spatial patterns in relation to process and function and often encompasses multiple ecosystems and large areas such as the Appalachian Mountains or Brazilian rainforests (Turner et al. 2001). Whereas ecosystems such as forests or grasslands are somewhat homogeneous, landscapes are heterogeneous (e.g., cropland, including surrounding forests). Landscape ecology was first used to improve human-based landscapes in Europe, and humans often strongly influence one or more of the ecosystems within a landscape.

Landscapes do not always encompass large areas. Current usage focuses more on relationships of heterogeneity and geospatial distribution to process and function than on scale. More important than size is the notion of an ecosystem or habitat (see **patches,** discussed shortly; fig. 6.2) submerged within a broader system or habitat—that is, a **matrix**. Like an ecosystem, a landscape has no clearly defined minimum size or acreage.

For example, Delibes-Mateos et al. (2018) evaluated natural and anthropogenic factors that encouraged the presence of European rabbits in Spain. The scientists selected media reports accessible on Google Advanced Search using the Spanish key words for "pest" and "rabbit" and analyzed the reports using a model that selected habitat features that favored the presence of rabbits. The researchers found that rabbits were concentrated in central-southern parts of Spain, that damage was greatest in cropland and habitats where vegetation was scarce, and that rabbit presence coincided with railways and highways. They used the findings to relate recent increases in crop damage by rabbits to infrastructure advances in Spain and suggested the information would help design management strategies.

DeVault et al. (2007) used landscape concepts to assess wildlife damage to corn and soybean fields in Indiana. Using global positioning satellite coordinates, the scientists related distances of damage to neighboring forests and human habitation, comparing actual distances with those randomly generated in a stratified sampling design. The researchers concluded that fields adjacent to forests would sustain the greatest damage and those near human habitations would sustain the least. Targeted removal of destructive species

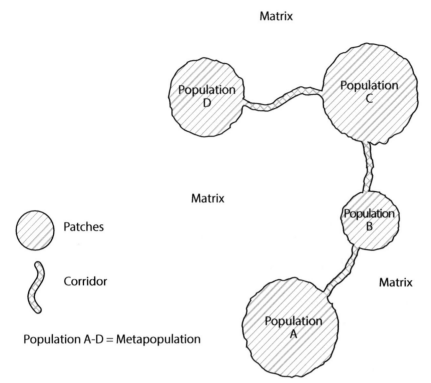

Figure 6.2 With metapopulations, populations reside in patches (such as meadows) connected via corridors within a general matrix (such as a forest). Genes can move between populations. Illustrations by Lamar Henderson, Wildhaven Creative LLC.

along crop/forest interfaces appeared cost-effective management. Ferraz et al. (2003) used a landscape approach to study damage by capybaras (*Hydrochoerus hydrochaeris*) to cornfields in Brazil. Fields closer to adjacent forests and water received the most damage. Tourenq et al. (2001) found that rice fields in France carried increased risk of damage by the greater flamingo (*Phoenicopterus ruber roseus*) when placed near wooded margins and natural marshes. As mitigation, the researchers suggested scaring devices and hedgerows placed strategically along the ecotones.

Landscape ecology has introduced other concepts and terminology, including patches, corridors, and habitat fragmentation. **Patches** are areas of relatively homogeneous habitat surrounded by a broader but different habitat, the **matrix** (see fig. 6.2). For example, a patch of forest may be surrounded by broad expanses of cornfields, the matrix. The patches may or may not be interconnected by thinner runs of similar habitat, or **corridors**. The notion of patches stems originally from that of island biogeography (MacArthur and Wilson 1967) in that patches can be visualized as "islands." **Habitat fragmentation** occurs when a habitat needed for wildlife is continually made smaller and smaller or is isolated from previously interconnected habitats; this occurs slowly in nature but is accelerated by human activity. Habitat fragmentation may relegate a

wildlife population to one or more patches. Often the patches themselves get smaller and smaller. As patches shrink, space along edges become relatively greater and interior space diminishes. If the patches are interconnected with corridors sufficient to allow movement and genetic exchange between the populations of wildlife, the interconnected individual populations can be thought of as a **metapopulation** (Levins 1969).

Landscape ecology has application in explaining and in managing wildlife problems associated with fragmented landscapes. Landscape fragmentation apparently increased contact between domestic and African wild dogs, allowing transfer of diseases and subsequent decline of the African wild dog (*Lycaon pictus*; Ward et al. 2009). Root et al. (2009) used genetic information from raccoon metapopulations in Pennsylvania to assess effectiveness of an **oral rabies vaccination** program. The researchers found isolation, i.e., restricted gene flow in some metapopulations and recommended that future programs adjust delivery of vaccines to ensure all metapopulations are reached. Losses of farm-raised channel catfish have been attributed to the eastern metapopulation of the American white pelican (*Pelecanus erythrorhynchos*; King and Anderson 2005) rather than the whole pelican population.

Sufficient deer metapopulations and corridors exist among fragmented forest habitat in some US states

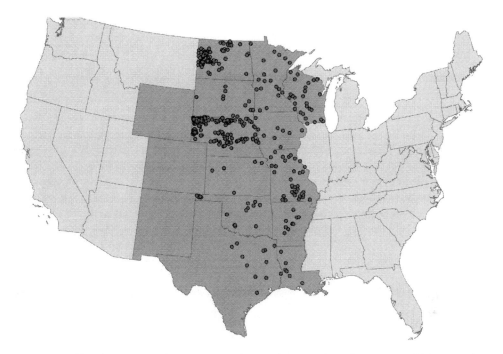

Figure 6.3 Locations of breeding mountain lion (*Puma concolor*) populations in the midwestern United States over the next 25 years, as predicted by LaRue and Nielsen (2016, 122). Improved prey base and suitable habitat will allow re-establishment of the northward range and long-term population stability. Reprinted from *Ecological Modelling*, vol. 321/121–129, 2016, with permission of Elsevier.

to set the ecological stage for reestablishment of mountain lions. Recent reintroductions or those underway—e.g., elk in at least seven states—will provide additional resources. Free-roaming mountain lions have been confirmed in at least 13 midwestern states. LaRue and Nielsen (2016) used landscape level modeling to postulate that mountain lions would re-establish in eight stable breeding populations in the midwestern United States over the next 25 years (fig. 6.3). Confirmed presence of reproducing mountain lion populations would represent a northward range extension from Louisiana and Arkansas, where established populations are already known or postulated.

Metapopulations have also served as the conceptual bases for megaparks in managing elephants and their damage in southern Africa (van Aarde and Jackson 2007).

Examples

Belant (1997) argued that gulls at landfills and airports should be managed at city or county, i.e., landscape, levels. He reviewed integrated methods of control, including architectural design and construction methods to reduce roof substrate, to manipulate turf height in loafing areas, to cover refuse landfill sites or compost facilities, to drain water, and to keep wildlife from water using wire grid systems. He argued that effective management was needed at all locations within the landscape because effective management at just one location, such as a landfill, would simply force birds to an unprotected site.

The importance of managing airports at the landscape level was reinforced by Pfeiffer et al. (2018). They evaluated landscape features for potential adverse effects on bird strike rates at 98 civil airports in the United States between 2009 and 2015. The researchers found adverse effects for proximity of wetlands, water, and cultivated crops within 3, 8, and 13 km radii from the airports. The findings are being used to encourage collaborative efforts among wildlife managers, airport planners, and landowners to reduce probabilities of airstrikes by minimizing crops (especially corn) and increasing distances between patches of open water.

COMMUNITY STRUCTURE AND DIVERSITY

Statement

Landscapes and ecosystems have structures and diversity that shape the nature, patterns, and extent of human-wildlife conflicts; understanding these factors can help predict damage and plan effective management strategies.

Explanation

Landscape or ecosystem structure is called its **physiognomy**. Structure exists in vertical and horizontal dimensions, and, in time, in both aquatic and terrestrial ecosystems. Vertical physiognomy includes structure and patterns of physical attributes such as soil strata (e.g., horizons) as well as grosser structures such as canyons, mountains, and rock faces. Horizontal physiognomy includes structure and patterns of physical attributes such as soil types, rocks, ponds, and streams or rivers.

Physiognomy also includes how biotic communities within landscapes and ecosystems are structured. Vertical structure might mean separation of the biotic community into mature tree canopies, young tree subcanopies, and shrub layers. Vertical structure is the basis for classifying vegetation types into standard physiognomic layers, which can then be used for mapping. Horizontal structure includes patterns exhibited by biotic communities. Physiognomy of the biotic community may change with time, as with succession (discussed in this chapter) or with changes in community composition due to season, such as migration or aestivation.

Abiotic physiognomy influences biotic physiognomy, as with altitudinal zonation (changes in structure and composition of living organisms as one advances up a mountain) or the distribution of trout in streams in relation to patterns of rocks used for hiding and predation. Biotic physiognomy can also influence the abiota, as, for instance, when invading species (i.e., the first species into a new area) break down rock with root hydraulics or alter the vertical structure of soil.

Structure often exists even though it may not be expected or obvious. For example, pelagic communities of seabirds display vertical structures while over-flying oceans, with the smaller mollusk and fish-catching birds such as shearwaters and petrels nearer the water and the larger, scavenging (and sometime stealing) albatrosses above them.

Landscape and ecosystem ecologists look for patterns in such structures. Ecologists use tools such as satellite imagery to capture images and then conduct geospatial analyses using specialized software that reveals such patterns. The ecologists then relate patterns to processes (e.g., energy flow) and functions (e.g., production of timber) of landscapes and ecosystems.

Ward et al. (2009) discussed the importance of landscape structure in understanding and managing wildlife diseases. The researchers noted, for example, that aerial delivery of rabies vaccines is less effective in hilly areas because density of delivery is reduced on slopes and also in urban and suburban areas; in these conditions, hand delivery or other approaches such as plastic bag baits may be needed. Features such as rivers and mountains can serve as natural barriers to movement of zoonoses or their vectors. Including such features in overall management strategies can improve effectiveness and efficiency of control. Ward et al. (2009) suggested that mathematical models, based on metapopulations in fragmented landscapes, can help plan details of zoonotic disease management, such as rabies vaccination strategies. K. Smith et al. (2005) used GIS mapping of rabies occurrence and mathematical models to design cordon sanitaires that minimize public exposure and serve as barriers to the spread of rabies. These scientists also suggested that landscape-based models improved the understanding of spatially transmitted foot and mouth disease on British farms, including aerial plumes from local farms and longer distance movement by farmers, veterinarians, and contaminated vehicles.

Horizontal distribution of species can be **random, clumped, even** (**uniform**), or some combination of these patterns (fig. 6.4). Distribution can be strongly influenced by available physical and biological re-

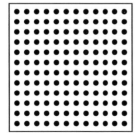

Uniform

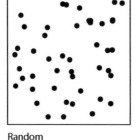

Random

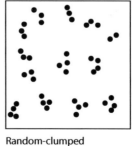

Random-clumped

Figure 6.4 Patterns in distribution of living organisms influence how one estimates and manages the population. Illustrations by Lamar Henderson, Wildhaven Creative LLC.

sources. Information on distribution can be gathered from ground studies, by aircraft, or by using global positioning systems that allow precise location of objects, and they can be analyzed using geographic information systems. Knowing patterns of damage can help determine where to focus management actions, such as setting traps or baits. Often information on species distribution is also needed to ensure statistical validity of sampling methods, such as in stratified sampling.

Biodiversity is related to the stability of ecosystems and landscapes. Some biotic communities are dominated by a single or a few species whereas diversity in other communities is spread more evenly among species. If the ecosystem has a particularly abundant species, such as a prairie dog (*Cynomys*), an elephant, or a wolf, that species may be a keystone species. **Keystone species** disproportionately dominate the structure, diversity, and function of the entire ecosystem (Paine 1966). A predator on top of a food chain is called an **apex predator** and is often also a keystone species. Actions by humans or nature that affect keystone species are of particular interest to managers because impacts on keystone species can destabilize the whole system with ensuing increases in human-wildlife conflicts.

A keystone species differs from a **foundation species**, a dominant primary producer in an ecosystem. For example, kelp is a foundation species in kelp forests. Here the sea otter is the keystone species because it controls populations of herbivores feeding on the kelp, and hence indirectly constrains the growth of kelp.

Roemer et al. (2007) argued that removal of larger predators, including such apex predators as the grizzly bear (*Ursus arctos horribilis*), the gray wolf, and the mountain lion (see chapter 2), caused a predator–prey disequilibrium in North America, resulting in some of today's conflicts with wildlife. In northern Yellowstone Park, for example, overpopulation of elk from lack of wolf predation led to heavy browsing of willows and loss of beaver wetlands. A similar disequilibrium involving overabundance of moose, willow loss, and reduced wildlife diversity occurred in the southern Greater Yellowstone Ecosystem. Evidence supporting this notion has been provided by some researchers (e.g., Beschta et al. 2018). Others (e.g., Kauffman et al. 2010) have presented evidence that contradict aspects of the theory. More generally in North America, mesopredators, especially the coyote, have replaced the wolf as apex predator, leading to a broad range of conflicts with people such as those already described for duck production in the prairie pothole region and for the fox in Illinois (see chapter 4).

Johnson et al. (2007) found that survival of small to mid-sized marsupials was greater in areas of Australia where dingoes were present as apex predators. These researchers argued that high densities of dingoes protected small ground-dwelling marsupials from extinction by preventing release of fox and cats, intense mesopredators of smaller marsupials. The authors argued conversely that removal of dingoes from wide areas of Australia since settlement indirectly facilitated the extinction of marsupials by releasing mesopredators, a consideration for future conservation strategies worldwide. Impacts of **mesopredator release** have been documented broadly in oceanic, freshwater, and terrestrial ecosystems globally for species, including lizards, rodents, birds, rabbits, fishes, and sea turtles (Prugh et al. 2009).

Species diversity is often measured with indices—e.g., Simpson's or Shannon's Diversity Indices—or functions that measure diversity, such as **dominance** (the relative importance of one or more species), **richness** (number of species and distribution of individuals among species), and **evenness** (also the distribution of individuals among species). These indices are also important in identifying damage, its distribution, and its management.

Examples

In the Muddy River drainage of the Mojave Desert, the exotic invasive plant saltcedar (*Tamarix ramosissima*) has taken over riparian habitats. Aggressive efforts have been made to eradicate saltcedar; those efforts, in turn, have reduced both structural and compositional diversity of riparian ecosystems. Fleishman et al. (2003) explored the diversity and vertical physiognomy of the affected habitats. They used species richness indices to assess both avian and floral diversity and measures of total vegetation volume to assess vertical physiognomy. These scientists found that avian diversity related more closely to vertical physiognomy than to species richness of flora. The species contributing to richness sometimes had overlapping functions, serving mostly as "insurance"—i.e., as backups—rather than in daily ecosystem function. The researchers found that birds used exotic saltcedar for nesting and protection, especially if native flora were nearby. They concluded that bird diversity did not suffer from presence of saltcedar. They suggested gradual removal of the saltcedar, allowing time for replacement with native vegetation, rather than aggressive removal and concomitant loss of the volume of vegetation needed by the birds.

ENERGY FLOW

Statement

Trophic level cycling needs to be considered when deciding on some management options, such as using pesticides; trophic interactions described by Lotka-Volterra equations can predict outcomes, useful in managing damage, of predation or competing species.

Explanation

Movement of energy through ecosystems begins with sunlight (most common), an energy subsidy from another ecosystem, or from chemically derived energy in a few benthic oceanic ecosystems. Energy is captured by producers using **photosynthesis** (or, rarely, chemosynthesis), which converts the sun's energy into a chemical form. Energy is then passed via food chains and webs to other consumers, including herbivores, omnivores, carnivores, and detrivores. **Herbivores** eat plants whereas **carnivores** eat animals. **Omnivores** consume both plants and animals. And **detrivores** (scavengers and decomposers) eat **detritus**, the dead remains of formerly living organisms, usually a mix of materials from the body and fecal matter.

By the **first law of thermodynamics**, energy is neither created nor destroyed but can be converted from one form, such as sunlight, to another, such as the chemical energy bound in adenosine triphosphate (ATP). ATP serves as currency for chemical energy exchanges in cells of all known life. By the **second law of thermodynamics**, however, some energy is lost as heat during each transfer. Major paths of energy flow through ecosystems include **grazing circuits** (food webs based directly on sunlight, such as prairies or savannas) and **detritus food circuits** (food webs based on consuming dead organisms).

Energy flows through organisms in communities or ecosystems from one trophic (feeding) level to the next; these collective interactions are described as food chains, food webs, and food pyramids. Constrained by the first two laws of thermodynamics, only about 10% of energy is transferred from one **trophic level** to the next; the rest is lost as heat. **Food chain concentration** or **biomagnification** can occur for some compounds, such as organochlorine pesticides (these are lipid soluble), second-generation **anticoagulant** rodenticides, and radioactive materials, usually concentrating by an order of magnitude at each level of transfer. Food chain concentration should be considered whenever pesticides are used in managing wildlife damage.

For example, strychnine baits may only be used to control pocket gophers in the United States. Baits must be placed in burrows to prevent hazards to nontarget animals in the food web that might access baits placed above ground. In California, concerns have included the threatened California red-legged frog (*Rana aurora draytonii*), the California tiger salamander (*Ambystoma californiense*), and the federally endangered San Joaquin kit fox (*Vulpes macrotis mutica*).

Predation and **parasitism** (wherein one species benefits at the expense of another) are specific types of trophic interactions. These are described by Gause's law (chapter 4) and Lotka-Volterra equations (box 6.1; fig. 6.5).

If one were to combine the zero isoclines (box 6.1) of two interacting species on the same graph, one could envision four possible combinations (see fig. 6.5b–e). In viewing the graphs, pay particular attention to the size of K relative to the size of K/α as they intercept each axis. In the first situation (see fig. 6.5c), K_1 is greater than K_2/α_{21} along the N_1 axis, indicating that species 2 is a relatively weak competitor. K_1/α_{12}, however, is greater than K_2, indicating that species 1 is a relatively strong competitor. Following the directions of the arrows, interactions will move along the isoclines until species 1 achieves its carrying capacity, and in this case species 2 becomes extinct; species 1 wins (circled isocline intercept with N_1).

In the second situation (see fig. 6.5d), just the opposite occurs. Here, carrying capacity for species 2 is greater than the competitive strength of species 1, indicating that species 1 is a relatively weak competitor. Following the arrows in this situation, species 2 achieves its carrying capacity and species 1 becomes extinct.

In the situation in figure 6.5e, both species have carrying capacities that exceed their relative competitive strengths. Here, an unstable equilibrium is achieved, with some arrows pointing toward the intercept of the isocline and some away from it. Eventually, one species will force the other to extinction, but it is not possible to predict the winner from this model.

In the last situation, (see fig. 6.5f), the competitive strengths of both species exceed their respective carrying capacities, and, as indicated by all the arrows moving toward the intercept of the isoclines, a stable equilibrium is achieved. This is what might occur, for example, when some niche differentiation has occurred.

Although the models presented here are relatively simple and only approximate actual interactions between species, more sophisticated versions are available with better predictive capabilities.

BOX 6.1 Lotka-Volterra Models

Lotka (1925) and Volterra (1926) independently derived models to describe relationships between two species. The models explain predator and prey and competitive relationships between species. The models are relatively easy to understand and are useful. They also can be derived from equations that model logistic growth of populations:

$$dN/dt = r_{max} N \, {}^{*}(K-N)/K \text{ (from chapter 5).}$$

The $(K-N)/K$ term is a mathematical way to model intraspecific competition—competition within the population or species. As N approaches K, intraspecific competition leads to a geometric decline in rate of growth, r. Lotka-Volterra replaced this term with one that models both intraspecific and interspecific (i.e., between species) competition. In a **Lotka-Volterra model**, N_1 represents the population size of one (i.e., the first) species, and N_2 represents the population size of the other (i.e., the second) species. In analogous ways, carrying capacities are represented by K_1 and K_2, and growth rates by r_1 and r_2.

Although one might expect competition to increase as N_1 and N_2 increase, one cannot expect the competitive effects of each species to be the same. For example, species 1 might have twice the competitive effect on species 2 as species 2 has on species 1, the total competitive effect being $(N_1 + 0.5 \, {}^{*} N_2)$. To account for this as a general factor, the Lotka-Volterra model adds a constant, α_{12} (the value 0.5 in the preceding example) to represent the effect of species 1 on species

2, and a constant α_{21} to represent the effect of species 2 on species 1. Thus,

$$dN/dt = r_{max1} \, {}^{*} N_1 \, {}^{*} (K_1 - (N_1 + \alpha_{12} \, {}^{*} N_2))/K_1,$$

represents the total impact of competition on growth rate on the first species, and

$$dN/dt = r_{max2} \, {}^{*} N_2 \, {}^{*} (K_2 - (N_2 + \alpha_{21} \, {}^{*} N_1))/K_2,$$

represents the total impact of competition on growth rate on the second species. The two equations are the Lotka-Volterra model.

Now, using the model, consider the range of pressures that are exerted on the sloped line (i.e., the zero **isocline** where $dN_1/dt = 0$; fig. 6.5a). An analogous zero isocline can be calculated for species 2: $dN_2/dt = 0$ (fig. 6.5b). The equations can be mathematically manipulated and reduced, so when $dN/dt = 0$,

$$N_1 = K_1 - \alpha_{12} \, {}^{*} N_2,$$

and

$$N_2 = K_2 - \alpha_{21} \, {}^{*} N_1.$$

To understand what is happening conceptually, it might help to do iterative (mental or actual) calculations using the Lotka-Volterra models, substituting some hypothetical values for the constants and variables. The arrows in figure 6.5a indicate the impact of size of one population on the other species: sometimes increasing or sometimes decreasing but always toward the sloping line.

Lotka-Volterra models can predict outcomes of the removal of predators for prey or for other predators (i.e., the mesopredator release effect). Examples are the outcomes of the removal of wolves for populations of other, lesser predators such as coyotes or skunks. Similarly, the models can predict outcomes of removing a species when it competes with another (i.e., the **competitor release effect**). The models have been used to help predict competition between kangaroos and livestock in Australia (Moloney and Hearne 2009), means of minimizing trapping strategies for management of beaver problems in North America (Bhat et al. 1993), impacts of white-tailed deer culling on management of Lyme disease in North America (Li and DeMasi 2009), the effects of introducing wolves

to the ecosystems of Yellowstone National Park and dingoes to Australia (Baker et al. 2017), and the effects of fear on predator–prey relations using Michaelis-Menten type prey harvesting (Lai et al. 2020).

Examples

FOOD CHAIN CONCENTRATION. Anticoagulant rodenticides are used globally to control rodents in fields, homes, and for conservation. The compounds, such as second-generation brodifacoum, have been particularly effective in eradicating rodents from islands where rats (*Rattus rattus*, *R. norvegicus*, and *R. exulans*) have caused the extinction or near extinction of seabirds and have impacted lizards, invertebrates, plants, and overall ecosystem functioning (box 6.2) (Duron

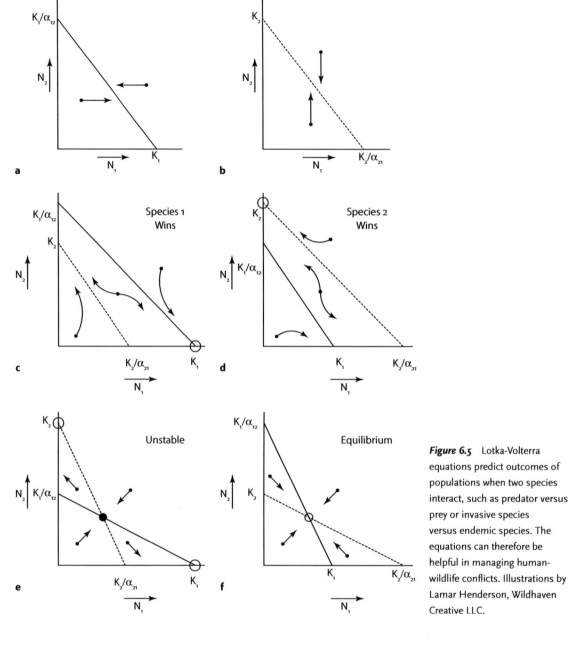

Figure 6.5 Lotka-Volterra equations predict outcomes of populations when two species interact, such as predator versus prey or invasive species versus endemic species. The equations can therefore be helpful in managing human-wildlife conflicts. Illustrations by Lamar Henderson, Wildhaven Creative LLC.

et al. 2017). But second-generation anticoagulants are more potent (provide a lethal dose in a single feeding) than the first-generation ones (require multiple feedings to provide a lethal dose), and their residues can be retained in carcasses for long periods of time so that compounds like brodifacoum can be passed on to predators and scavengers. Although secondary poisoning may not be a concern on remote islands that lack predators or scavengers, such poisoning can occur on mainlands or on islands, given the persistence

and movement of the residues in both aquatic and terrestrial food webs (Regnery et al. 2019).

Exposure to nontarget wildlife is widespread. López-Perea and Mateo (2018) summarized studies documenting exposure of a broad range of nontargeted wildlife to anticoagulant rodenticides in Europe and the United States. The list included predatory birds such as hawks, eagles, falcons, and owls and mammals such as fox, lynx (*Lynx lynx*), bobcat, mountain lion, European badger (*Meles meles*), stoat,

BOX 6.2 Conservation of Island Species

Since Darwin's accounts of his discoveries on the Galápagos Archipelago, islands have been seen as safe havens for some of the world's most diverse and unique wildlife. Isolated from unceasing evolutionary pressures of continents, species finding **sweepstakes routes** (i.e., extremely uncommon routes, as unlikely as winning a sweepstakes) onto these natural refuges were free to evolve new lifeforms at leisurely paces counted in eons. Birds, sea reptiles, and other oceanic species found islands and used them as safe areas for reproduction and other activities. Because defenses against common diseases, herbivory, and predation were mostly unnecessary, island species invested their energies in more unique evolutionary paths. Islands, representing only about 5% of the earth's land mass, became home to about 39% of the world's critically endangered species (Keitt et al. 2019).

Humans changed all of that by compressing travel to islands into days or years rather than the eons it took for some organisms formerly using sweepstakes routes. Sometimes accidentally, sometimes intentionally, people brought herbivores, omnivores, predators, and diseases to many of these natural refuges (see chapter 2). Once there, rats, mice, foxes, feral pigs, goats, cats, and dogs efficiently removed unprepared wildlife from the Galápagos Islands and thousands of other islands. Three species of rat colonized over 80% of the world's archipelagos, a process that continues today. These rodents attacked seabirds sitting on nests, and, because of their omnivorous habits, broadly affected whole ecosystems. The three species alone have accounted for the extinction of at least 50 species on 40 different islands (H. Jones et al. 2008). Because of their broadly based herbivory, feral goats were identified in the 1970s as threatening insular plant species (Lucas and Synge 1978). Here, however, seed banks and restricted accessibility of localized plants made extinctions unlikely. Instead, broad-scale impacts on vegetation were common, with cascading effects on other wildlife.

Recognizing their roles in the cause of this problem, and the importance of island flora and fauna to global biodiversity, humans have more recently attempted to eradicate invasive species from critical islands and put an end to the extinctions. To date, there have been over 1,200 eradication attempts on over 700 islands. Observed benefits of successful eradications have included measurable increases in growth of seabird and other resident wildlife populations, assisted and unassisted recolonization by oceanic wildlife, and preparation of islands for later conservation introductions.

Much remains to be done. Only about 100 species of terrestrial vertebrates, or about 12% of the 860 critically threatened species that occur on islands, have benefited from invasive mammal eradications. By contrast, about 47% of similarly endangered seabird species have benefited. Rates of island eradications are increasing. And with improved technologies, a better understanding of the human dimensions surrounding these issues, and improving funding strategies, one can envision a brighter future for many of these wildlife species (Keitt et al. 2019).

American mink (*Mustela vison*), and raccoon. The researchers also developed a model for bioaccumulation of anticoagulant residues in predators and illustrated it using barn owls (*Tyto alba*).

Direct impacts can be substantial. For example, Molenaar et al. (2017) found that 19 of 110 red kites (*Milvus milvus*) studied had been poisoned by anticoagulant rodenticides. The kites were part of a group found dead and tested for pesticides from areas in England and Scotland where reintroductions had been made earlier. Indirect effects of exposure can also occur. Rattner et al. (2020) found prolonged effects on prothrombin clotting time when American kestrels (*Falco sparverius*) were exposed to brodifacoum. Consequently, the kestrels were more sensitive to brodi-

facoum when exposed to it a second time. Serieys et al. (2018) found that a combination of urban factors along with chronic exposure to anticoagulant rodenticides increased the susceptibility of bobcats to mange.

Concerns for anticoagulant intoxication of wildlife have led to restrictions on use of the compounds. For example, anticoagulant rodenticides are not classified as biocidal in the European Union because they persist, bioaccumulate, and are toxic. Although the European Union recognizes the need for continued use of these compounds, they have encouraged use of rodenticides with lowered dosages of second-generation anticoagulants until suitable alternative control methods are found. In 2014, California prohibited the use of second-generation anticoagulants except by certified pest control operators, allowing only the use of first-generation anticoagulants by landowners and home-owners. Despite the prohibition, second-generation residues continue to be found in mountain lion and other California wildlife (Rudd et al. 2018). Gabriel et al. (2015) found toxicity poisoning the cause for 10% of 167 fisher (*Pekania pennanti*) deaths they investigated and suggested the continued impacts were due to illegal use of the anticoagulants for marijuana production in the forests. Gabriel et al. (2018) found that northern spotted owls (*Strix occidentalis caurina*) and barred owls (*Strix varia*) were also exposed to these anticoagulants in areas with illegal marijuana fields.

LOTKA-VOLTERRA. Caut et al. (2007) used population data and a generalized Lotka-Volterra population model to explain the irruption of mice as a competitor release effect. These scientists used one equation to model rat population dynamics and a second to model population dynamics of the mice. The researchers added a third equation to simulate prey species common to both the rats and the mice. From the equations, they found four points of equilibria: when both species, rats and mice, were eradicated; when the rat was eradicated, allowing the mouse to achieve its maximum population size as a point of equilibrium; when the rat was controlled but not eradicated, triggering competitor release; and intense control but not eradication of either, also triggering competitor release under some conditions.

The study suggests possible irruption of minor competitors during periods when both pests are being controlled. Such irruptions would occur when the control methods are beginning to impact the superior competitor but before they impact the inferior one. Consequences could be serious, in that the irruptions could lead to unintended effects on endangered flora and fauna. Prior knowledge of such releases might help to design management strategies that accommodate responses from both species.

SUCCESSION

Statement

Wildlife found in earlier successional stages tend to have characteristics that lead to conflicts with people; pest irruptions sometimes follow efforts to push back ecological succession.

Explanation

Invader species—the first species into an area—eventually alter their environment so they can no lon-ger survive in it. These species are then replaced by others, replacement continuing in an orderly manner until a **climax community** (energetically balanced) emerges, a process called **succession**. Each successional community is called a **seral stage**, or **sere**. **Climatic climax communities** are mature seres for major biomes or regions. **Edaphic climaxes** are seres made stable for prolonged periods by factors such as recurrent fires or floods. **Primary succession** occurs when organisms enter an area previously devoid of life. **Secondary succession** occurs when an ecosystem is pushed back to an earlier sere by some factor, such as fire, herbicide use, or intensive grazing. Earlier seres are more productive than later seres but tend to have less diversity and stability.

Species characterizing early seres are sometimes called **R-growth species**. They tend to reach reproductive age early, produce large numbers of offspring but invest little in their care, and have short generation times (i.e., be short-lived). They are often opportunistic. Species characterizing mature seres are called **K-growth species**. They tend to produce fewer young but invest more in parental care, reproduce late in life, and have long generation times.

Human-wildlife conflicts are sometimes the consequence of pushing an ecosystem to an earlier sere, simpler and more productive but also characterized by R-growth species. People use various energy subsidies to do this, from slash-and-burn, to herbicides, brush-hogs, and "chaining." People remove fencerow vegetation or consolidate smaller farms into larger ones, all cropped in a single high-yield variety to boost production. Although such systems are productive, they are also less stable than mature systems. Species may be released from the complex community interrelationships that had kept them in check, such as predation, parasitism, or abiotic limiting factors associated with nutrient cycling.

Activity of some wildlife can also push ecosystems to earlier seres. Rooting by feral pigs has this effect. Overabundant populations of herbivores such as deer, goats, or rabbits push ecosystems to earlier seres. Browsing by overabundant deer can eliminate the regenerating capability of a forest (e.g., Bradshaw and Waller 2016).

Example

In the late 1950s, a portion of the island of Cotabato, Philippines, was set aside for human settlement as part of a governmental transmigration program. The new settlers cleared large areas and began growing crops. An outbreak of ricefield rats (*Rattus argentiventer*) ensued in which the rodents destroyed much of the harvest; international assistance was required to provide food relief and prevent massive human starvation. Similar outbreaks continue to occur for analogous reasons throughout many parts of the globe.

Summary

- Some practitioners view ecosystems and landscapes as fundamental units for managing wildlife conflicts.
- Ecotones provide habitat for species such as deer, coyotes, and parasitic cowbirds that often conflict with human interests.
- Cities are ecosystems that support generalist wildlife such as starlings, house sparrows, commensal rodents, and feral cats and dogs. Cities require major energy subsidies that provide economic incentives to push agriculture into intense production, exacerbating conflicts with wildlife.
- Climates impact conflicts with wildlife. Global warming portends even more dramatic conflicts.
- Landscape concepts—such as ecosystems within matrices, habitat fragmentation, and metapopulations—are useful in understanding human-wildlife conflicts and planning management strategies.
- **Community physiognomy** greatly influences patterns and distribution of wildlife damage in ecosystems.
- Species interact in ecosystems through food chains and webs; interactions of predators and competitors can be described by Lotka-Volterra models.
- Food chain concentration should be considered before using any pesticides; Lotka-Volterra models can predict whether removal of predators or competitors will cause ecological problems such as the mesopredator release effect.
- Some human-wildlife conflicts are explained by efforts to push communities from mature successional stages to earlier ones, as is done with most agriculture; early communities tend to have move irruptive species and simplified interactions.

Review and Discussion Questions

1. It has been argued that diversity equals stability in communities and ecosystems. How might this notion apply to wildlife damage and its management?
2. Landscape ecology, habitat fragmentation, metapopulations, and corridors are often concepts used in conservation biology. How might they be applied in the understanding and management of human-wildlife conflicts? How might these concepts point to ties between conservation biology and wildlife damage management?
3. Choose three characteristics of seral stages, and contrast early versus late seres for each characteristic. Then relate each characteristic and stage to the likelihood of wildlife damage. Be specific.
4. Choose a specific study where Lotka-Volterra equations were applied to wildlife damage management. Assess the contribution of the models to the study. Were they helpful? Give some specific examples of other applications to wildlife damage management.
5. Contrast zinc phosphide and brodifacoum, both vertebrate pesticides, in terms of food chain concentration. How do these compounds compare to organochlorine pesticides such as DDT (dichloro-diphenyl-trichloroethane)?

PART III · **SURVEYS OF DAMAGING SPECIES**

In this part, I look at damage of which wildlife are the proximate cause, both globally and in North America. Included are wildlife diseases and zoonoses.

7

Global Conflicts: Invasive and Endemic Species

To achieve global notoriety as an invasive, the species must have attained sufficient distribution as an exotic and then served as the proximate cause of sufficient conflicts with people to attract their interest. Endemic species just need to serve as the proximate cause of damage.

Statement

Most successful invasions involve exotics of little interest to humans. Some contribute to overall diversity at local scales and may be otherwise beneficial. Most agricultural species are exotics. A relatively few exotic species serve as the proximate causes of harm to people or their interests and are termed **invasive species**.

Explanation

Some invasive species, such as the European starling, have seen management actions for years. Only in recent decades, however, has the full extent of invasive species come to the forefront and been added to the responsibilities already within the purview of those managing human-wildlife conflicts. Successful invaders are found among all types of wildlife, not just vertebrate species. Invertebrates such as zebra mussels (*Dreissena polymorpha*) in the United States and American crayfish (e.g., the signal crayfish *Pacifastacus leniusculus*) in Europe have come under management scrutiny, as have invasive plants such as the blackberry (*Rubus*) and the mimosa (*Mimosa*).

All species continually attempt to invade new areas. A species of wildlife leaving an area is an emigrant, but it is an **exotic** when it appears in the new habitat. Most species either fail to make it to new lands or, once there, are unable to become established. Failures occur for many reasons, but all lead to an inability of the exotic species to complete a necessary life cycle and spread in the new area. For example, **protists** (unicellular or multicellular organisms without specialized tissues) causing avian malaria were unable to colonize the Hawaiian Islands until an appropriate vector, the house mosquito (*Culex quinquefasciatus*), was introduced.

The likelihood of a successful invasion depends partly on the number of individuals that emigrate and partly on the persistence (frequency) of invasions, called **propagule pressure**. Propagule pressure, combined with factors such as available vectors and pathways for invasion, suitable characteristics of the invading species (e.g., wings facilitate invasions by birds), and available resources at the new location (e.g., Britton-Simmons et al. 2008) underwrite successful invasions (Fine 2002; Lockwood et al. 2005).

Successful invasions occur in at least three stages: transport and introduction to the new area; establishment of one or more colonies via naturalization (adapting to or altering environments to reliably complete life cycles); and invasion of surrounding areas from the colonies (after Di Castri 1990; Blackburn et al. 2015). Lockwood et al. (2013) include an important fourth stage for exotic invasive species, the stage at which the invaders have caused sufficient impact to attract the attention of humans.

Invaders tend to be generalists that can survive for long periods of time with limited resources. They are often small, mobile, agile, and inconspicuous. Invaders tend to be R-growth species (see chapter 6) with high fecundity and growth rate, limited genetic variation, and simple life cycles. They are effective in dispersing seeds or young either themselves or with the help of environmental factors (phoresis) such as wind or other organisms. Though experience with a similar environment helps, the presence of fewer competitors, predators, or pathogens is probably more important to success (Di Castri 1990). Although these attributes characterize some successful invaders, there is sufficient case-by-case variation that the attributes have little predictive value.

Environmental conditions can facilitate successful invasions. Open, simple, cleared areas seem particularly vulnerable, whereas more complex areas with a diverse array of organisms seem less vulnerable. For instance, areas settled by humans or having recent geological or evolutionary disturbance are more likely than undisturbed areas to allow successful invasions (Lockwood et al. 2013). Thus, a recent volcanic eruption or flood that clears land can make it accessible to invasive species. Areas where humans transport people and products for trade, colonization, or war are also vulnerable to invasion (Di Castri 1990). Rapid transit, such as use of aircraft, particularly facilitates the transfer of exotic wildlife diseases and zoonoses because the speed of transit is sometimes faster than the incubation times of the diseases.

At least two notions have been proposed to explain vulnerability to invasive wildlife: the biotic resistance hypothesis and the biotic acceptance hypothesis. The biotic resistance hypothesis is based on the relationship between diversity and stability of an ecosystem; ecosystems with the greatest species diversity will also be the most resistant to invasions by exotic species. Conversely, ecosystems with less diversity will be more susceptible. If an ecosystem has many niches but they are already filled by diverse organisms, none are available for newcomers. This hypothesis has been used to explain isolation (i.e., some niches remain unoccupied) as a factor in vulnerability to invasions of islands, such as New Zealand, Australia, and the Hawaiian Islands, and some peninsulas (e.g., Fagerstone 2007). But the hypothesis is not supported by some investigations, for example, the successful lionfish (*Pterois volitans*) invasions in two protected marine areas in the Caribbean with differing biological diversities (Cobián-Rojas et al. 2018).

The biotic acceptance hypothesis supposes that some ecosystems are sufficiently diverse to allow for new niches for newcomers (Stohlgren et al. 2006). The new niche might be split from an existing one, or perhaps not all niches are fully used. Biotic resistance and biotic acceptance are opposing hypotheses, and which is most useful in predicting and managing invasions is still being debated. I do not favor one over the other, but I present both because they help conceptualize how successful invasions may occur.

A third theory on successful invasions has been proposed: the human activity hypothesis. Human activity affects propagule pressure. For example, invasion of a new plant species on the Hawaiian Islands might occur naturally every 20,000 or 30,000 years. Assuming wildlife can hitch a ride, human transport by boat and airplanes might allow repeated introductions within a year, thereby increasing propagule pressure by a factor of thousands. Hence, human transport could greatly facilitate invasive species completing their first stage, moving to the new area.

Once there, human disturbance of habitat might help by simplifying ecosystems making them susceptible to invasion (i.e., human release hypothesis; Zimmermann et al. 2014). Studies (e.g., Leprieur et al. 2008; Gallardo and Aldridge 2013) have shown that indicators of human activity relate directly to the abundance of invasive species. Examples include human economic indicators such as real estate sales and demographic factors such as human population density in a geographic area or percentage of the area that is urbanized (e.g., Taylor and Irwin 2004; Pyšek and Richardson 2006).

The economic costs of invasive species can be astronomical. For example, invasive insects alone cost humans at least $70 billion per year in goods and services with an additional $6.9 billion in health-related expenses (Bradshaw and Waller 2016 in Diagne et al. 2020). Diagne et al. (2020) have developed a public database, called InvaCost, that addresses economic costs of invasive species on a global scale. The database included over 2,400 cost estimates at the time of their publication. The developers described the database as "the most up-to-date, comprehensive, harmonised and robust compilation and description of economic cost estimates associated with biological invasions worldwide" (1).

Table 7.1 lists 100 of the world's worst exotic invasive species, as proposed and presented by the Global Invasive Species Database (http://www.iucngisd.org /gisd/search.php) of the International Union for the Conservation of Nature (IUCN). A primary purpose of the list is to bring attention to issues surrounding invasive species so that the issues are resolved, and the species can be removed. For example, **rinderpest** virus (*Rinderpest virus*) was declared eradicated from the globe and removed from the list in 2013. Giant salvinia (*Salvinia molesta*), a free-floating complex of closely related ferns native to Brazil, was added (Luque et al. 2014). Widely distributed as an ornamental, the ferns have escaped and infested lakes and waterways throughout the globe. The plants form thick mats that impede boat traffic and block sunlight from aquatic ecosystems.

The listing does not rank species. Rather, it chooses species that represent a balance of taxa causing problems, limiting any genus to one example even though there may be other species within the genus that also cause problems for humans. For instance, the western mosquitofish (*Gambusia affinis*) is included on the list whereas the eastern mosquitofish (*Gambusia holbrooki*), similar in population size and distribution and also concern for people, is not. The list guided my choice of examples.

Given the broad scope and opportunities for discovery and observation, the study of invasive species benefits from involvement of citizen scientists. Johnson et al. (2020) described 26 such initiatives whose data were used in scientific publications related to invasive species.

Examples

I start with plants, progress to the smallest animals, and move upward in size and complexity.

PLANTS. Many successful plant invasions are the consequences of human-facilitated transport (e.g., Pauchard and Shea 2006) and characteristics of the plants. Baker (1965) saw an "ideal" plant invader as a plastic perennial that germinates under most conditions, grows fast, flowers early, self-pollinates, produces an abundance of seeds that disperse widely, reproduces vegetatively as well, and competes effectively. He thought that only some of these characteristics were needed for the invasion to be successful. Plants invading natural areas tend to be aquatic or semiaquatic, nitrogen-fixing legumes, grasses, climbers, or clonal trees, different (about 25% overlap) species than those invading agroecosystems (Daehler 1998).

Water hyacinths (*Eichhornia crassipes*), originally from South America, are now found throughout the globe. They grow and spread quickly, choking waterways and blocking the penetration of sunlight into water, thereby reducing the diversity of aquatic ecosystems (Lowe et al. 2000) (fig. 7.1). Water hyacinths can become dense enough and cover sufficiently to disrupt waterflow and the production of hydroelectric power, as occurred at the Owen Falls Dam facility in Uganda in the late 1980s (Kateregga and Sterner 2007). In China, the aquatic plants have formed heavy mats in the watershed of the Three Gorges River Dam and affected water quality (Ding et al. 2008; Xiong et al. 2018). Commercial sale and trade of water hyacinths and other aquatic invasive weeds are important pathways for their spread. Humair et al. (2015) found that over 500 invasive seed plants are being traded daily on the internet.

Miconia (*Miconia calvescens*), South American tropical plants with huge red and purple leaves, are sold as ornamentals throughout the tropics. Fruit-eating birds helped them become established in the wild in Tahiti beginning around 1937. Miconia have now invaded over half of the island and are dominant canopy trees in many areas. Several endemic species are either threatened or extinct because of it. Miconia were introduced as ornamentals into Hawaii in the 1960s, are on other Pacific islands (Lowe et al. 2000), and have now invaded the rainforests of Australia, New Caledonia, and Sri Lanka (González-Muñoz et al. 2015).

INSECTS, ARTHROPODS, AMPHIBIA, AND REPTILES. Exotic invasive arthropods, amphibia, and reptiles tend to be omnivores, have high reproductive rates, are capable of high population numbers and densities, and are hard to detect because of their behavior or small size (Pitt et al. 2005). Some, such as the cane toad, are poisonous to people, domesticated animals, and other wildlife. Others, such as the Caribbean tree frog (*Eleutherodactylus coqui*), make loud, incessant noises, reducing value of real estate. In sufficient numbers, released exotic arthropods, amphibia, and reptiles can outcompete their native counterparts, dramatically altering both aquatic and terrestrial ecosystems.

TABLE 7.1   *One hundred of the world's worst invasive alien species*

MICROORGANISM

Avian malaria	*Plasmodium relictum*
Banana bunchy top virus	*Banana bunchy top virus*

MACRO-FUNGI

Chestnut blight	*Cryphonectria parasitica*
Crayfish plague	*Aphanomyces astaci*
Dutch elm disease	*Ophiostoma ulmi*
Frog chytrid fungus	*Batrachochytrium dendrobatidis*
Phytophthora root rot	*Phytophthora cinnamomi*

AQUATIC PLANT

Caulerpa seaweed	*Caulerpa taxifolia*
Common cordgrass	*Spartina anglica*
Giant salvinia	*Salvinia molesta*
Wakame seaweed	*Undaria pinnatifida*
Water hyacinth	*Eichhornia crassipes*

LAND PLANT

African tulip tree	*Spathodea campanulata*
Black wattle	*Acacia mearnsii*
Brazilian pepper tree	*Schinus terebinthifolius*
Cogon grass	*Imperata cylindrica*
Cluster pine	*Pinus pinaster*
Erect pricklypear	*Opuntia stricta*
Fire tree	*Morella faya*
Giant reed	*Arundo donax*
Gorse	*Ulex europaeus*
Hiptage	*Hiptage benghalensis*
Japanese knotweed	*Fallopia japonica*
Kahili ginger	*Hedychium gardnerianum*
Koster's curse	*Clidemia hirta*
Kudzu	*Pueraria montana var. lobata*
Lantana	*Lantana camara*
Leafy spurge	*Euphorbia esula*
Leucaena	*Leucaena leucocephala*
Melaleuca	*Melaleuca quinquenervia*
Mesquite	*Prosopis glandulosa*
Miconia	*Miconia calvescens*
Mile-a-minute weed	*Mikania micrantha*
Mimosa	*Mimosa pigra*
Privet	*Ligustrum robustum*
Pumpwood	*Cecropia peltata*
Purple loosestrife	*Lythrum salicaria*
Quinine tree	*Cinchona pubescens*
Shoebutton ardisia	*Ardisia elliptica*
Siam weed	*Chromolaena odorata*
Strawberry guava	*Psidium cattleianum*
Tamarisk	*Tamarix ramosissima*
Wedelia	*Sphagneticola trilobata*
Yellow Himalayan raspberry	*Rubus ellipticus*

AQUATIC INVERTEBRATE

Chinese mitten crab	*Eriocheir sinensis*
Comb jelly	*Mnemiopsis leidyi*
Fishhook flea	*Cercopagis pengoi*
Golden apple snail	*Pomacea canaliculata*
Green crab	*Carcinus maenas*
Marine clam	*Potamocorbula amurensis*
Mediterranean mussel	*Mytilus galloprovincialis*
Northern Pacific seastar	*Asterias amurensis*
Zebra mussel	*Dreissena polymorpha*

LAND INVERTEBRATE

Argentine ant	*Linepithema humile*
Asian long-horned beetle	*Anoplophora glabripennis*
Asian tiger mosquito	*Aedes albopictus*
Big-headed ant	*Pheidole megacephala*
Common malaria mosquito	*Anopheles quadrimaculatus*
Common wasp	*Vespula vulgaris*
Crazy ant	*Anoplolepis gracilipes*
Cypress aphid	*Cinara cupressi*
Flatworm	*Platydemus manokwari*
Formosan subterranean termite	*Coptotermes formosanus shiraki*
Giant African snail	*Achatina fulica*
Khapra beetle	*Trogoderma granarium*
Little fire ant	*Wasmannia auropunctata*
Red imported fire ant	*Solenopsis invicta*
Rosy wolf snail	*Euglandina rosea*
Spongy (gypsy) moth	*Lymantria dispar*
Sweet potato whitefly	*Bemisia tabaci*

AMPHIBIAN

Bullfrog	*Lithobates catesbeianus*
Cane toad	*Bufo marinus*
Caribbean tree frog	*Eleutherodactylus coqui*

FISH

Brown trout	*Salmo trutta*
Carp	*Cyprinus carpio*
Largemouth bass	*Micropterus salmoides*
Mozambique tilapia	*Oreochromis mossambicus*
Nile perch	*Lates niloticus*
Rainbow trout	*Oncorhynchus mykiss*
Walking catfish	*Clarias batrachus*
Western mosquito fish	*Gambusia affinis*

BIRD

Indian myna bird	*Acridotheres tristis*
Red-vented bulbul	*Pycnonotus cafer*
Starling	*Sturnus vulgaris*

REPTILE

Brown tree snake	*Boiga irregularis*
Red-eared slider	*Trachemys scripta*

MAMMAL

Brushtail possum	*Trichosurus vulpecula*
Domestic cat	*Felis catus*
Goat	*Capra hircus*
Gray squirrel	*Sciurus carolinensis*
Macaque monkey	*Macaca fascicularis*
Mouse	*Mus musculus*
Nutria	*Myocastor coypus*
Pig	*Sus scrofa*
Rabbit	*Oryctolagus cuniculus*
Red deer	*Cervus elaphus*
Red fox	*Vulpes vulpes*
Ship rat	*Rattus rattus*
Small Indian mongoose	*Herpestes javanicus*
Stoat	*Mustela erminea*

Source: After Lowe et al. 2000, with permission.

Figure 7.1 Water hyacinth have invaded aquatic ecosystems throughout the globe over the last six decades, partly a response to untreated sewage water and other nutrients that feed the plants. These are Landsat images of Valsequillo reservoir that provides water to the Puebla municipality in central Mexico. Top: Image taken on January 10, 2000. Bottom: Image taken on January 9, 2020. Water hyacinth (gray) now covers almost half of the reservoir. Images by Lauren Dauphin, using Landsat data from the US Geological Survey. Description from story by Kasha Patel.

Humans often provide the initial transportation within or between land masses for invasive arthropods, amphibia, and reptiles, sometimes intentionally and sometimes not. Thus, the American bullfrog (*Lithobates catesbeianus*) was transported throughout the globe for its value as a food and for sport. Introductions of Burmese pythons (*Python molurus bivittatus*) or cobra species into the Florida Everglades probably followed their transport into the United States as pets. The release of brown tree snakes in Guam was accidental, but again humans probably provided the transportation.

Crazy ants have caused extensive damage to ecosystems on islands (McGlynn 1999). The ants form colonies in the canopies of tropical forests, allowing multiple colonies with multiple queens. Supercolonies with 300 queens have been observed on the Christmas Islands, with infestations covering about 28% of its 10,000 hectares of rainforest (Abbott 2005). Crazy ants are omnivores with both predatory and scavenging habits. They eat grains, seeds, and detritus. They "farm" scale insects and aphids by both overseeing crawlers and protecting them from predators. Scale and aphids can consequently also invade some island forests, resulting in mold growth, canopy dieback, and death of canopy trees.

Crazy ants have destroyed red land crab (*Gecarcoidea natalis*) populations on Christmas Island. The ants spray the crabs with lethal doses of formic acid as their paths cross. Crab carcasses then become a high-protein food for the ants (O'Dowd et al. 2003). Fifteen to twenty million crabs have been killed since 1989, eliminating the crab as a keystone species on parts of the island. In those areas, litter cover has doubled as has seedling species richness. Changes in seedling densities and canopy holes have cascaded through the food web, reducing the presence of some endemic species. Similar impacts have been observed on other islands. Presence of the ant also favors introduction of exotic rats and cats.

The ants have also affected reproductive and foraging behavior of the Christmas Island thrush (*Turdus poliocephalus*) and the emerald dove (*Chalcophaps indica*; Davis et al. 2008), affected the roosting habits of the now extinct Christmas Island pipistrelle (*Pipistrellus murrayi*), and are affecting the avoidance and resting behavior of the one remaining endemic mammal, the Christmas Island flying fox (*Pteropus natalis*; Dorrestein et al. 2019).

Cane toads were introduced from Central America into sugarcane-growing parts of the world to control beetles. The beetles avoided the toads by climbing to higher parts of plants; the toads then preyed on other wildlife. From egg to adult, cane toads contain toxic bufadienolides, so predatory wildlife and pets, and occasionally people, are poisoned. Cane toads eat some threatened species, outcompete other native frogs for breeding sites, and transmit diseases such as *Salmonella* (Shanmuganathan et al. 2010).

About 100 cane toads were introduced into Australia in 1935 to control sugarcane beetles. Today the cane toads are found throughout Australia's tropics and subtropics and have reached western Australia. Models of global warming predict a further extension southward. Following the toad's arrival at the Kakadu National Park, native predators, including the quolls

(*Dasyurus*) and large goannas (monitor lizards; *Varanus*), declined markedly. Flavor aversion learning was used to train quolls to avoid cane toads, and the "smart" quolls were then released into the park. The quolls avoided the toads but were attacked by dingoes and other predators (Jolly et al. 2018).

American bullfrogs are endemic to the eastern United States but were introduced over the last century into other states and to at least 40 other countries in Asia, Europe, and Central and South America (Lever 2003). Populations now occur in Argentina, Brazil, Costa Rica, Ecuador, Uruguay, and Venezuela in South American and in Belgium, France, Germany, Greece, and Italy in Europe (Ficetola et al. 2007). Tadpoles of these bullfrogs are large and outcompete the larvae of native species; the adults prey broadly on native species, including other amphibia (Kats and Ferrer 2003; Snow and Witmer 2010). Blaustein and Kiesecker (2002) found that bullfrog tadpoles can affect how native tadpoles use their environment, making them more vulnerable to predation by fish. Bullfrogs carry frog chytrid fungus (*Batrachochytrium dendrobatidis*; see chapter 9) and are partly responsible for distributing this emerging infectious disease (Snow and Witmer 2010).

Established wild populations of the red-eared slider turtle, native to the Mississippi Valley region of the United States, are now found in Australia, France, Italy, Spain, Japan, and Taiwan (Ficetola et al. 2009). These turtles are shipped worldwide as popular reptilian pets (Connor 1992). Sales peaked in the late 1980s and early 1990s along with popularity of the *Teenage Mutant Ninja Turtles* television series and movies. Since 1975, the sale of red-eared slider turtle eggs and turtles under 4 inches (small enough for a child to put in his or her mouth) as pets has been prohibited in the United States because the reptiles can transmit *Salmonella* bacteria (e.g., Nagano et al. 2006). Sliders are still raised for sale in other countries, both as pets and for other purposes (T. Williams 1999).

Docile hatchlings and young turtles become aggressive adults, growing up to 13 inches long and living over 30 years. After becoming unmanageable, the pets are discarded in local ponds and waterways by former turtle enthusiasts. Increased numbers of turtles coincided with decreased numbers of birds in ponds of several parks in London after the "Ninja Turtle" craze subsided. Discarded turtles can outcompete local wildlife for basking sites (fig. 7.2). These turtles also consume local plants and algae, invertebrates, fish, frog eggs and tadpoles, and aquatic snakes. Cadi and Joly (2004) described their impact on European pond turtles (*Emys orbicularis*). In addition, the turtles can hybridize with others such as the indigenous yellow-bellied slider (*Trachemys scripta scripta*) in Florida or the Big Bend slider (*T. gaigeae*) in New Mexico (Stuart 2000). Further, Meyer et al. (2015) found genetic evidence that parasite "switching" could occur between the invasive slider and the endemic Mediterranean pond turtle (*Mauremys leprosa*) in France and Spain, raising concerns that trade in these sliders might facilitate the transfer of unwanted parasites.

FISH. Most exotic fish are introduced as food sources, for sport angling or commercial fishing, or for aquariums or ornamental ponds. Aquaculture is a growing industry worldwide and contributes to introductions of exotic fish and diseases into native ecosystems. Predatory fish such as bass (*Micropterus*) and trout (*Oncorhynchus*) compete with native species. Carp, introduced worldwide for sport and food, muddy the water and otherwise degrade aquatic habitat. Diseases carried by introduced fish can broadly affect native species.

Leprieur et al. (2008) found six major drainage basins worldwide where exotic fish represented more

Figure 7.2   Red-eared sliders (*Trachemys scripta*) competing with a mallard (*Anas platyrhynchos*) for a basking site. Photo by Mbz1.

than a quarter of the total number of species per basin: the Pacific Coast of North and Central America; southern South America; western and southern Europe; central Eurasia; South Africa and Madagascar; and southern Australia and New Zealand. They note that these areas, based on the World Conservation Union Red List, also have the highest proportions of fish species facing extinction. The Northern Hemisphere has the greatest number of exotic fish. Leprieur et al. found significant correlations between the presence of exotic fish and human activity but not with either biotic acceptance or biotic resistance. Among human factors, gross domestic product of river basins best predicted the presence and impacts of exotics.

Common carp have received global distribution partly because they are an inexpensive source of protein from fish that are resistant to stress from handling and can survive in waters with low oxygen concentrations (Arlinghaus and Mehner 2003). According to the Food and Agricultural Organization of the United Nations, common carp were the tenth most farmed item in world aquaculture in 2017, accounting for 3.7% of the world's quantity of all species produced (Cai et al. 2019). They also have been distributed worldwide as ornamentals (e.g., as nishikigoi or koi), often being put in ponds or local waterways and sometimes used as baitfish (Aguirre and Poss 2000). In addition, carp are introduced to clear waterways of aquatic weeds. Once in a waterway, the carp are able to negotiate turbid waters and leap over obstacles a yard high (Koehn 2004), moving into unintended areas and becoming invasive.

Carp destroy vegetation and increase water turbidity by rooting and dislodging plants. They may uproot or consume aquatic macrophytes or prevent light from reaching them by increasing turbidity (Vilizzi et al. 2015). Carp prey on eggs of other fish and may have been responsible for the decline of the razorback sucker (*Xyrauchen texanus*) in the Colorado River basin (Taylor et al. 1984).

Brown trout (*Salmo trutta*) are native to Europe and Asia, but pure native populations probably now exist only in a few places such as in Corsica. They were introduced for angling worldwide, including North and South America, Africa, Asia, Australia, and New Zealand. First introductions in the United States were into the Pere Marquette River (now Baldwin River) in Michigan in 1883 (Courtenay et al. 1984). Because of heavy angling and low or nonexistent natural reproduction in some areas of the United States, brown trout are sometimes propagated in fisheries and restocked in streams and ponds annually.

These trout are aggressive predators that strongly impact other fish by **displacement behavior**, competition, and predation (Budy and Gaeta 2017). They displaced native adult brook trout (*Salvelinus fontinalis*) throughout the northeastern United States (Fausch and White 1981), replaced cutthroat trout (*Oncorhynchus clarkii*) in some large rivers (Behnke 1992), and contributed to the decline of golden trout (*O. aguabonita*) in the Kern River and Lahontan cutthroat trout (*O. clarkii henshawi*) in Lake Tahoe. In upland waters of Australia and New Zealand, brown trout prey on indigenous endangered or vulnerable fish or exclude them competitively so they become fragmented populations. Brown trout may have been involved in extirpation of an endemic New Zealand grayling (*Prototroctes oxyrhynchus*; McDowall 1990). In Japan, introduced brown trout have displaced native white-spotted charr (*Salvelinus leucomaenis*) in a part of the Ishikari River in Hokkaido (Takami et al. 2002) and are aggressive invasive piscivores in other lakes.

Mozambique tilapia (*Oreochromis mossambicus*) are native to southern Africa. They have become popular for mosquito control and are relatively easy to raise, growing uniformly, quickly, and surviving under a broad range of environmental conditions. These tilapia have been introduced broadly and are now found in many subtropical and tropical habitats throughout the world (Russell et al. 2012). Once released, they compete indirectly with native fish for food and nesting habitat and directly by preying on smaller fish. They reduce plant growth because of a rooting behavior that increases water turbidity (Russell et al. 2012). Striped mullet (*Mugil cephalus*) are threatened in Hawaii because of competition with these tilapia, as are desert pupfish (*Cyprinodon macularius*) in California's Salton Sea. The fish have caused havoc in Australian streams since their introduction for mosquito and weed control in the 1970s. Tilapia were found in El Junco, a volcanic lagoon in a volcanic mountain in the Galápagos, in 2006. The population of tilapia was expanding and consuming a copepod important for eating algae and preventing blooms in the lagoon. Rotenone, a piscicide, was used to eradicate the fish and allow the lagoon to restore itself. Ironically, the Mozambique tilapia may themselves become threatened in their native range, where they are hybridizing with an introduced fish, the Nile tilapia (*O. niloticus*).

Nile perch (*Lates niloticus*) are native to the Nile River system, Lake Mariout, and some West African river systems and occur in the Zaire (Congo) River system and in the lakes Albert and Turkana. They were introduced into Lake Victoria in 1954 and, after several decades, exploded in biomass as an apex predator. In the process, about 200 species of smaller fish, including many endemic and unique haplochromine cichlids (*Haplochromini*), were either greatly repressed in numbers or disappeared (Ligtvoet et al. 1991; Kaufman 1992). These perch became the basis of a major export industry to Europe, United States, Australia, and New Zealand as well as a local food source. Because the flesh is more oily than other local species, the fish require a long smoking time, increasing the demand for firewood with consequent effects on coastal forests.

Aggressive fish, Nile perch can grow over 6 feet in length and weigh over 400 pounds. Their presence in Lake Victoria has supported not only a commercial but also a growing tourist sport fisheries industry, so ecological impacts are considered in relation to economic benefits by governments of countries that adjoin the lake (Ogutu-Ohwayo 2004; Aloo et al. 2017). These perch have been introduced into lakes in countries around the world. Attempted introductions into reservoirs in Texas in the 1970s and 1980s have been unsuccessful (Howells and Garrett 1992).

Mosquitofish are small fish native to eastern and southern United States. They were introduced into many waterways during the early 1900s to control mosquitoes and are now believed to be the most widely spread freshwater fish in the world, often associated with degraded waterways (Lee et al. 2018). An opportunistic omnivore, they eat mosquito larvae but also a broad range of algae, crustaceans, insects, and amphibians, including the larvae and eggs of indigenous species. Mosquitofish influence trophic structure, including rare fish and invertebrates. Impacts on invertebrates may be a particular concern when these fish were introduced for mosquito control into normally fishless waters. Ironically, mosquitofish are probably no more efficient at eating mosquito larva than local predators (Haas et al. 2003).

BIRDS. Birds are among the most common natural invaders of new areas. They are often found on islands because their ability to fly facilitates moving within and between land masses. Even small birds such as the arctic tern (*Sterna paradisaea*) or hummingbirds (*Trochilidae*) routinely fly great distances as part of their migratory life cycles. Occasionally storms and winds move birds away from their accustomed migratory paths and into new territories where they might succeed as invasive species.

People have also transported birds. Some were taken to newly colonized lands as reminders of homelands. Others were imported to solve problems such as controlling insects that damage crops. Many were imported because they are exotic, aesthetically pleasing, and serve as pets. Once in the new land, flight can help birds find areas suitable for colonization and for further invasions.

As with other invasive wildlife, successful bird species tend to be generalists with broad diets that can adapt to many types of environments, including those disturbed or inhabited by humans. Some are aggressive, outcompeting local species for food, nesting space, and cover. Some can have several broods of young in good seasons and make large investments in parental care. Three bird species are listed among the 100 world's worst invasive pests: the European starling, the red-vented bulbul (*Pycnonotus cafer*), and the Indian myna bird (see table 7.1).

The European starling originated in Europe, southwest Asia, and North Africa. Starlings, often introduced for aesthetic reasons, are now distributed globally except in the Neotropics. These birds were introduced into New Zealand partly to control agricultural insects, and they subsequently became pests in both New Zealand and Australia. They were introduced in the United States as a bird referenced by Shakespeare and have since spread throughout the country and into Canada and Mexico.

Starlings congregate in large flocks and can cause economic damage to crops, including fruit and grain. These birds, particularly females during the reproductive season, feed on a broad range of insects, including some that attack corn and other crops. Although beneficial in this context, starlings are generally seen as more damaging than beneficial in most cropland settings and as particularly damaging to fruit crops (Anderson et al. 2013). The bills and eye placement of starlings are adapted for foraging on ground invertebrates (Linz et al. 2018), and they have the evolutionary advantage over frugivorous birds of being able to probe with an open bill (GISD database). At livestock feeding lots and dairy farms, starlings take livestock feed, and they are a particular concern during winter months when other food sources are not available (Linz et al. 2018). Further, starlings carry diseases transmissible to livestock such as *Salmonella*, Shiga toxin–producing *Escherichia coli*, and *Yersinia*. (Linz et al. 2018). Because they often travel between farms and mingle with livestock, starlings can transfer diseases between farms (Gough and Beyer 1981).

This species competes aggressively for nesting cavities, either natural or those constructed by other species. Ingold (1998) found that starlings usurped nest-site cavities to the detriment of northern flickers (*Colaptes auratus*) and red-bellied woodpeckers (*Melanerpes caroli-nus*); Frei et al. (2015) found this competition the strongest factor affecting nest survival of the threatened red-headed woodpeckers in southern Ontario.

Starlings damage property with nest construction and are messy defecators, defacing properties where they nest. Their droppings can serve as a medium for the fungus histoplasmosis (*Histoplasma capsulatum*; Linz et al. 2018). Because starlings often forage in flocks near airports, they are a concern for collisions with aircraft during takeoffs and landings (Dolbeer et al. 2000), although this is a species that also encounters aircraft beyond airport boundaries (DeVault et al. 2016).

Red-vented bulbuls originated on the South Indian subcontinent but have become established throughout the Pacific region, including Fiji, Samoa, Tonga, the Hawaiian Islands, and New Zealand, as well as in Dubai, United Arab Emirates. They have been found more recently in Europe (Malaga, Spain) and mainland United States (Houston, Texas; Thibault et al. 2018). Bulbuls inhabit dry scrub, plains, and cultivated lands in their native habitat. Bulbuls prefer dry lowlands where they were introduced, becoming an agricultural pest by damaging fruit, flowers, beans, tomatoes, peas, and ripening soft fruit such as bananas. Bulbuls facilitate the dispersal of seeds of some invasive plant species such as the invasive tree miconia in Tahiti (Meyer and Florence 1996). On Oahu (Hawaii), the birds can cause economic damage to orchids and anthuriums (Thibault et al. 2018).

Common or Indian myna are native to India. They were introduced throughout the tropical world to control agricultural insect pests. Myna are now found on all continents except Antarctica and many of the Pacific Islands. As with starlings, mynas compete with native species for nest sites in cavities (e.g., Charter et al. 2016). The birds also destroy young chicks and eggs and evict small mammals (e.g., Tindall et al. 2007). They sometimes "mob" other birds or mammals, e.g., native possums in Australia. They can carry diseases and parasites and amplify those in the invaded ecosystem, affecting ecosystem health (Chalkowski et al. 2018). The mynas damage grape and other fruit crops, including apricots, apples, pears, strawberries, and gooseberries. Their droppings can carry diseases such as histoplasmosis.

MAMMALS. Globally invasive mammals seem particularly tied to human activity. These mammals can be grouped into four categories. First are commensal mammals, including the ubiquitous house mouse and Norway rat, but also four other rat species that have traveled with humans. Second are feral mammals, released from being agricultural animals, livestock, or pets. Third are game mammals, introduced into new lands because they offered sport. Finally, there are species introduced for other reasons, such as pest control, research, medicine, pets, or aesthetics. For example, rhesus monkeys (*Macaca mulatta*) have been raised and used for research but have become a problem on some islands where they reside.

Once established, exotic invasive mammals can serve as the proximate source of conflicts with humans and their activities, their livestock, and pets. They can damage both agroecosystems and natural ones, sometimes destroying functions of ecosystems and endangering other wildlife. Possibly because of their long ties with humans, commensal mammals are also among the most difficult to manage.

Norway rats probably originated in China but are now found worldwide. This species is among the largest of the three common commensals and is fossorial (burrowing). Norway rats live with humans, thriving in urban areas. Strains have been domesticated for research and as pets. These rats are euryphagic and can adapt to local resources. For example, the rats eat fish when living near fisheries and can dive for mollusks in the Po River of Italy (Galef 1980). They eat seeds, seedlings, and agricultural crops and damage property. They can eat sufficient seeds and seedlings at times to reduce the ranges of desirable plants. These rodents consume and contaminate stored foods and spread diseases such as trichinosis, rat bite fever, viral hemorrhagic fever, and hantavirus (*Orthohantavirus*). Norway rats have caused or contributed to the extinction or reduced ranges of many birds, mammals, reptiles, and invertebrate species through predation and competition, particularly on islands (see Box 6.2).

Next to humans, the house mouse is probably the world's most widely distributed mammalian species. Originally from Eurasia and northern Africa, house mice have traveled with humans for at least 8,000 years (Cucchi et al. 2005; Cucchi and Vigne 2006). They have been pets as well as commensal for at least 3,000 years (García-García 2020). These mice damage crops and consume or contaminate stored foods and can be subject to periodic irruptions (fig. 7.3). House mice have contributed to the decline of some species, including some albatrosses (*Diomedeidae*) and petrels (*Procellariidae*; Cuthbert and Hilton 2004). The mice carry parasites and diseases such as leptospirosis, plague, *Salmonella*, lymphocytic choriomeningitis, and toxoplasmosis (Witmer and Jojola 2006; see chapter 9).

Figure 7.3   House mouse (*Mus musculus*) irruptions, called plagues, occur periodically in grain growing regions of Australia, disrupting communities and causing massive losses to farmers. This one, in 2021, is said to have grown to "Biblical" proportions. Photo by Grant Singleton, CSIRO.

Another invasive mammal, red deer are among the largest deer species. Native to southwestern Asia, Europe, and northern Africa, red deer were introduced into Argentina and Chile, Australia, and New Zealand for hunting and food. These deer can impact native flora and fauna. For example, Forsyth et al. (2015) used models to predict profound long-term effects of the combination of red deer herbivory and rodent seed predation on forests in New Zealand. Red deer compete with native ungulates such as guanaco (*Lama guanicoe*) and the Patagonian huemul (*Hippocamelus bisulcus*) in Chile and Argentina (Flueck et al. 2003). The deer species also competes with livestock for forage (Lowe et al. 2000).

Pigs (fig. 7.4) are native to Europe and Asia as far south and east as the Malaysian Peninsula, Sumatra, and Java. They have traveled along with humans as

Figure 7.4   Feral pigs in a field in the United States. Photo by USDA/APHIS/Wildlife Services, National Wildlife Research Center.

domesticated species and have reached virtually all regions of the world, including many islands. In much of the world, pigs have also been released and established themselves in feral populations. Feral pigs root, form wallows, compete with native wildlife for food, damage streams, prey on ground-nesting native wildlife, alter seed banks by seed predation, and change soil temperature and leaching characteristics; thus, they have major impacts on ecosystems by reducing overall vegetation. Rooting by feral pigs, for example, has slowed oak regeneration in parts of eastern North America. Digging and rooting can sometimes be sufficiently intense to move an ecosystem to an earlier sere, sometimes exacerbating exotic plant invasions. Diminished numbers or even extinction of native species is sometimes a consequence.

In Hawaii the introduction of feral pigs may have remained unremarkable until earthworms (*Oligochaeta*) were also released and became feral (see, e.g., Nogueira-Filho et al. 2009). With this hypothesis, earthworms became a staple food for the feral pig, allowing an increase of populations and their ensuing damage. The pigs formed wallows that serve as **microhabitat** for insect vectors of diseases such as avian malaria and avian pox, indirectly contributing to the decline or extinction of native Hawaiian birds.

Feral pigs are aggressive euryphagic omnivores, eating young land tortoises, sea turtles, and sea birds, thereby contributing directly to the demise of these populations on both continents and islands (see chapter 5). For example, feral pigs were believed to play a major role in wildlife extinctions on the Galápagos Islands (Loope et al. 1988). Concerns were sufficient that intensive efforts continued for 30 years until feral pigs were eradicated from Santiago Island, one of the protected Galápagos Islands (Cruz et al. 2005; box 6.2).

In addition to their impacts on wildlife and natural ecosystem, feral pigs can negatively affect agricultural crops, timber, and pastures. The crops they damage in the United States include hay, small grains, corn, and peanuts as well as some vegetable crops, watermelons, soybeans, cotton, and tree fruits, and conifer seedlings (West et al. 2009). Feral pigs directly damage infrastructure, including fences, roads, dikes, and irrigation canals. Equipment can be damaged and operators injured from holes made by the feral pigs. These pigs prey on young livestock and can carry diseases such as brucellosis, pseudorabies, leptospirosis, and foot-and-mouth disease.

Another worldwide invasive animal, long-tailed macaques (*Macaca fascicularis*) come from Southeast Asia but have been introduced into parts of Indonesia, Mauritius, Palau, and Hong Kong. Macaques have few enemies in the new areas and are considered invasive or potentially invasive. They do well in disturbed habitats and compete with local avifauna for fruit and seeds. Macaques disperse seeds of exotic plants and damage agricultural crops in areas where they have become established. These monkeys may carry Ebola virus (*Ebolavirus*), monkeypox, and a form of malaria that can also infect humans.

ENDEMIC SPECIES WORLDWIDE

Statement
Although this chapter emphasizes conflicts with invasive species from a global perspective, those surrounding endemic species are also a global concern that warrants attention.

Explanation
In his review of human-wildlife conflict and coexistence at global and regional scales, Nyhus (2016) aggregates conflicts into those with large terrestrial and amphibious species, abundant agricultural pests, feral animals, marine species, and those responsible for disease transmission. As examples of terrestrial and amphibious species, Nyhus further subaggregates species into carnivores, herbivores and omnivores, and reptilian species.

As examples of carnivores, the researcher includes "wolves in Asia, North America, and Europe, jaguars (*Panthera onca*) in the Americas, lions (*Panthera leo*) and wild dogs in Africa, and tigers (*Panthera tigris*) in Asia"

(148) and suggests that overall about 75% of the world's felid species are affected. As examples of herbivores, Nyhus includes elephants, deer, and hippopotamuses, and he includes swine and all bear species as examples of omnivores. Crocodilians (*Crocodylidae*), including alligators (*Alligator mississippiensis*), crocodiles, and caimans (*Alligatoridae*), and hundreds of snake species, serve as examples of reptilian species that conflict with people, although some snake species are also globally invasive. For feral animals, Nyhus points particularly to feral cats, dogs, and horses. And as examples of marine wildlife often in conflict with people, the researcher points to shark (*Selachimorpha*) attacks on people and boat collisions with fin whales (*Balaenoptera physalus*), right whales (*Eubalaena*), humpback whales (*Megaptera novaeangliae*), and sperm whales (*Physeter macrocephalus*).

Nyhus (2016) also points out that many of these species are large-sized, a characteristic that can underlie serious damage. Many are also endemic and charismatic, flagship species for the conservation movement and sometimes for hunting. **Charismatic megafauna**— large animals with widespread popular appeal, such as elephants, tigers, lions, leopards (*Panthera*), bears, hawks, and eagles—often live along with humans in multiple-use areas, not in wildlife safe areas. For example, over 80% of African and Asian elephants range outside protected areas (Woodroffe et al. 2005b). Amur (Siberian) tigers (*Panthera tigris*), wolves, bears, wolverines (*Gulo gulo*), and lynx also range near people, and it is reasonable to believe the overlap will increase in the future (see chapter 17). Many of these charismatic megafaunas are listed as threatened or endangered. Conservation of these species depends on their survival in areas where they conflict with people, sometimes as the proximate causes of serious damage to property or human injury. The goals and methods of management must therefore support both reduction of injuries to people as well as the survival of the species. Or, in the words of Woodroffe et al. (2005a), "if these species cannot be conserved in multiple-use landscapes, there is a very real probability that they cannot be conserved at all" (389).

Although current methods are the purview of section IV of this book, it should be noted here that acceptable and effective methods are particularly needed to manage these charismatic megafaunas. I point to two general approaches as examples. Woodroffe et al. (2005b) and others (e.g., Ivașcu and Biro 2020) have suggested that management of these species might benefit by revisiting past methods, some of them ancient, an idea that seems to have gained recent traction. These researchers, for example, suggested reconsideration of shepherding and husbandry practices. Shepherding was mostly a lost art in North America and Europe where livestock were left to roam unattended over wide areas that were mostly free of predators. Predator populations are recovering, however, as exemplified by the reestablishment of coyotes in the High Plateau of Texas, the return of mountain lions and wolves to much of North America's mainland, and the increased numbers of lynx, bear, wolves, and wolverine throughout much of Europe (Chapron et al. 2014). Attesting the value of revisiting ancient methods of management, guard animals are again being used successfully in livestock protection in many parts of the globe. In a major meta-study comparing the effectiveness of a broad spectrum of methods for protecting livestock from predation, Eeden et al. (2018) found that nonlethal methods could be as effective, or more effective, than lethal ones, and that livestock guardian animals were the most effective among the nonlethal methods. Use and improvements in methods such as fladry are being investigated. Range riders also are being reintroduced to parts of the American West (e.g., S. Wilson et al. 2017).

Zoning offers another approach. Linnell et al. (2005) described zoning practices as they relate to carnivores, especially those with large home ranges, but the basic concepts apply to other species targeted for conservation. As with zoning laws in cities where some areas are set aside for residents, other areas for commerce, and some for both, **zoning** in conservation sets aside some areas for human activity, others for wildlife, and some for both. Under the zoning concept, for example, a wildlife refuge would constitute one zone, with others set aside for just human activity or for combined activities of wildlife and humans (such as buffer zones around refuges; Linnell et al. 2005).

With zoning, incompatible human or wildlife activities are prevented by excluding people or wildlife from some zones, and compatible activities allow both humans and wildlife in other zones. One benefit is that zoning extends areas suitable for wildlife beyond safe areas such as refuges into those habited by humans. This is particularly important for carnivores because their large home ranges often take them beyond the boundaries of refuges. Wildlife densities, mitigation of damage, and human activity can be managed in each zone to achieve its designated function. For example, targeted wildlife might be removed from a zone where damage is not tolerated, and humans could be compensated should such damage occur anyway. By contrast, most human activity could be made illegal and laws

strictly enforced in zones designated for conservation of the carnivores. Examples include fencing dingoes from the sheep-farming areas of Australia; introduction of wolves into Yellowstone National Park with their concomitant designation as experimental wildlife rather than endangered outside of the boundaries of the Park; management of bears near the border of Croatia in Slovenia; and use of a region-specific quota system for cougar management in New Mexico, thereby enhancing their value as game species while conserving them and keeping them from bighorn sheep (*Ovis canadensis*, Linnell et al. 2005).

The effectiveness of zoning often depends heavily on public policies and their enforcement. Trouwborst (2018) recently reviewed two major directives, the Bern Convention and the European Union Habitats Directive, to assess their applicability to zoning practices, especially for carnivores that are now recovering in Europe. He concluded that the two legal instruments "do allow—and partly call for—geographically differentiated management of large carnivores" (318). He noted further that most zones where large carnivores are treated favorably will be in conformity with European obligations, whereas those wherein the carnivores receive less favorable treatment may or may not be in conformity.

Summary

- All species continually attempt to invade new areas, but most attempts fail. The intensity of the effort at invasion can be measured as propagule pressure.
- Successful invasions occur in three stages: movement to the new area, establishment of colonies, and invasion of surrounding areas from the colonies. A fourth stage occurs when impacts are great enough to garner human attention.
- Likelihood of success may be facilitated by at least three factors: biotic resistance, biotic acceptance, and human activity. Human activity can be measured using economic indicators such as real estate sales or demographic factors such as area of human habitation and human population density; it is probably a major contributor to the success of invasions.
- Most successful invasions are of little concern to humans, and some benefit them.
- Few species are both successful and damaging in the eyes of humans. These species are termed invasive species.
- Invasive species are found among all taxa and are often generalists; many have characteristics of

species in early successional seres. They are viewed by humans as among the most damaging of wildlife.

- Examples of some of the most damaging exotic invasive species are provided in table 7.1.
- Endemic species are also a concern, including many large-sized animals that can cause serious damage but are also charismatic and have garnered public interest.

Review and Discussion Questions

1. Review the list of 100 worst exotic invasive species. Can you find one that has only negative characteristics? If this is the list of "worst" invasive species, can we assume that all species have at least some positive values in the eyes of humans?

2. Do you think that, at the global scale, the story of damaging wildlife is really the story of exotic invasive species? What about endemic wildlife with broad regional or global distribution such as wolves, mountain lions, bobcats, or elephants?

3. Why might an island or peninsula be more susceptible than a continent to successful invasion by an exotic species? Cite some literature to support your view.

4. From the list of 100 worst exotic invasive species, choose the 10 worst and rank them in severity as sources of proximate damage for peoples. What criteria did you use for ranking? How evenly were these distributed among taxa?

5. Choose a species not discussed in chapter 7 but that is nonetheless an invasive species or an endemic charismatic species. Find the species on a database such as that provided by Invasive Species Specialist Group or on the internet. Summarize the information you find, including common and scientific name, origin, present distribution, and problems associated with it.

8

North American Conflicts

Here I highlight some of the wildlife in North America (Mexico, Canada, and the United States) that are the proximate causes of human-wildlife conflicts. I avoid duplication of species listed in chapter 7, while recognizing that many are a concern in North America.

Statement
Human-wildlife conflicts occur in North America with endemic and invasive species, including plants and invertebrate and vertebrate animals, including reptiles, birds, and mammals.

Explanation
The Center for Invasive Species and Ecosystem Health (CISEH, https://www.bugwood.org/) lists over 2,000 plants, 201 pathogens, 494 insects, 16 crustacea, 37 mollusks, 77 fishes, 6 amphibia, 72 reptiles, 26 birds, and 22 mammals as exotic and invasive in North America. For example, Asian giant hornets (*Vespa mandarinia*) are listed as detected in the United States. They were found in the Pacific Northwest in late 2019 and 2020 in two locations. With additional queens found elsewhere, entomologists are concerned the hornet may become invasive in North America. Using ecological niche modeling, Zhu et al. (2020) predicted that an established population could spread broadly along the western coast of the Pacific Northwest over a period of 20 years, and, with human assistance, could also become established along the eastern coast.

Sometimes called the "murder hornet," this social wasp uses spiked, fin-shaped mandibles to decapitate honeybees and other prey. The hornets can destroy an entire beehive in hours, then take the thoraxes of the bees to feed their young. The hornets have stingers long enough to puncture beekeepers' suits, and they use them to inject a potent venom that causes excruciating pain. The hornets kill about 50 people a year in Japan.

Native species can also be invasive. Osage orange trees (*Maclura pomifera*) are an example. These trees had a historical range mostly within the Blackland Prairies, Chiso Mountains, and the Red River drainage of Arkansas, Texas, and Oklahoma. Because these trees have densely packed, sharp thorns when severely pruned, they were planted as hedgerows during the settlement of the western United States to keep free-ranging cattle from gardens and crops and to form windbreaks under the "Great Plains Shelterbelt" Works Project beginning in 1934. Osage orange trees became widely distributed throughout the central plains states and have since moved beyond the plains into other states, state parks, and other protected areas. The trees shade areas beneath them, facilitating erosion. Osage oranges are now reported as invasive in at least 25 states, from Washington to Massachusetts.

Coyotes (*Canis latrans*) are also invasive in parts of North America. For example, these mammals are listed as invasive in Florida, where they have occurred by human introduction and range expansion since at least 1925. Native to western North America, coyotes replaced wolves, gradually expanding their ranges to include most of Canada, the United States, and Mexico. Coyotes have also moved from natural areas into suburbs and cities (Morey et al. 2007).

Species are sometimes seen as problematic within their native ranges. For example, Hygnstrom et al. (1994), in their manual for wildlife damage practitioners, list many native species as sometimes conflicting with human interests. Examples include crayfish, alligators, nonpoisonous snakes and rattlesnakes, eagles, gulls, waterfowl, woodpeckers, beavers, mountain beavers, muskrats, porcupines, prairie dogs, foxes, raccoons, skunks, bats, and deer.

One hundred of the most serious and persistent damaging species of North America are included in table 8.1. In creating this list, I aimed for a balanced approach that included representative taxa rather than a comprehensive listing or ranking.

Examples

I start with plants, continue to the smallest animals, and then move upward in size and complexity to mammals.

PLANTS. Most damaging plant species in North America are exotic and invasive. Thousands of plants have been imported and sold as ornamentals, used in agriculture, or brought for purposes such as marsh draining. Many find their way from backyards or agricultural fields into natural ecosystems. Plants can cause ecological and economic problems. The ecological impacts include alteration of fire regimens (e.g., cheatgrass [*Bromus tectorum*]; and buffelgrass [*Pennisetum ciliare*] in the Sonoran Desert, McDonald and McPherson 2013), nutrient cycling (smooth cordgrass, *Spartina alterniflora*; e.g., Osgood and Zieman 1998), and water dynamics (tamarisk, saltcedar; Mack et al. 2000). **Genetic pollution** (mixing genes and causing

hybridization or introgression and thereby affecting the survivability of a species) can also impact plants, sometimes endangered ones. The economic costs for damage can also be staggering. For example, Soltani et al. (2017) estimated that farm gate value of soybeans would be reduced by about $16.2 billion annually in the United States and about $1.0 billion annually in Canada between 2007 and 2013 if weeds were left unmanaged.

Purple loosestrife (*Lythrum salicaria*) are herbaceous perennials that probably arrived in North America by the early 1800s in the ballast water of ships from Europe, Asia, or Africa. By the late 1800s they had spread throughout the northeastern United States and southeastern Canada. They invaded estuaries throughout this region, becoming prominent problems in the 1930s when they took over floodplain pastures along the St. Lawrence River. These plants crowd out native vegetation along riverbanks and waterways, and they may replace monotypic stands of cattails with another exotic monospecific community that also has limited value to wildlife (Thompson et al. 1987).

Paperbark trees (*Melaleuca quinquenervia*) originated in Australia, where Aboriginal people used them for their antibacterial and antifungal properties. They were first offered for sale as ornamental trees in Florida in 1887. These plants were also used to help dry up shallow water basins. They grow aggressively, crowd out other vegetation, and catch fire readily. The species now occupies over 202,000 hectares of flatwoods, marshes, and cypress swamps in southern Florida (Turner et al. 1997). Control efforts, including mixes of physical, chemical, and biological methods that are repeated annually, have been intense and expensive. The methods have been effective in some areas of Florida, constraining the growth of these invaders to maintenance levels (Rodgers et al. 2017).

Kudzu (*Pueraria montana*) are pea family vines that came to North America from Japan in 1876 for the Philadelphia Centennial Exposition. They were planted in the eastern United States for control of erosion during the depression years. Nicknamed "the vine that ate the South," the plants quickly spread over millions of acres of farmlands and forests, covering and smothering native vegetation wherever they gained roothold, and leaving a "desert of vines." Their vines can extend over 100 feet and cover entire canopies, telephone poles, or abandoned houses (fig. 8.1). They grow from runners at nodes and crowns but rarely from seeds. Kudzu plants carry soybean rust fungus, a fungus that may have arrived in the United States via hurricane in 2004. Used as forage for livestock, kudzu may also have some

TABLE 8.1   *One hundred of North America's damaging species*

AQUATIC PLANTS

Giant salvia	*Salvinia molesta*
Hydrilla	*Hydrilla verticillata*

TERRESTRIAL PLANTS

Brazilian pepper tree	*Schinus terebinthifolius*
Common broom	*Cytisus scoparius*
Garlic mustard	*Alliaria petiolata*
Japanese barberry	*Berberis thunbergii*
Kudzu	*Pueraria montana*
Leafy spurge	*Euphorbia esula*
Osage orange	*Maclura pomifera*
Paperbark tree	*Melaleuca quinquenervia*
Purple loosestrife	*Lythrum salicaria*
Spotted knapweed	*Centaurea stoebe*

ARACHNIDS

Honey bee mite	*Acarapis woodi*
Honey bee varroa mite	*Varroa destructor*
Reptilian mite	*Amblyomma sparsum*

INSECTS

Africanized honey bee	*Apis mellifera scutellata*
Emerald ash borer	*Agrilus planipennis*
Formosan subterranean termite	*Coptotermes formosanus*
Large elm beetle	*Lymantria dispar*
Pine shoot beetle	*Scolytus scolytus*
Red imported fire ant	*Tomicus piniperda*
Small poplar borer	*Solenopsis invicta*
Spongy (gypsy) moth	*Saperda populnea*

CRUSTACEA

Fishhook waterflea	*Cercopagis pengoi*
Rusty crayfish	*Orconectes rusticus*
Spiny waterflea	*Bythotrephes cederstroemi*

MOLLUSCS

African giant snail	*Achatina fulica*
Green mussel	*Perna viridis*
Zebra mussel	*Dreissena polymorpha*

AMPHIBIA

American bullfrog	*Lithobates catesbeianus*
Cane toad	*Bufo marinus*
Coqui	*Eleutherodactylus coqui*

REPTILES

American alligator	*Alligator mississippiensis*
Burmese python	*Python molurus*
Curious skink	*Carlia ailanpalai*

FISH

Black carp	*Mylopharyngodon piceus*
Green sunfish	*Lepomis cyanellus*
Northern snakehead	*Channa argus*
Red shiner	*Cyprinella lutrensis*
Sea lamprey	*Petromyzon marinus*
Silver "flying" carp	*Hypophthalmichthys molitrix*

BIRDS

Acorn woodpecker	*Melanerpes formicivorus*
American crow	*Corvus brachyrhynchos*
American white pelican	*Pelecanus erythrorhynchos*
Black-billed magpie	*Pica hudsonia*
Black vulture	*Coragyps atratus*
Brown-headed cowbird	*Molothrus ater*
Canada goose	*Branta canadensis*
Common grackle	*Quiscalus quiscula*
Common pigeon	*Columba livia*
Double-crested cormorant	*Phalacrocorax auritus*
European starling	*Sturnus vulgaris*
Golden eagle	*Aquila chrysaetos*
Great blue heron	*Ardea herodias*
Great horned owl	*Bubo virginianus*
House sparrow	*Passer domesticus*
Laughing gull	*Leucophaeus atricilla*
Mallard	*Anas platyrhynchos*
Northern flicker	*Colaptes auratus*
Red-tailed hawk	*Buteo jamaicensis*
Red-winged blackbird	*Agelaius phoeniceus*
Sandhill crane	*Grus canadensis*
Snow goose	*Chen caerulescens*
Turkey vulture	*Cathartes aura*
Yellow-bellied sapsucker	*Sphyrapicus varius*

MAMMALS

Badger	*Taxidea taxus*
Beaver	*Castor canadensis*
Black bear	*Ursus americanus*
Black-tailed prairie dog	*Cynomys ludovicianus*
Bobcat	*Lynx rufus*
California ground squirrel	*Otospermophilus beecheyi*
Canada lynx	*Lynx canadensis*
Cotton rat	*Sigmodon hispidus*
Coyote	*Canis latrans*
Eastern chipmunk	*Tamias striatus*
Eastern cottontail	*Sylvilagus floridanus*
Eastern gray squirrel	*Sciurus carolinensis*
Eastern mole	*Scalopus aquaticus*
Feral cat	*Felis catus*
Feral dog	*Canis familiaris*
Feral pig	*Sus scrofa*
Little brown bat	*Myotis lucifugus*
Long-tailed weasel	*Mustela frenata*
Meadow vole	*Microtus pennsylvanicus*
Moose	*Alces alces*
Mountain beaver	*Aplodontia rufa*
Mountain lion	*Puma concolor*
Muskrat	*Ondatra zibethicus*
Nine-banded armadillo	*Dasypus novemcinctus*
Northern pocket gopher	*Thomomys talpoides*
Norway rat	*Rattus norvegicus*
Nutria	*Myocastor coypus*
Opossum	*Didelphis virginiana*
Porcupine	*Erethizon dorsatum*
Racoon	*Procyon lotor*
Red fox	*Vulpes vulpes*
Rocky mountain elk	*Cervus elaphus nelsoni*
Striped skunk	*Mephitis mephitis*
White-footed deer mouse	*Peromyscus leucopus*
White-tailed deer	*Odocoileus virginianus*
Woodchuck	*Marmota monax*

Note: Selected examples. Any genus is represented by only one species, although several species may cause damage.

Figure 8.1   Kudzu, "the vine that ate the South," suffocates native vegetation in a field near Port Gibson, Mississippi. Photo by Gsmith.

medicinal uses, and they contain a starch used for food in parts of Asia. They have been utilized in the United States to make soaps, lotions, jelly, and compost. A small patch of kudzu was discovered in 2009 near Leamington, Ontario, Canada.

"Kudzu bugs" (*Megacopta cribraria*) arrived in Atlanta, Georgia, in 2009 on a plane from Asia and quickly made kudzu their favorite food. Probably facilitated by the presence of kudzu, the bugs have since invaded much of the southeastern United States and could help to manage the invasive plant. Unfortunately, the bugs also eat soybeans, posing agricultural and economic risks (Dhammi et al. 2016).

INSECTS AND OTHER INVERTEBRATES. Insects, many of them exotic and invaders, affect agriculture, agroforestry, and natural forest ecosystems. Invaders include the emerald ash borer (*Agrilus planipennis*), the Africanized honeybee (*Apis mellifera*), the brown marmorated stink bug (*Halyomorpha halys*), the spotted lantern fly (*Lycorma delicatula*), the spongy moth (*Lymantria dispar*, formerly known as the gypsy moth), the pine shoot beetle (*Tomicus piniperda*), and the Asian giant hornet (table 8.1).

Most damaging aquatic invertebrates have moved into new areas where they outcompete local fauna, disrupting food production such as commercial fishing and aquaculture. For example, spiny water fleas (*Bythotrephes cederstroemi*, a crustacean; see table 8.1) were first found in 1984 in Lake Huron, probably having arrived in water ballast from Great Britain or northern Europe. The water fleas have since spread to all the Great Lakes and to many other lakes throughout the region. They were found most recently in Lake Champlain, Vermont, in 2014. The fleas' full impacts on the lakes' ecosystems are unknown. They foul fishing gear, compete with native fish for plankton, and may be a factor in the decline of desirable fish such as alewife (*Alosa pseudoharengus*) in Lakes Ontario, Erie, Huron, and Michigan. They serve as food for some fish, but their long spines damage the guts of others.

Given that about 9.4 million reptilian and 14.0 million small animal pets were kept in the United States in the year 2020, many imported, it is unsurprising that exotic reptilian ticks and mites are arachnids of concern. Some reptilian ticks (e.g., *Amblyomma sparsum* and *A. marmoreum*) are known vectors of the bacteria

(*Ehrlichia ruminantium*) that cause heartwater in both domestic and wild ruminants (Burridge 2005) and may occur in humans. Mendoza-Roldan et al. (2020) summarized the parasites and other vectors found on reptiles that can transfer diseases to humans. Two mites, the honey bee mite (*Acarapis woodi*) and the honey bee varroa mite (*Varroa destructor*; see table 8.1) are known problems for both wild and domesticated honey bees.

Spongy moths (aka gypsy moths) are among North America's most devastating forest pests. The moths were imported to Boston, Massachusetts, from Europe in 1868 or 1869 by a silk producer to improve the resistance of silkworms to disease. Feral populations emerged within 10 years. State and federal eradication efforts began in 1890. The moths had moved into most of the northeastern United States by 1994 and are expected to go as far west as Iowa and as far south as South Carolina by 2025. Within their range, spongy moths erupt in cycles that are difficult to predict. During a high cycle, the moths defoliate trees, especially oak trees (*Quercus*). With repeated defoliation, up to 20% of the oaks die, with occasional very heavy die-offs. The long-term impacts on forest vegetation are unknown, but replacement of oaks by less susceptible species is likely. Aggressive efforts are made to monitor for the presence of spongy moths in new areas and to eradicate the moths where they are found. Between 1980 and 2020, state and federal governments spent over $282 million to suppress damage by these moths (Coleman 2020). GYPEK, a species-specific and effective insecticide containing the spongy moth nuclear polyhedrosis virus, was manufactured and used by the US Forest Service on federal lands. Its production, which required infecting live caterpillars, was labor intensive and expensive, and was stopped in 2019. *Bacillus thuringiensis* var. *kurstaki* is used effectively, once caterpillars are established and feeding, as are more traditional insecticides.

Many invasive mollusks are exotic, transported to North America for use in aquariums, on plants, or in ship ballast. All are prolific breeders, and some have already caused severe problems in natural and agricultural ecosystems. Each also has its own story. For example, giant African land snails (*Achatina* or *Lissachatina fulica*, *A. achatina*, and *Archachatina marginata*) have been widely spread in North America as pets and for study in schools. Green mussels (*Perna viridis*) were first reported in 1999 in US coastal waters following an accidental release, probably from ballast water, with a subsequent occurrence near St. Augustine, Florida, in 2002. These mussels are believed to be ag-gressive competitors with local species; according to Power et al. (2004), they may become the marine equivalent of the freshwater zebra mussel. Veined rapa whelk (*Rapana venosa*), carnivorous gastropods that have spread rapidly into parts of Europe, were discovered in the Chesapeake Bay in 1998; these predatory whelks could seriously harm the shell and oyster industry there.

Rusty crayfish (*Orconectes rusticus*) are native to the Ohio River drainage system of the United States but have moved beyond their historical range and have established in Illinois, Wisconsin, Minnesota, parts of 17 other states, and central Canada (Durland Donahou et al. 2019). Rusty crayfish have large chelae (pinchers) and use their size advantage to outcompete other species. They reduce vegetation so that habitats for other native species are lost. Predators tend to take less-threatening crayfish, so rusty crayfish prevail, having more and more dramatic impacts on the aquatic ecosystems that they invade.

Twenty years after the first zebra mussels were found, Strayer (2009) reviewed their status in North America. Zebra mussels probably first arrived in ballast water of ships from Russia and were first seen in the Great Lakes in 1988. Early predictions were correct that zebra mussels would spread rapidly throughout major waterways of North America. Zebra mussels moved from the Great Lakes into many major waterways of the eastern half and midwestern part of the United States. Models now predict spread of the mussels into waters with sufficient soluble calcium (i.e., not the Pacific Northwest or New England), except for waters that are very cold (e.g., northern Canada) or very warm (e.g., southwestern United States and Mexico). The spread of zebra mussels in the Great Lakes has slowed due to **displacement** by the quagga mussel (*Dreissena rostriformis*; Matthews et al. 2015), another invasive species. This shift in dominance may reduce predator risks to certain heavy metals that are bioaccumulated in zebra mussels (Matthews et al. 2015).

Zebra mussels absorb and retain nutrients such as phosphorus and calcium. They take the nutrients along as they recede to the lower benthos, in this manner clearing waterways such as the Hudson River. On this river, the zebra mussels enriched the littoral or benthic zone as they moved into lower water, reducing the productivity of the upper level photic (planktonic) and pelagic systems. Plankton-based food chains often disappeared, and pelagic fish were reduced by 28%, whereas systems based on deeper rooted plants and attached algae flourished. Littoral fish increased by 97% (Strayer et al. 1999).

AMPHIBIA AND REPTILES. Pitt et al. (2005) suggested that while many amphibians and reptiles were globally declining, a few were expanding rapidly and could be problematic. Species of concern were euryphagic, had high reproductive rates, and could achieve high population sizes. Many would be undetected in transit. A mix of exotic and native amphibians and reptiles are a concern in North America. For example, imported pet snakes such as boa constrictors and pythons can cause problems when they get into natural ecosystems.

About 7,000 to 8,000 people are bitten by venomous snakes in the United States each year, and about six people die from these bites (Forrester et al. 2018). Children are bitten by rattlesnakes, especially when playing in grass. Livestock and pets, including hunting dogs, may also be bitten by snakes, sometimes causing death. Invasive snakes can cause bites, but native species are a primary concern. For example, Ruha et al. (2017) evaluated 450 snake bites reported to the North American Snakebite Registry between January 1, 2013, and December 31, 2015. The researchers found that native species comprised 99% of cases, mostly rattlesnakes and copperheads. Most (69.3%) victims were male, and 28.2% were children aged 12 years or younger.

Alligator attacks are increasing in the United States. Langley (2010) found that there were 567 injuries and 24 deaths between 1928 and January 1, 2009. In the 2005 report, Florida had over 334 documented attacks, with 14 fatalities (and over 17,000 nuisance complaints per year); Texas had 15 attacks; and Georgia had nine attacks, with one fatality. Most people injured were attempting to capture (or pick up or exhibit) an alligator (17.4%), swimming (16.7%), fishing (9.9%), or retrieving golf balls (9.5%). Alligators usually seize an appendage and twist it off by spinning. Alligators also serve as vectors for at least one zoonotic disease. West Nile virus infection was confirmed in three infected American alligators following an epizootic event at an alligator farm in Florida in 2002 (Jacobson et al. 2005).

Curious skinks (Carlia ailanpalai) are terrestrial lizards that were introduced into Guam in the 1960s. In areas without snakes, these skink have achieved population densities approaching 10,000 per acre, both in forested areas and near human habitation (Pitt et al. 2005). By numbers alone, these skinks are a force outcompeting native lizards. They have also become a food source for brown tree snakes, furthering the population growth of the snakes. The skinks also serve as prey for the yellow bittern (Ixobrychus sinensis), birds whose numbers around airports now pose risks for collisions with aircraft (Pitt et al. 2005).

Burmese pythons are native to Southeast Asia. They were accidently and intentionally released in the Everglades National Park, Florida, and established breeding populations there by the 1990s. Often, pet snakes become too large for their tanks and are released in the wild by their owners. Other snakes, such as Indochinese cobras (Naja siamensis), have also become established. These snakes compete with native species and prey on native birds, including the federally endangered wood stork (Mycteria americana; Dove et al. 2011) and mammals, including Key Largo wood rats (Neotoma floridana smalli), round-tailed muskrats (Neofiber alleni), and some reptiles, including alligators. Both alligators and pythons prey on the threatened and endangered Key deer (Odocoileus virginianus). Pythons preying on larger mammals, such as deer, raccoons, and opossums (Didelphis virginiana), have also reduced their availability as secondary hosts for mosquitoes. As a consequence, there has been an over 400% increase in feeding on the blood of the primary host, cotton rats (Sigmodon hispidus), by mosquitoes that carry Everglades virus. The shift, based on dilution effect theory, prompted Hoyer et al. (2017) to predict an increase of this zoonotic disease in people.

Because of their large size, pythons can frighten Everglades Park visitors (fig. 8.2). Based on surveys, increased numbers of road kills and numbers removed, and reduced prey populations, the population now lies between 100,000 and 300,000 individuals and appears to be expanding. Traps, hand capture, detection with dogs, and Judas pythons (Pitt et al. 2005; B. Smith et al. 2016) are being used in attempts to control the snakes in the park.

FISH. Aquatic ecosystems are particularly susceptible to human influence. People have often moved fish between bodies of water. Some species, including largemouth bass, rainbow trout, and common carp, have been introduced by public agencies to improve sport fishing or for biological control of weeds. Butterfly peacock bass (Cichla ocellaris) were introduced into Florida for both sport and control of tilapia. Often placed there for aesthetic value, goldfish (Carassius auratus) are widespread in water bodies of North America. Bighead (Hypophthalmichthys nobilis), black (Mylopharyngodon piceus), and silver carp (H. molitrix) were brought into aquaculture facilities because they might meet a niche food market or control aquatic weeds. Many, such as the red-bellied Pacu (Colossoma bidens) and some tilapia, were pets that escaped or were released. Some introductions occurred with fish used as baitfish or contaminants of baitfish.

Figure 8.2 An invasive Burmese python (*Python molurus*) meets a native American alligator (*Alligator mississippiensis*) (top). Burmese pythons can become impressively long; this one was 17 feet, 7 inches long and had 87 eggs (bottom). Top photo by Lori Oberhofer, National Park Service. Bottom photo by Catherine Puckett, USGS, Public domain.

Exotic and invasive fish can carry diseases. For example, stocking rainbow trout has introduced whirling disease in about 20 states (Fuller et al. 2019b). Introduced fish may outcompete native ones, such as rainbow trout forcing the decline of cutthroat trout. Exotic fish may be a source of genetic pollution. For example, blue catfish (*Ictalurus furcatus*), by hybridizing, contributed to the decline of the threatened Yaqui catfish (*I. pricei*) in Mexico (Fuller and Neilson 2021). Rainbow trout likewise hybridized with native cutthroat trout, golden trout, and redband trout (*O. mykiss*

subspecies; e.g., Fuller et al. 2019b). Exotic fish often alter habitat, as with the effects of carp or suckers on water turbidity, sometimes dramatically changing community composition and dynamics.

Sea lampreys (*Petromyzon marinus*; fig. 8.3) are jawless fish native to the Eastern Seaboard of the United States. They are usually marine but move into fresh water to spawn. They were probably introduced into the Great Lakes region from the Finger Lakes or Lake Champlain in New York and Vermont. By the 1930s or 1940s, sea lampreys had spread to Lakes Michigan,

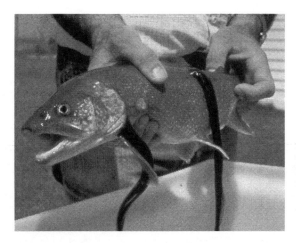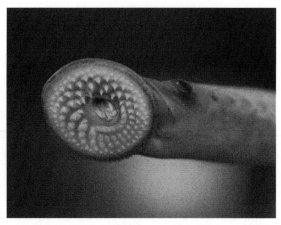

Figure 8.3 Mouth of a sea lamprey (right). Sea lampreys (*Petromyzon marinus*) have parasitized lake trout (*Salvelinus namaycush*) and other fish since they invaded the Great Lakes in the early 1900s (left). They have dramatically altered the Great Lakes ecosystems. Left photo courtesy of the US Geological Survey. Right photo by T. Lawrence, Great Lakes Fishery Commission.

Huron, and Superior, where they parasitized fish, including lake trout (*Salvelinus namaycush*), lake whitefish (*Coregonus clupeaformis*), chub, and lake herring (*C. artedi*). Sea lampreys have had major ecological influences on the fisheries of the Great Lakes. Millions of dollars are spent each year to control lampreys and to restore fish to the lakes. Costs to the fishery industry are estimated in the billions of dollars (Hansen 2010).

Like sea lampreys, white perch (*Morone americana*) found access from drainage areas along the eastern United States into the Great Lakes and the waterways of the upper midwestern states, this time via the Erie Barge Canal in the 1930s. White perch are now found in about 20 states outside of their native range. The perch eat fish eggs during spring, preying on eggs of the walleye (*Stizotedion vitreum*), white bass (*Morone chrysops*), and others (Fuller et al. 2019c). They also feed on eastern shiners (*Notropis*; Fuller et al., 2019c) and may contribute to the decline of some of these species (Fuller et al. 2019c). White bass are sympatric with other white bass species and hybridize with them in western Lake Erie (Fuller et al. 2019c).

Brook stickleback (*Culaea inconstans*) and red shiners (*Cyprinella lutrensis*) were moved into new areas as baitfish or as contaminants in the shipment of bait or stocking fishes. Native to the northern United States, brook stickleback have been reported in at least 14 additional states. Bait-bucket releases have probably introduced red shiners into at least 15 states outside of the Mississippi River Basin. Shiners are extremely aggressive and may have carried the Asian tapeworm into the Virgin River, Utah, where it infected the endangered woundfin (*Plagopterus argentissimus*; Nico et al. 2021). Due to

their predation, competition, hybridization, and introduction of parasites, red shiners are viewed by some professionals as second only to mosquitofish in their potential to negatively impact native fish.

At least four species of snakehead (*Channa*) have been introduced into North American waters. Snakeheads originated in Asia, where they are important commercial fish, grown in aquaculture, for sport, and for aquariums. Introductions in the United States are sometimes accidental from aquariums and sometimes intentional, as potential food fishes. Blotched snakeheads (*C. maculata*) have been established on Oahu, Hawaii, since before 1900. Giant snakeheads (*C. micropeltes*) and northern snakeheads (*C. argus*), sometimes called "frankenfish" in popular media, have been reported in at least 6 and 16 states, respectively. A reproducing population of northern snakehead was found in Crofton Pond, Maryland, and in the Potomac River (Courtenay and Williams 2004). Bullseye snakeheads (*C. marulius*) have been illegally stocked in residential lakes and canals in Broward County, Florida. Snakeheads are predatory fish and may harm competing native fish and crustaceans (Fuller et al. 2019a). The overall ecological impacts are unknown, but scientific concerns center around their aggressive predatory nature and their potential impacts on native fishes and ecosystems.

Silver carp were introduced in 1973 by a fish farmer in Arkansas who thought the fish could improve water quality at sewage treatment lagoons and aquaculture facilities and serve a niche market for food (Conover et al. 2007). The carp were evaluated by a state agency and a university in Arkansas and were tested in at least four lagoons and at Mallard Lake. The present population in the Mississippi River Basin may have come from one of these locations. Feral populations have been confirmed in at least 10 states outside the Basin (Nico et al. 2019), with a general movement toward the Great Lakes. Silver carp jump (fig. 8.4), are difficult to handle during production, are not cultured in the United States, and are listed as injurious under the Lacey Act (see chapter 15). The commercial harvest of silver carp has increased in the Mississippi and Illinois Rivers with efforts by the industry to establish more markets. Silver carp are plankton eaters, affecting mussels, larval fish, and some adult fish, and compete with gizzard shad (*Dorosoma cepedianum*) and bigmouth buffalo (*Ictiobus cyprinellus*) in these rivers. In addition, silver carp might be a source of Asian tapeworm.

BIRDS. Flight and gregarious or predatory behavior underlie many of the conflicts with birds in North America. Flight allows gregarious birds, such as many **granivorous** or omnivorous species, to move great distances daily and to congregate in large flocks for feeding, and then to assemble or reassemble for loafing or roosting. Predatory birds such as raptors tend toward individual rather than gregarious lifestyles, but with these birds the ability to fly enhances the accessibility of livestock such as poultry and young calves. Individual birds also cause damage, such as woodpeckers damaging house siding or utility poles. Birds can carry diseases such as West Nile virus or avian influenza. Invasive birds such as European starlings compete with native cavity nesting birds and otherwise harm natural ecosystems.

Red-winged blackbirds are one of a group of about ten blackbird species in North America, including starlings since their introduction from Europe in the late 1800s. Included also are common grackles (*Quiscalus quiscula*), great-tailed grackles (*Q. mexicanus*), yellow-headed blackbirds (*Xanthocephalus xanthocephalus*), brown-headed cowbirds, and Brewer's blackbirds (*Euphagus cyanocephalus*). Blackbirds are the most abundant birds in North America. They damage ripening corn, sunflowers, and rice and can damage sprouting rice in localized areas (Linz et al. 2017). Starlings contaminate feed at cattle feedlots and take feed. They damage fruit, including cherries, grapes, and apples. Blackbirds congregate in large flocks and roosts, especially in the winter, making noise and defecating (Linz et al. 2017) (fig. 8.5). The resulting guano can host the fungus that causes histoplasmosis. Linz et al. (2017) list eight zoonotic diseases associated with flocking blackbirds, including red-winged blackbirds. In addition to histoplasmosis, these include encephalitis (St. Louis and western equine), Lyme disease, avian salmonellosis, and shiga toxin-producing *Escherichia coli*.

Aircraft also collide with blackbirds. Of the 33 birds most frequently identified as struck by civil aircraft in the United States between 1990 and 2019, the European starling ranked sixth (5,118 strikes) overall and seventh (307 strikes) for 2019 only (Dolbeer et al. 2021). Because of their small body masses, less than 3% of these collisions resulted in damage, but costs for repairs remain high.

Double-crested cormorants are part of a group of herons, egrets, bitterns, pelicans, and other cormorants that feed on fish in aquaculture facilities, including publicly operated hatcheries and commercial fish farms where they can cause serious economic

Figure 8.4 Silver carp, invasive from Asia and now found in North American waterways, often respond to speeding boats by jumping. This survey team is electrofishing. Photo by Ryan Hagerty, US Fish and Wildlife Service.

Figure 8.5 Blackbirds form large flocks, often damaging crops and becoming a nuisance and a health hazard in urban and rural settings. Photo by Edibobb.

losses (King 2005). Cormorants recovered from a precipitous decline in the 1950s caused partly by use of organochlorine pesticides such as dichloro-diphenyl-trichloroethane (DDT). Since then, cormorant populations have increased dramatically, adjusting both habits and breeding locations to take advantage of the aquaculture industry in states such as Mississippi and Arkansas. For example, many previously migratory cormorants now remain near southern aquaculture facilities year-round, causing substantial economic losses (Engle et al. 2021).

Cormorants roost in rookeries in numbers sometimes over 10,000, sizes sufficient to depredate recreational fisheries, lakes, and reservoirs. Guano accumulation in permanent roosts causes secondary problems such as odor, destruction of vegetation, including large mature trees, erosion, and sometimes loss of desirable habitat for tourism and recreation, such as at Lake Guntersville in Alabama (Barras 2004). In the Great Lakes, cormorants damage forest island ecosystems, reducing woody vegetation and opening areas for ground-nesting colonial waterbirds such as gulls and American white pelicans. These species perpetuate damage to soil and woody vegetation, prompting efforts to manage tree damage caused by the cormorants (Dorr and Fielder 2017).

Red-tailed hawks (*Buteo jamaicensis*) are raptors that, along with goshawks and great-horned owls (*Bubo virginianus*), sometimes take poultry or other livestock (Washburn 2016). Raptors generally leave puncture wounds in the back and breast and often pluck the feathers; owls often also decapitate the prey. Raptors usually take only one prey animal daily, unlike some mammalian predators. Golden eagles (*Aquila chrysaetos*) sometimes take lambs or kids on open ranges, leaving puncture wounds that are about 1 to 2 inches apart from their three front talons, and a wound 4 to 6 inches behind from the hallux (rear toe). Black vultures (*Coragyps atratus*) also kill or injure vulnerable livestock, often newborn calves, piglets, or lambs (Avery and Lowney 2016). The relatively large size of hawks and vultures is a concern for damage to aircraft from collisions. For example, turkey vultures ranked number 22 (901 strikes) in bird species struck by aircraft in the United States from 1990 to 2019, but almost half (49.5%) of the collisions resulted in damage to the aircraft (Dolbeer et al. 2021).

Acorn woodpeckers (*Melanerpes formicivorus*) are one of several woodpeckers that peck holes in wooden siding and similar structures to find insects or store acorns. House siding sometimes amplifies their pecking sounds, aiding the woodpecker in establishing a territory. A particularly notable example was the holes pecked in foam insulation at Launch Pad 39B, Cape Canaveral, Florida, in June 1995. The woodpeckers, probably flickers (*Colaptes*), pecked about 200 holes in the insulation, some 4 inches deep. Launch of the space shuttle Discovery was delayed by five days until the insulation could be replaced and pyrotechnics deployed. Woodpecker spotters were utilized to watch the launch pads around the clock. Sapsuckers (*Sphyrapicus*) damage ornamental or commercial trees, such as Christmas trees, by feeding on sap, bark, or insects in the tree trunks.

MAMMALS. Mammals are often crepuscular or nocturnal and are less easily connected than birds to the damage they cause. Nonetheless, the damage can be extensive.

White-tailed deer are mammals that, along with other ungulates, including moose and elk, browse agricultural crops and trees such as ornamentals, fruit, Christmas trees, and forests. Ungulates can also cause damage by rubbing and trampling. Deer and elk damage agricultural crops such as alfalfa, oats, and winter wheat; elk raid haystacks; and deer may consume stored crops (VerCauteren et al. 2003). In addition to cattle grazing, the ungulates can affect long-term availability of forage and reforestation efforts. Moser and Witmer (2000) found reduced shrub and mammal biodiversity in areas of the Blue Mountains of eastern Oregon where elk and cattle were present in high densities. Deer feed on buds during the winter, affecting the following year's production, and take the lead growth of smaller Christmas trees, often eliminating their commercial value. Deer rub trees with diameters of about one-half inch to 4 inches, affecting ornamental, forest, and Christmas tree industries.

When overabundant, deer and elk can affect natural ecosystems by selectively foraging, thereby modifying patterns and distribution of vegetation. The effects cascade to other species, including insects, birds, and other mammals (e.g., Côté et al. 2004). Aircraft, automobiles, and other vehicles collide with deer, elk, and other ungulates, causing economic losses that are estimated at over a billion dollars annually and injuring and killing both humans and wildlife (fig. 8.6) (Huijser et al. 2007). Deer are vectors of diseases such as Lyme disease, serving as hosts for the tick that transfers the disease to humans, and they can transfer other diseases such as tuberculosis to livestock.

Nutria (*Myocastor coypus*), exotic invaders introduced in 1899 for fur farming, feed on aquatic plants and crops close to waterways, damage wooden structures and floating docks, and eat and damage young woody vegetation, creating major problems in bottomland forest regeneration efforts. They can be a particular problem in southern states with crops such as rice and sugarcane and in wildlife areas such as refuges, where their eradication is often desired. Muskrats, too, damage wetlands and ponds, whether natural or manmade, throughout North America

Figure 8.6 A deer found in the wheel well of a military aircraft. Major efforts are made at military and commercial airports worldwide to minimize collisions with wildlife on takeoffs and landings. Collisions are dangerous to wildlife and crew and costly to aircraft. Photo by Air Combat Command, US Department of Defense.

(J. E. Miller 2018). They burrow extensively into dams and levees, adjusting burrow design and height according to fluctuating water levels. They can have important impacts on the structure and species composition of wetland marshes, both natural ones and those constructed for water treatment, changing dense vegetation into patchworks of open areas (Miller 2018). Muskrat burrows can collapse under the pressure of livestock, sometimes injuring the livestock but also stimulating additional burrow construction or reconstruction. Thus, the damage is often exacerbated at ponds, stock tanks, and waterholes for livestock.

Beavers cut saplings and cut or sometimes girdle large trees, preferring trees such as cottonwood, sweet gum, loblolly pine, and aspen, and use them to build dams that can block water flow and cause flooding (Taylor et al. 2017). They sometimes plug road culverts with sticks, build dams under highway bridges and railroad grades, and dig into levees and man-made dams, causing damage. Significant in riparian ecosystems, they can cut and remove numbers of trees and flood and kill large tracts of mature timber. For example, beavers cut and remove Fremont cottonwood (*Populus deltoides*) saplings, trees that are declining in western arid riparian habitat. Breck et al. (2003) showed that access to trees was enhanced in the Green River, Utah, by controlled flooding, increasing their vulnerability to damage by beavers. Beavers sometimes carry diseases such as the protozoan infection, *Giardia*, which can be transferred to humans, causing diarrhea and gastroenteritis (Taylor et al. 2017).

Red foxes, another invasive mammal, are one of a group of canids—including also coyotes, wolves, and dogs (domestic and feral)—that collectively prey on animals from small reptiles and rodents to large cervids. These animals take livestock, with foxes being a particular problem for poultry. Foxes usually attack the throats of lambs and poultry, sometimes biting the neck or back, and carry the carcass to a den. They open eggs and lick out the contents, leaving the shells beside the nest, and they consume first the legs then the breast of poultry, partially burying the remains. Domestic and feral dog attacks are often characterized by inefficient mutilation of the prey, with broad damage to the flanks, hindquarters, and head. Coyotes often attack the throat and suffocate the prey. Less frequently, they attack from the rear or attempt to pull the prey down from the underbelly (Sterner and Crane 2000). If the prey is small enough, coyotes can bite through the skull or the neck. Coyote attacks on pets and small children have increased as humans and coyotes have increasingly shared suburban and urban space. Coyotes are euryphagic, opportunistic foragers and will damage crops such as watermelons, cantaloupes, grapes, and other fruit. They can carry rabies, sarcoptic manges, and other diseases transmissible to pet dogs and livestock, and they readily prey on game species of wildlife, including wild turkey, antelope, and especially fawns and young poults.

Wolves attack caribou (*Rangifer*), moose, elk, and cattle, often by damaging muscles and ligaments of the back legs. The anticipated impacts on livestock led ranchers to oppose reintroduction of the wolves into Yellowstone National Park. Wolves are spreading rapidly from the upper midwest and expanding ranges in the west (Mech 2017). With reintroductions underway, many efforts in North America now are focused on minimizing livestock-wolf conflicts. Although infrequent, wolves will attack people, hunting dogs, and domestic pets (Penteriani et al. 2016; Mech 2017). McNay and Hicks (2002) summarized 80 encounters between wolves and people in Alaska and Canada in which the wolves appeared unafraid of people. Of these attacks, 12 wolves were known or suspected to be rabid, and 39 aggressions were by wolves known to be healthy. Sixteen people were bitten by healthy wolves; none of the injuries were life-threatening.

Mountain lions attack deer, elk, and livestock, including sheep (*Ovis aries*), goats, cattle, and horses, in western North America. They are opportunistic and will take available smaller prey. Mountain lions use their powerful jaws to puncture skulls and can break the neck and back of even large prey. These cats often feed on the front quarter of prey first, also dragging the carcass to a feeding site where it is cached and to which the lion will return repeatedly to feed. Although still rare, contacts between mountain lions and humans are increasing in North America and Europe, as are the number of fatal attacks on both humans and their pets (Penteriani et al. 2016). Penteriani et al. (2016) have suggested that the increasing numbers of attacks by carnivores on humans are associated with increasingly risky human behavior, such as leaving children unattended.

Summary

- Human-wildlife conflicts occur widely in North America, caused by both native and exotic species that include plants and animals, invertebrate and vertebrate.
- Most damaging plants are exotic and invasive; they cause ecosystem problems by altering fire regimes, impacting nutrient and water cycling, and creating genetic pollution.
- Most damaging insects and other invertebrates, such as the spongy moth and the zebra mussel, are also exotic and invasive; they disrupt food production and commercial fishing in aquatic systems and can have widespread impacts on forests and agroecosystems.
- Damaging amphibians and reptiles are both exotic, such as the Burmese python, and native, such as the American alligator. Exotic snakes have altered the Everglades ecosystem in Florida, and alligator attacks on people are increasing.
- Damaging fish, such as the sea lamprey, snakehead, and carp, also include both exotic and native species. Often moved by people between bodies of water, they can parasitize other fishes or outcompete local species, thereby dominating ecosystems and carrying diseases.

- Flight and their gregarious nature underlie the damage caused by many birds, such as the exotic starling or the red-winged blackbird, allowing them to travel great distances daily and then congregate in large groups to feed on crops, at cattle feedlots or hatcheries, or to attack livestock and roost at large roosts or rookeries in urban areas or parks.
- Mammals, such as white-tailed deer and coyotes, often cause damage at night or at dusk. Damage to crops and livestock can sometimes be extensive, with attacks on pets and people increasing as they encroach on wildlife habitat and as more wildlife move into suburban and urban settings.

Review and Discussion Questions

1. An important part of wildlife management is restoration of wildlife and its habitat. Often this involves the translocation of wildlife species. For example, elk have been reintroduced in several eastern states in the United States in recent years. What lessons might be drawn from this chapter that should be considered before taking such management actions?

2. What does this chapter suggest regarding relationships between wildlife problems and movement of boats from one body of water to another, e.g., from oceans into lakes of North America? Or from one lake to another in North America? Or from pond to pond?

3. What do you see as the long-term impacts of damaging species on ecosystems of North America? For example, how about impacts of black or silver carp on the Mississippi River system? How about impacts of invasive weeds on the grasslands and livestock industry of Saskatchewan?

4. The US Department of Agriculture's Animal and Plant Health Inspection Service has recently told teachers and public schools to no longer use giant snails in their classrooms. Why? What environmental impact is this federal agency trying to avoid?

5. Describe some of the effects of feral pigs on natural ecosystems in Texas or Mississippi. How about their impact on endangered bird species in Hawaii or other islands of North America?

9

Zoonoses and Wildlife Diseases

I explore wildlife diseases including zoonoses and their sources, symptoms, and means of transmission; how humans facilitate the spread of the diseases; and how the diseases affect humans and their interests, including natural ecosystems.

Statement

Recent increases in wildlife diseases and zoonotic epidemics may be the consequence of increased human transport of wildlife and wildlife pathogens, the movement of people into more intimate contact with wildlife, and natural or human-caused perturbations of ecosystems, with consequent changes in new and old diseases so they can invade new hosts and areas.

Explanation

Diseases are abnormal conditions that impair the functions of organisms. They are usually associated with sets of signs and symptoms. If affected organisms are nondomesticated, it is a wildlife disease. If a disease can be transmitted from other organisms (either wild or domesticated) to humans or vice versa, it is also a zoonosis.

Diseases are caused by agents. **Agents** are often biotic but might be abiotic, such as radiation or chemicals like heavy metals, toxins, or **prions** (abnormally folded protein bodies, possibly the cause of bovine spongiform encephalopathy or "mad cow disease" in cattle, Creutzfeldt-Jakob disease in humans, chronic wasting disease in cervids, and scrapie in sheep; Barnes 2005). Biological agents are called **pathogens**. Many are also **parasites** in that they are smaller than organisms they infect and benefit at the expense of the infected organisms. Most parasitic pathogens are also **endoparasites** because they live inside infected organisms. Pathogens include **microparasites** (microscopic parasites that can complete life cycles within one organism—e.g., viruses, fungi, obligate intracellular bacterial microparasites called **Rickettsia**, and other bacteria) and **macroparasites** (parasites that can be seen with the naked eye—e.g., protozoa and helminths) that need to move from host to host to complete their life cycles. Examples include *Rickettsia* that cause Rocky Mountain spotted fever and *Yersinia pestis*, bacteria that cause plague.

Some pathogens are fungal, such as *Pseudogymnoascus destructans*, which causes white-nose syndrome and has killed more than 6 million bats in 33 states and seven Canadian provinces since it was first discovered in New York State in 2006–2007.

Hosts are organisms that harbor pathogens. Long-term hosts may be considered **reservoirs** or **niduses** ("nests" or loci) for the disease (see tables 9.1–9.2). Diseases are transmitted by organisms called **vectors**, or carriers. Vectors may also be **ectoparasites** if they are smaller than the host and live on the outside of it. These carriers do not cause diseases but rather convey agents from hosts (or reservoirs) to hosts. Vectors are often arthropods or insects such as mosquitoes, flies (single-winged *Dipterans* including biting **midges** or "no-see-ums" and **sand flies**), **sand fleas** (small decapod crustaceans), **lice** (*Phthiraptera*; wingless obligate parasites, 3,000 spp.), **fleas** (*Siphonaptera*), ticks (*Ixodidae*), or other **mites** (*Acarina*). Movement of vectors can be facilitated by birds or mammals. At times, hosts may themselves serve as vectors. For example, the spread of the mosquito-borne disease West Nile virus along the eastern coast of the United States coincided with migratory routes of birds.

Vectors carry diseases in various ways. Flies may simply have bodies contaminated by feces or saliva of infected hosts (reservoirs) or by soil. Infected mosquitoes and ticks can inject pathogens directly into bloodstreams of new hosts. Pathogens may accumulate in gelatinous masses in guts, making it difficult for vectors such as mosquitoes to ingest second meals. Mosquitoes may then proactively expulse (regurgitate) the masses into bloodstreams of hosts to clear their guts for more meals (Barnes 2005).

Pathogens and vectors have often coevolved so that pathogens do not harm their vectors. Vectors that carry parasitic pathogens for prolonged periods of time are called **secondary** or **intermediate hosts**. Final targeted organisms are then called definitive hosts (Barnes 2005). Definitive hosts are often, but not always, where parasites reach maturity and reproduce. Sometimes, definitive hosts are simply ones considered most important by humans. For example, sludge worms (*Tubifex*) are considered to be secondary hosts for **whirling disease** (a protozoan disease that infects salmon and trout species) even though the protozoa sometimes reproduce inside the worms. In this case, fish are considered definitive hosts.

Movement of a disease between hosts is called its **biological life cycle**. The natural portion of the cycle, between wildlife as hosts and vectors, is sometimes called the **sylvatic** (or **enzootic**) **cycle**. Sylvatic cycles

may allow diseases to persist in the wild and hosts to serve as reservoirs. If a disease has only a sylvatic cycle, it is a wildlife disease. If the disease "spills over" to humans, even if the infection "dead-ends" without further transmission, the disease is called a "zoonosis." Some zoonoses are transmitted between humans.

Biological life cycles can be straightforward or complex. I illustrate using **Lyme disease** (fig. 9.1). Lyme disease is caused by the spirochete *Borrelia burgdorferi* and occurs in North America, Europe, South Africa, and Australia. This disease appeared or reappeared along with reforestation and movement of people into suburban settings. The disease was first identified in the United States in Old Lyme, Connecticut, in 1981, but the skin reaction was described in Sweden in 1909.

With Lyme disease, vectors are *Ixodes* ticks—e.g., *I. scapularis*. Tick larval and nymphal stages feed on small animals, preferring the white-footed deer mouse in North America. Pinhead-sized larvae emerge from eggs in the soil during summer, attach to a host, feed, and then drop to the ground and molt. The only slightly larger nymphs emerge the next summer, find a host, feed, and drop to the soil to molt. Adults emerge at the end of summer and choose larger mammals such as white-tailed deer as hosts. Adult males die shortly after mating; females die after laying eggs and completing the life cycle (see fig. 9.1) (Barnes 2005).

If any host is infected with *B. burgdorferi*, the tick becomes a vector, passing the bacteria to its next hosts. Spillover to humans occurs as early as the nymphal stage. The tiny nymphs are hardly visible to the human eye and are difficult to detect on the skin. The spirochetes only begin to replicate in the guts of ticks after the ticks begin ingesting blood, and they must move to salivary glands to be injected, so the infected ticks are not contagious for 36 to 48 hours after attaching to humans or other hosts (see fig. 9.1) (Barnes 2005).

Lyme disease symptoms sometimes include a ring-shaped rash that grows to several inches in diameter at the site of the bite. As the ring grows, the center clears, giving the appearance of a target. Initial symptoms are flulike and include fever, chills, headache, fatigue, and joint and muscle aches. Aggressive treatment with antibiotics is often needed to deal with the disease. Untreated, Lyme disease can lead to chronic arthritis, cardiac and neurologic problems, and in some individuals death.

Although we think of zoonoses as being transmitted from wildlife to humans, reverse zoonoses also happen. For example, molecular typing studies of *Giardia*, caused by parasitic protozoa (*Giardia lamblia*)

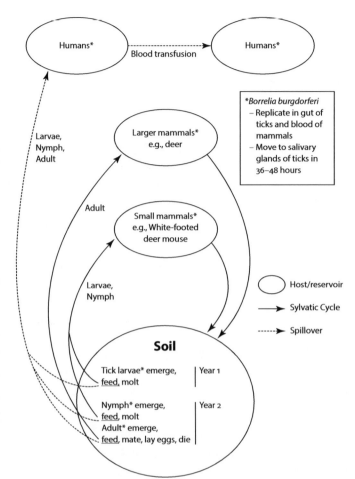

Larvae, Nymph, Adult

Borrelia burgdorferi
– Replicate in gut of ticks and blood of mammals
– Move to salivary glands of ticks in 36–48 hours

Larger mammals*
e.g., deer

Adult

Small mammals*
e.g., White-footed deer mouse

Larvae, Nymph

Host/reservoir

Sylvatic Cycle

Spillover

Soil

Tick larvae* emerge, <u>feed,</u> molt | Year 1

Nymph* emerge, <u>feed,</u> molt | Year 2
Adult* emerge, <u>feed,</u> mate, lay eggs, die

Figure 9.1   Life cycle of Lyme disease. Illustration by Lamar Henderson, Wildhaven Creative LLC.

carried by beavers, indicate that it sometimes moves along with human sewage back to beavers. A concern, of course, is that infected beaver populations can then reinfect humans (Thompson et al. 2009).

Outbreaks happen when infectious diseases occur at higher frequencies than expected. If the disease is unknown or has not occurred for a long time, a single case might be considered an outbreak. **Epidemics** occur when infectious diseases spread rapidly among large numbers of people (e.g., the severe acute respiratory syndrome [SARS] epidemic in Asia during 2003). **Pandemics** are even larger; pandemics occur when spread of a disease is potentially global. Past avian influenza pandemics and SARS-CoV-2 (COVID-19) are examples.

Some wildlife diseases have been known since ancient times and are called **lingering**. This does not imply that the diseases are static and relict. Rather, such diseases are dynamic, having survived by changing along with environmental conditions. Lingering diseases re-emerge in changed forms when conditions

are right for them. Examples include brucellosis and dog rabies. Other wildlife diseases are new and emerging. In a joint consultation held in 2004 the World Health Organization (WHO), the Food and Agriculture Organization, and the World Organization for Animal Health defined an **emerging zoonosis** as "a pathogen that is newly recognized or newly evolved, or that has occurred previously but shows an increase in incidence or expansion in geographical, host, or vector range." In addition to avian influenza, as described below, other emerging zoonoses include bovine spongiform encephalopathy, Nipah virus, and possibly toxoplasmosis (box 9.1).

Diseases respond to environmental changes partly because of the genetic bases of pathogens. Viruses composed of ribonucleic acids (RNAs) mutate frequently and are therefore particularly labile, but deoxyribonucleic acid (DNA)-based pathogens also mutate. Here again the RNA-based COVID-19 serves as an example. Mutated forms developed quickly in regions where the virus first became established, such as in New York

BOX 9.1 Toxoplasmosis

Toxoplasmosis is a concern for people, pets, livestock, and wildlife. It infects many people globally, including about 1.1 million people in the United States each year, and is the second leading cause of death due to foodborne illnesses in the United States (Aguirre et al. 2019). It can be particularly dangerous to pregnant women and to those with compromised immune systems. A retinal form is recognized as a leading form of blindness throughout the world.

Toxoplasmosis is a parasitic infection caused by the protozoan *Toxoplasma gondii*. Felids, including domestic and wild cats, are definitive hosts while many warm-blooded animals serve as intermediate hosts (Frenkel et al. 1970). The characteristics and status of the disease were also reviewed by Aguirre et al. (2019), both summarized partly here.

The disease has three stages: tachyzoites, bradyzoites, and oocysts. Each stage can infect either intermediate or definitive hosts. A tachyzoite penetrates a host's cell then forms a vacuole that protects the tachyzoite from the host's defenses. It multiplies until the cell ruptures. After repeated divisions, the tachyzoite enters a stage where it forms a cyst. The cyst grows until it contains up to several hundred bradyzoites. The cyst is usually formed in muscular or skeletal tissue including the brain, eye, or cardiac muscle. The cysts remain dormant without causing problems in most intermediate hosts. When ingested by a felid, however, the cyst is digested by intestinal enzymes, releasing the bradyzoites. They invade extracellular intestinal tissues forming at least five different asexual types of *T. gondii*, eventually also forming gamonts that begin a sexual cycle. Oocysts are produced and shed from the infected felid for several months.

Until recently, toxoplasmosis was viewed mostly as a wildlife disease transmitted by carnivores eating contaminated prey. The disease was also seen as zoonotic but with a "spillover" transmitted to limited numbers of humans due mostly to infected domestic cats. J. L. Jones and Dubey (2010), however, estimated that, with 78 million domestic cats and 73 million feral ones in the United States, and with 1 million oocysts per shedding cat (estimated at about 30% of the total), there are at least 50 billion oocysts in the environment of the United States at any time. Thousands of cougars and millions of bobcats add to the environmental burden. With live prey as well as eviscerated tissues from infected hunted deer and bear, Jones and Dubey (2010) suggested that there was a well-formed sylvatic cycle for the disease.

The disease is now recognized to occur globally in humans, pets, livestock, and wildlife, often having both terrestrial and aquatic components. As specific examples, an outbreak of toxoplasmosis occurred in a town in Canada in 1994 that was traced to contaminated drinking water. Infected cats, both wild (*Felis concolor*) and domestic, were found near the water supply (Aramini et al. 1998). An outbreak of toxicosis also caused high mortality in sea otters along the coast of California. The source of the protozoa was believed to be contamination of seawater from oocysts in runoff contaminated with the urine and feces from domestic and wild cats (Dabritz et al. 2007). Sea otters probably got infected by feeding on marine invertebrates (M. A. Miller et al. 2008).

Aguirre et al. (2019) have summarized the complex human dimensions surrounding resolution of the disease. They suggest that effective solutions will require broad cooperation across disciplines and recommend a One Health approach.

City and California in the United States. A few of the mutations, such as the British, South African, and Delta forms, offered advantage to the virus, and these forms then spread globally.

Pathogens also recombine their nucleic acids to form new types, take nucleic acids from other pathogens already in the host, or sometimes take nucleic acids from the host itself. Pigs and birds have particular bents as "melting pots" where pathogens can meet and exchange strands. Although most new combinations fail, a few become emergent pathogens, surviving in new hosts, vectors, or environments. Rapid expansion

of the disease may follow. For example, the "swine flu" virus, H1N1, originated in North America in April 2009 and quickly became the 2009 Influenza A/H1N1 pandemic. From April 12, 2009, to April 10, 2010, this influenza infected an estimated 60.8 million people in the United States and killed between 152,000 and 575,000 people worldwide (US Centers for Disease Control and Prevention [CDC] 2019). H1N1 comprised two swine, one human, and one avian strain of influenza (Belshe 2009).

The frequency of wildlife diseases (including zoonoses) is increasing (e.g., McMahon et al. 2018). The factors facilitating diseases appear to be the same as those already described for other categories of human-wildlife conflicts: a burgeoning human population with increased food demands, intensified agriculture, expanded human habitation and contact with wildlife, and increased human densities in urban areas (e.g., McMichael 2004); increasing transport of people, organisms, and products around the world for business or recreation, and legal and illegal wildlife trade; and a changing climate that facilitates mutations of pathogens and alters population dynamics of pathogens, vectors, and hosts (e.g., McMahon et al. 2018).

For example, intensified agriculture changed how animal feed was processed in the United Kingdom, allowing prion-contaminated feed to be given to cows and leading to the emergence of a previously unknown bovine spongiform encephalopathy in cattle (Mahy and Brown 2000). The transport of infected rodents from Ghana into the United States, and their subsequent commingling in pet shops, allowed monkeypox virus to "species jump" (box 9.2) into pet prairie dogs, subsequently infecting over 70 people (Hutson et al. 2007).

Native pathogens operate in all ecosystems and seres, serving critical functions in detritus circuits, killing susceptible organisms, and breaking down dead organisms to release and recirculate minerals and nutrients. Ecosystems have coevolved with native pathogens, resulting in pathogens that are neutral or beneficial. For example, native pathogens can facilitate tree decline and therefore the tree composition and overall diversity of forest ecosystems. Pathogens work in concert with abiotic factors such as mineral deficiencies, wind, and lightning to remove susceptible trees. Such pathogens might include viruses, bacteria, fungi, nematodes, parasites such as mistletoe, and insect defoliators, as well as bark and wood borers. Hagle et al. (2003), among others, has pointed to over 1,300 pathogens, diseases, and insect pests of conifers in the northern and central Rocky Mountains, most of which are native and of little or no concern. Native pathogens usually become problematic when conditions change, favoring new pathogenic forms.

Exotic pathogens, however, can disrupt normal ecosystem function by affecting individual species and organisms. There is not enough time for coevolution of defenses against new pathogens, and such defenses may not already exist. Susceptibility to American chestnut blight and Dutch elm disease (*Ophiostoma ulmi*) are examples.

The effects of exotic pathogens on ecosystems can be both subtle and complex but have profound conse-

---

BOX 9.2 Species Jump

A **species jump** occurs when a pathogen overcomes a species barrier and infects a new organism. Woolhouse et al. (2005) provide examples and describe underlying epidemiological theory. The likely importance of an outbreak is based partly on its **basic reproduction number**, R_0. R_0 is the ratio of secondary cases (number of individuals in the new population that are infected from primary cases) to primary cases (number of individuals infected by the jump). Suppose, for example, a pathogen jumped a species barrier and infected two individuals (i.e., two primary cases). Suppose the primary cases subsequently infected six more individuals. The R_0 would then be 6:2, or 3.

Woolhouse et al. (2005) stated that if $R_0 < 1$, the transmission potential is low, and the disease is unlikely to pose a great threat to the new hosts. Examples include Ebola, monkeypox, and most avian influenza viruses. If, however, $R_0 > 1$, there is risk for a major epidemic or pandemic. Examples include avian influenza type A and SARS-CoV viruses such as COVID-19. Woolhouse et al. (2005) point out that when $R_0 > 1$, most new infections come from within the new population. This is the situation that led to the rapid spread of COVID-19 throughout the globe. Once a person carrying the disease arrived at a new location, that person spread the disease rapidly to others who in turn spread it to others, and so on. When $R_0 < 1$, however, spread of the disease into the new population depends on repeated and continued species jumps, which is unlikely.

---

quences. For example, Borer et al. (2007) suggested that generalist invasive viral pathogens (e.g., barley and cereal yellow dwarf viruses) contribute to the invasion and domination of California's perennial grasslands by exotic annual grasses. Over 20 million acres of California's native perennial grasses have been replaced by exotic annual grasses and forbs since European settlement. The exotic annual grasses have been successful even though they are inferior to the native perennial species in competing for resources.

Borer et al. (2007), modifying a Lotka-Volterra competition model to accommodate diseases and seasons, found that disease-free native perennials could not be successfully invaded by healthy annuals. However, the presence of viruses changed susceptibility. Because the native perennials are a necessary reservoir for the viruses, the simulations predicted an equilibrium in which native perennials were dominated by invasive annuals. Aphids serve as vectors for reinfection of perennials, and annual immigration of aphids from perennial lawn grasses or irrigated grass crops such as barley are needed for the equilibrium to persist.

The reader should explore resources available through the internet where one can get current information on wildlife diseases and zoonoses. For example, the Center for Invasive Species and Ecosystem Health (https://www.invasive.org) provides information on invasive species—including invasive and exotic pathogens in North America—affecting both terrestrial and aquatic species. The National Wildlife Health Center, a part of the US Geological Survey, maintains a database on emerging wildlife diseases such as avian influenza, coral diseases, snake fungal diseases, and sylvatic plague. The Wildlife Health Information Sharing Partnership (WHISPers) provides a database for wildlife mortality events across North America that are verified by trained biologists. The National Center for Emerging and Zoonotic Infectious Diseases (NCEZID), a part of the US Centers for Disease Control and Prevention (CDC), is a national resource on both national and global zoonotic diseases. Conover and Vail (2014) provide interesting and useful information on past and emerging zoonoses. To highlight some wildlife diseases and zoonoses, I provide selected listings in tables 9.1 and 9.2 and examples below.

Examples

PLANT DISEASES. Chestnut blight (*Cryphonectria parasitica*) was first observed in 1904 infecting chestnut trees in the Bronx Zoo, New York. Released from ecosystems that kept it in check in Asia, the fungus spread rapidly throughout forests of the eastern United States so that by 1940 the tree was no longer a dominant species in the forest canopy. Chestnut trees have since been mostly replaced by oak and hickory. Chestnut blight also spread over Europe; however, trees there showed resistance and recovered. *C. parasitica* infects upper branches of trees but does not kill root collars or root systems. Consequently, native chestnut trees still survive in parts of the United States as new saplings emerging from old root systems. Most become infected and die before they can produce viable seeds.

Diamond et al. (2000) analyzed **mast** (edible seeds and fruits of woody plants) production in the eastern United States before and after the presence of chestnut blight. Even though chestnut trees were replaced by two other mast producing species, the amounts and quality of mast were sufficiently reduced to impact animals. The researchers pointed particularly to white-tailed deer, black bear, wild turkey, gray squirrel, and ruffed grouse (*Bonasa umbellus*) as wildlife that depend on mast production for survival.

Westbrook et al. (2020) described two approaches whose long-term goal is to restore blight-tolerant American chestnut trees (*Castanea dentata*) to the eastern deciduous forests. With one, the American Chestnut Foundation is backcrossing blight-tolerant Chinese chestnut trees (*Castanea mollssima*) with wild-type susceptible American chestnut trees, then selecting hybrid trees that are morphologically indistinguishable from the American chestnut tree but intermediate in resistance to the blight. With the help of volunteer citizen scientists, the Foundation has applied pollen from the backcrossed trees to flowering wild American chestnut trees in at least 40 trials and mine land restoration projects across the eastern United States. The authors summarized a second approach in which transgenic American chestnut trees would be distributed in eastern forests, again with the help of citizen scientists. "Darling 58," one of two transgenic American chestnut trees made resistant to chestnut blight by inserting a wheat oxalate oxidase gene into its genome using CRISPR technology, has been developed, and a plan for long-term distribution of its progeny now awaits federal consideration and approval (Steiner et al. 2017). Westbrook et al. (2020) suggest that a sufficient population of transgenic blight-tolerant American chestnut trees could be available for large-scale forest restoration within 20 to 35 years after federal approval.

Dogwood anthracnose (*Discula destructiva*) is a fungal infection of flowering dogwoods (*Cornus florida*) and the Pacific dogwood (*C. nuttallii*) in North America. It may have been introduced into the United States on infected horticultural stock from Asia in the 1970s. Symptoms appeared in dogwoods in Washington, DC, New York State, and Connecticut at about the same time. According to the CABI Invasive Species Compendium (2021), dogwood anthracnose is now present in localized areas of British Columbia and Ontario, Canada, and is widespread in the United States in localized areas from the eastern to the western seaboards. It has also been found in a few areas in Germany, is localized in Italy and Switzerland, and is widespread in the United Kingdom. Infected plants form girdling cankers at leaf nodes and then exhibit

TABLE 9.1   *Selected wildlife diseases*

Disease	Agent	Vector/transmission	Host/reservoir
PLANTS			
Chestnut blight	Fungus	Spores in wind	Chestnut trees
Dogwood anthracnose	Fungus	Spores in wind, rain	Dogwoods
Dutch elm disease	Fungus	Elm bark beetle	Dutch elm trees
Eelgrass wasting disease	Fungus	Unknown, warm temperatures?	Eelgrass
Florida torreya mycosis	Fungus	Unknown	Florida torreya
Oak wilt	Fungus	Insects, root transfer	Oaks in United States
Pondberry stem dieback	Fungus	Unknown, beetle?	Pondberry
Powdery mildew, grape	Fungus	Spores in wind	European horse chestnut, Indian beam tree
Root fungus	Fungus	Infected soil	Different species, e.g., jarrah
Sudden oak death syndrome	Fungus	Infected air, water, plants	Some oaks, rhododendron, huckleberry
ANIMALS			
Anaplasmosis	Bacteria	Tick	Cattle, deer, sheep, antelope, wildebeest, blesbuck, duiker deer
Avian botulism, type C	Bacteria	Contaminated carcass, maggots	Waterfowl, shorebirds, gulls
Avian cholera	Bacteria	Direct, indirect contact	Waterfowl
Furunculosis, fish	Bacteria	Direct contact in water	Trout, other fish
Mycoplasma conjunctivitis	Bacteria	Direct contact, aerosol, eggs	Various passerine species
Upper respiratory tract disease, tortoise	Bacteria	Direct contact	Desert tortoises
Aspergillosis	Fungus	Inhalation	Birds worldwide
Bat white-nose syndrome	Fungus	Direct, human facilitated?	Bats
Chytridiomycosis, frog chytrid fungus	Fungus	Distributed by infected pets	Frogs, other amphibia
Chronic wasting disease	Prion	Contaminated food, water, soil	Cervids
Avian malaria	Protozoa	Mosquitoes	Hawaiian honeycreepers, other birds, crayfish
Crayfish plague	Protozoa	Infected crayfish	Crayfish
Whirling disease	Protozoa	Infected fish, birds, anglers	Sludge worms, salmon, trout, whitefish
African swine fever	Virus	Soft ticks	Wild boar
Aujeszky's disease (pseudorabies)	Virus	Oral, nasal contact	Wild boar
Avian pox	Virus	Mosquitoes, direct contact	Hawaiian honeycreepers
Bass virus	Virus	Contaminated water	Largemouth bass
Blue tongue	Virus	Biting flies	Wild ruminants, livestock
Bovine viral diarrhea, alpha	Virus	Direct, indirect contact, fluids	Cervids
Herpesvirus, catarrhal fever	Virus	Contact with birthing sheep	Cattle, ungulates
Canine distemper	Virus	Contaminated fluids, aerosol	Many carnivores
Duck plague	Virus	Direct contact; in water	Waterfowl
Epizootic hemorrhagic disease (EHD)	Virus	Biting flies	Deer, blesbuck, bighorn sheep, elk, pronghorn
Hemorrhagic septicemia, type IV-b (VHS)	Virus	Contaminated water	Finfish
Hemorrhagic disease	Virus	Direct contact, wind, dry air	Rabbit
Myxomatosis	Virus	Insects, direct contact	Rabbit
Salmon anemia, infectious	Virus	Direct contact, sea louse?	Atlantic salmon
Swine fever, classical	Virus	Direct, indirect contact	Wild boar

twig dieback. After successive years of dieback, shoots develop at the plant base. The shoots are susceptible to fungus, and cankers form around them and sometimes around entire trunks, eventually killing the dogwood.

Dogwood trees remove large amounts of calcium from soil. When their leaves drop, calcium becomes part of the litter zone. Dogwoods are, therefore, considered a primary calcium pump for deciduous forest ecosystems (e.g., Borer et al. 2013). Jenkins et al. (2007) studied loss of calcium in several ecosystem types affected by dogwood anthracnoses. Oak-hickory forests appear most likely to be affected, with a 42% decline in the amount of calcium cycled, losses not completely offset by overstory increases in cycling. Nation (2007) found calcium-dependent land snails at higher densities in the litter from flowering dogwood than American beech (*Fagus grandifolia*) and blackgum (*Nyssa sylvatica*), indicating impacts of the fungus may cascade beyond dogwoods to other forest species.

TABLE 9.2 *Selected zoonoses*

Disease	Agent	Vector/transmission	Host/reservoir
Avian influenza, highly pathogenic (HPAI), (H5N1)	Virus	Direct contact	Poultry, wild birds, including geese, swans, raptors, gulls, herons; pigs, cats, tigers
Colorado tick fever	Virus	Ticks	Porcupines, chipmunks
Ebola	Virus	Body fluids	Bats
Encephalitis, European	Virus	Tick, contaminated cow/goat milk	Tick, rodents, cows, goats
Encephalitis, Japanese	Virus	Mosquitoes	Water birds, pigs
Encephalitis, La Crosse	Virus	Mosquitoes	Chipmunks, tree squirrels
Encephalitis, Powassan	Virus	Ticks	Spotted skunks
Encephalitis, Russian	Virus	Ticks, contaminated cow/goat milk	Ticks, rodents, cows, goats
Equine encephalitis, eastern	Virus	Mosquitoes	Riparian birds; horses; dogs
Equine encephalitis, western	Virus	Mosquitoes	Passerines
Equine encephalitis, Venezuelan	Virus	Mosquitoes	Horses, rodents
Hantavirus (cardio)pulmonary syndrome (HPS, HCPS)	Virus	Contact with wildlife, contaminated material	Mice, rats
Hemorrhagic fever with renal syndrome (HFRS)	Virus	Contact with wildlife, contaminated material	Mice, rats, voles
Hendra virus	Virus	Contact with horses	Fruit bats (Pteropid), horses
Monkeypox	Virus	Animal blood or bite	Black-tailed prairie dogs, rodents, primates
Newcastle disease	Virus	Direct contact	Pigeons, waterfowls, other birds
Nipah virus	Virus	Contact with wildlife, contaminated material	Fruit bats (Pteropid), pigs
Rabies	Virus	Saliva of infected animals	Raccoons, raccoon dogs, wolves, foxes, bats, dogs, skunks, mongooses
Severe acute respiratory syndrome (SARS)	Virus	Aerosol, contact with infected animal	Horseshoe bats (Rhinolophidae)
Swine influenza, H1N1	Virus	Contact with pigs	Swine
West Nile encephalitis	Virus	Mosquitoes	Birds
West Nile virus	Virus	Mosquitoes	Birds, other vertebrate
Anaplasmosis, human monocytic	Bacteria	Ticks	Small mammals
Anthrax	Bacteria	Contact with hoofed animals	Hoofed animals
Avian botulism, A, and B	Bacteria	Ingest toxins, uncooked meat	Waterfowls, shorebirds, gulls
Avian tuberculosis	Bacteria	Contaminated feces, food, water	Birds, mammals
Bovine tuberculosis	Bacteria	Contact	Cervids, canids, bovids, others
Brucellosis	Bacteria	Ingest infected material	Elks, mule deer, buffalos, cattle, canids, goats, pigs
Cholera	Bacteria	Contaminated water	Contaminated water, shellfish
Ehrlichiosis, human monocytic	Bacteria	Ticks	White tailed deer
Leptospirosis	Bacteria	Contaminated water, food, soil	Livestock, commensal rodents, dogs, many wildlife
Lyme disease	Bacteria	Ticks	Mice, deer
Plague	Bacteria	Fleas	Rodents, prairie dogs, ground squirrels
Q-fever	Bacteria	Ticks	Many wildlife
Rocky mountain spotted fever	Bacteria	Ticks	Small mammals
Salmonellosis	Bacteria	Contaminated food, water, soil	Many reptiles, birds, mammals
Sylvatic plague	Bacteria	Direct contact or fleas	Rodents, canids
Tularemia	Bacteria	Ticks, biting flies	Birds, mammals, rabbits (type A), rodents (type B), hares (both), prairie dogs (pets)
Babesiosis	Protozoa	Ticks	Many mammals
Giardiasis	Protozoa	Fecal contaminated water	Beavers
Malaria	Protozoa	Mosquitoes	Vertebrates
Toxoplasmosis	Protozoa	Contaminated food or water	Felids, sea otter, birds, mammals
Visceral leishmaniasis	Protozoa	Sandfly	Dogs, other mammals
Aspergillosis	Fungus	Inhalation	Birds worldwide
Coccidioidomycosis	Fungus	Inhalation of fungus	Sea otters; soil
Histoplasmosis	Fungus	Inhalation	Bat, bird guano enriched soil
Acanthocephaliasis	Parasite	Infected water	Sea otters, birds, mammals
Chagas disease	Parasite	Blood-sucking insects	Armadillos, opossums, other mammals
Schistosomiasis	Parasite	Infected water	Snails
Trichinosis	Parasite	Infected meat	Many birds, mammals, including rodents, pigs, mountain lions, bears
Bovine spongiform encephalopathy	Prion	Infected food	Cattle

Oak wilt (*Ceratocystis fagacearum*; now *Bretziella fagacearum*; de Beer et al. 2017) is a fungal infection found only within the United States. It was first recognized in Wisconsin in 1942. Oak wilt may be the most destructive disease of oak in the United States, having particularly severe impacts on trees in the Edward's Plateau and other parts of Texas. The disease is an economic concern for both forest industries and individual homeowners. In one example, Haight et al. (2011) used cost for removal of trees as the single metric for an analysis of potential economic impacts of the disease on residents of Anoke County, Minnesota. They predicted a cost of $18–60 million for removal of 76,000 to 266,000 infected trees and suggested this was the lower end of potential costs. Wilson (2005) reported cases in Texas where individual removal and treatment of oak costs up to $20,000 per incident. Additional costs are associated with restrictions on export of wood and wood products. The European Union, for example, has restricted import of American oak logs because of concerns for oak wilt. The path being considered to allow such imports includes pesticidal treatment of the logs and their sealed containment during shipment (Bragard et al. 2020).

Oak wilt may have originated in Central America, South America, or Mexico. Infected trees wilt and die quickly. Patches of infected trees can be seen in forested areas by observing discoloration of canopy tops. The disease is transferred by spores carried from tree to tree by insect vectors. For example, sap beetles are attracted to aggregation pheromones of other beetles and fruity smells of mats of fungal spores for feeding and breeding. Contaminated beetles then pass spores onto wounds of other trees. Intertwined root systems of trees also transfer the disease, even between species of oak. Tree removal is often needed. Trenching around trees sometimes prevents transfer of the disease between root systems.

Japanese oak wilt disease (*Raffaelea quercivora*), analogous to its counterpart in North America, is viewed as one of the most serious threats to Japanese oak. The effects of the disease are sometimes exacerbated by deer damage in Japanese forests (Nagashima et al. 2019).

OTHER WILDLIFE DISEASES. Crayfish plague (*Aphanomyces astaci*) arrived in Italy in 1860, probably in the ballast of a North American ship. This disease is believed to be caused by a protist that only infects crayfish. It is endemic to North America, infecting signal crayfish, Louisiana swamp crayfish (*Procambarus clarkii*), and spiny cheek crayfish (*Orconectes limosus*). On exposure to the oomycetes that cause the disease, North American species respond vigorously with immune defenses, often showing no symptoms (Unestam and Weiss 1970).

In Europe, the noble crayfish (*Astacus astacus*), the white-clawed crayfish (*Austropotamobius pallipes*), the narrow-clawed or Turkish crayfish (*Astacus leptodactylus*), and other European crayfish are popular foods and important prey for fish such as trout, pike, perch, and chub (Westman and Savolainen 2001). Because these European crayfish offered little immunity, plague spread rapidly through crayfish populations of northern Italy, quickly decimating them, as it moved on to crayfish populations in France and Germany and then to other European countries. The plague created both economic loss of crayfish as a valued food and the fishery loss of important species in food chains.

European responses included replacing noble crayfish with the larger and comparably flavored signal crayfish from North America. Because they were asymptomatic, signal crayfish were assumed to be free of plague. Signal crayfish harbored plague, however, and vectored further impacts on European crayfish populations. Infected populations of crayfish continue to be a problem in Europe, partly due to North American crayfish as carriers but also due to illegal releases of contaminated aquarium water or use of live fish baiting or garden pond escapes, and to European crayfish serving as latent carriers of the disease. Chinese mitten crabs (*Eriocheir sinensis*) introduced illegally in some European waters, may also carry the disease as may leftover parts from the nearly 600,000 tons of North American crayfish that are sold for culinary purposes each year (Svoboda et al. 2017).

The disease has also negatively impacted native crayfish populations in Japan, where it was imported in 1927 along with two North American crayfish species. This undesirable consequence of a well-intended introduction underscores the risks of introducing asymptomatic but infected crayfish into areas with susceptible species (Martín-Torrijos et al. 2018), such as Australia and the Pacific Islands (Unestam 1975).

Frog chytridiomycosis, a highly virulent amphibian fungal disease, has contributed to the decline of at least 301 amphibian species; 90 are believed extinct in the wild, and another 124 have declined by over 90% (Scheele et al. 2019). This disease is caused by *Batrachochytrium dendrobatidis* and *B. salamandrivorans*. According to Scheele et al. (2019), "this represents the greatest documented loss of biodiversity attributable to a pathogen and places *B. dendrobatidis* among the most destructive invasive species, comparable to rodents (threatening 420 species) and cats (threatening 430 species)" (1459).

Once believed to have originated in Africa, it is now thought to have come from Asia (Doherty-Bone et al. 2020). Trade and human development have broken natural barriers, facilitating transfer of the disease. Frog chytrid fungus affects particularly larger-sized animals, populations located at higher elevations, and populations located in perennially wet environments, factors favorable for fungal infections. These factors might also help to explain the strong impact of the fungus on some North American and Australian species (Scheele et al. 2019). The disease peaked in the 1980s after several decades of explosive global spread. Recovery has been slow and incomplete for most surviving species, and the potential continues for additional impacts.

Research has intensified on potential cures for the disease. In a possible breakthrough, Savage et al. (2020) studied survivors and nonsurvivors of experimentally infected lowland leopard frogs (*Lithobates yavapaiensis*) and found that frogs producing large numbers of B- and T-type lymphocytes were more likely to die from the disease than those that provided a less intense immune response. The fungus destroys B and T cells with impunity. Savage et al. (2020) hypothesized that overproduction of these defensive cells might waste energy without contributing to a cure, facilitating the death of the weakened frog.

Viral hemorrhagic septicemia (VHS) is a reportable finfish disease caused by viral hemorrhagic septicemia virus (VHSv; Escobar et al. 2018). The disease was first identified from wild rainbow trout in Europe in the 1930s and was thought to be endemic in and limited to Europe but since has been found throughout the globe in a broad range of cool to cold freshwater and marine ecosystems and in a broad range of finfish. For example, VHS has been found responsible for substantial fish die-offs in both wild and cultured fish in both the Pacific and Atlantic Oceans. Anthropogenic factors such as high fish density and transport of fish further exacerbate the spread of VHS and other transmissible diseases in fish populations (Lafferty and Hofmann 2016). Given its broad distribution and its ability to infect a wide range of fishes and the likelihood of its continued spread through human activity, VHS is now considered a major concern for long-term natural and cultured fish production in Europe (Kim and Faisal 2011) and for natural fish populations in the Great Lakes regions (Rodger 2016).

VHSv is an RNA virus in the family *Rhabdoviridae*. Based on the composition of its nucleoproteins and glycoproteins, VHS is classified into four types (I, II, III, and IV) that are loosely associated with four geographic regions. The disease is transmitted through water and affects young fry, sometimes killing entire cohort populations. Signs of the disease can range from extensive hemorrhaging to lethargy and congregating near tank ponds or sides and erratic swimming. Survivors develop antibodies that protect them the rest of their adult lives.

The virus was found in the Great Lakes region for the first time in 2005. Within two years, the virus, labeled type IVb, was found in 25 species of fish in five of the lakes, in the Saint Lawrence River and in inland lakes in New York, Michigan, and Wisconsin. VHS has caused major fish die-offs there in yellow perch, muskellunge, freshwater drum, round goby, emerald shiners, and gizzard shad. The concern now is that this particularly virulent strain might move into new populations of freshwater fishes and in different geographic regions (Winton et al. 2008).

Efforts are ongoing to develop a vaccine, but none is yet available. Some gains in disease resistance have been made by selection of resistant stock and by intergeneric hybridization. Control currently relies on surveillance programs coupled with policies that include quarantine and eradication procedures. The programs have resulted in eradication of the disease from some parts of Europe (Rodger 2016) and in some North American hatcheries (Håstein et al. 1999).

Plasmodium relictum is a filarial protist that causes avian malaria. The disease infects erythrocytes of birds, damaging them and causing anemia. Because high levels of oxygen are important for bird metabolism and flight, the impacts are often lethal unless the birds have developed some tolerance to the disease. The protist is not spread directly between birds but requires vectors such as mosquitoes. *P. relictum* has been found in birds in England, France, New Zealand, and several mainland US states. It has been considered as a possible cause of the dramatic decline in populations of house sparrows in London (Dadam et al. 2019).

P. relictum spread quickly through Hawaii's bird fauna after the southern house mosquito (*Culex quinquefasciatus*) arrived in a ship's water barrels in 1826. The disease infected birds, including honeycreepers (*Carduelinae*) and Hawaiian crows (*Corvus hawaiiensis*), at lower elevations where mosquitoes could survive, and it became a factor in the extinctions of at least 10 bird species (Atkinson et al. 1995). Some honeycreepers in forests at lower elevations are developing tolerance to the disease, so the reservoir of avian malaria is increasing. Freed et al. (2005) have argued that this increases pressure for transfer of the disease to birds at higher elevations. Climate change is increasing ambient temperatures, exacerbating this pressure.

Facilitated adaptation using gene-edited malaria-resistant birds (CRISPR technology) is among the approaches being considered as a permanent means of repopulating Hawaiian forests with honeycreepers (Samuel et al. 2020). Another is the *Wolbachia*-mediated Incompatible Insect Technique being used to reduce mosquito-borne diseases such as malaria and dengue in Africa and other continents (Berio Fortini et al. 2020).

Chronic wasting disease (CWD) is a transmissible spongiform encephalopathy found broadly in cervids, including mule deer (*Odocoileus hemionus*), white-tailed deer, elk, and moose (fig. 9.2). The agent is a misfolded isoform of a host protein called a prion. The misfolded prion replicates and causes a cascade of developmental changes as it spreads between cells. The protein can be readily passed between individuals (Rivera et al. 2019). The syndrome was first identified in 1967 in a farm-raised mule deer in Colorado and subsequently in free-ranging deer and elk in Colorado and Wyoming. CWD may be endemic to Colorado, Wyoming, and Nebraska. It has since been found in free-ranging deer in at least 26 other states, three provinces in Canada, three European countries, and captive cervids in South Korea (Rivera et al. 2019).

Infected deer lose weight over weeks or months and experience excessive salivation, difficulty in swallowing, excessive thirst, and production of excessive urine. Most die within months, sometimes of aspiration pneumonia. The disease is highly transmissible, probably from animal to animal by direct contact or through contaminated food or water, and it is always fatal.

Hunters and others who eat venison are concerned that CWD might be transmitted to humans. Ranchers are worried that infected free-ranging deer or elk might contact or contaminate food or water that would transmit the disease to livestock. Experimental studies show that some transmission may be possible but only under laboratory conditions. Transmission to noncervid mammals has not been observed under natural conditions and seems unlikely. There appears to be a barrier at the molecular level that limits transfer to humans, cattle, and sheep (Rivera et al. 2019). The US CDC recommends that venison from CWD infected cervids not be eaten and that all venison be well cooked.

ZOONOSES. COVID-19, the highly contagious zoonotic disease caused by severe acute respiratory syndrome coronavirus 2 (SARS-CoV-2; fig. 9.3), was first identified in Wuhan, China, in December 2019. The spread to humans may have been the consequence of a species jump from bats via pangolins or perhaps other wildlife to humans. A wet market in Wuhan, where many species of wildlife are sold, and a research laboratory in Wuhan where bats are kept, have both been implicated. An ongoing pandemic followed, resulting in over 6.07 million human deaths globally at the time of this writing, of which over 1 million (https://coronavirus.jhu.edu/) occurred in the United States.

Symptoms include fever, cough, fatigue, difficulty breathing, and loss of sense of taste and smell. For

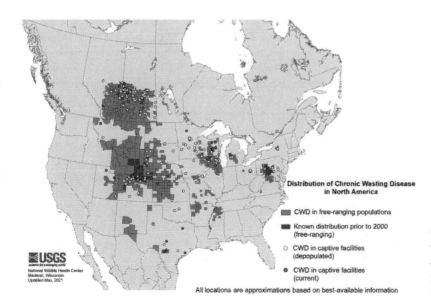

Distribution of Chronic Wasting Disease in North America

▓ CWD in free-ranging populations

▓ Known distribution prior to 2000 (free-ranging)

○ CWD in captive facilities (depopulated)

● CWD in captive facilities (current)

All locations are approximations based on best-available information

≋USGS
science for a changing world
National Wildlife Health Center
Madison, Wisconsin
Updated May, 2021

Figure 9.2 Distribution of chronic wasting disease in North America as of May 2021. Map provided by the US Geological Survey.

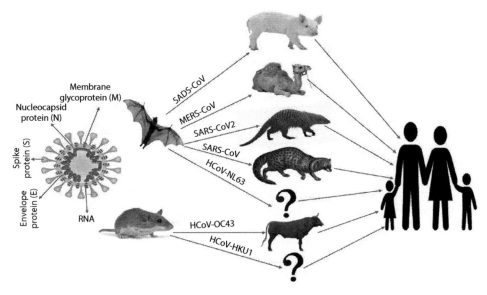

Figure 9.3 Transmission of coronavirus (SARS-CoV), showing animal origins of human coronaviruses. Diagram from Swelum et al. (2020). Open Access.

about one-third of people with the disease, no symptoms are noticed. For most of the remainder, the symptoms are mild to moderate. For 5% or so, the symptoms are severe and critical, often leading to hospitalization and death.

Infection occurs when people are near enough to each other for long enough; the transmission occurs when a person inhales contaminated droplets of air from an infected person's breath due to coughing, sneezing, singing, or speech. Scientifically based prevention such as social distancing, wearing masks, and washing hands with soap and water were found early on to be highly effective and were recommended by scientific authorities. The human dimensions surrounding these solutions have been found, however, to be much more complex. In some instances, political leaders have provided disinformation that encouraged group norms based on conspiracy and other unproven or inaccurate theories that opposed rather than supported the scientifically based preventive measures. In other instances, masses of people have simply chosen to ignore these preventive measures in favor of large group gatherings without preventive measures for weddings, political rallies, or parties or other social gatherings. The consequence has been unnecessary human deaths, which some estimate in the hundreds of thousands.

Vaccination for the disease along with preventive measures, now underway globally, appears at the time of this writing to be the most effective means of managing the disease in the long term. This approach, however, also has complex human dimensions in that some individuals oppose vaccinations, and many who would take the vaccination find it inaccessible to them.

The technologies and regulatory procedures used to develop, test, and approve this plethora of effective vaccinations in such a short time are impressive. For over 30 years, scientists explored the possibility of using mRNA rather than injecting an impaired virus or bacteria, or a fragment of a virus or bacteria, to induce immunity. With the mRNA approach, once injected into the bloodstream and taken up by cells, the mRNA produces the protein encoded in the RNA (for COVID-19, the segment of mRNA was selected that produces the protein spikes on the surface of the virus). The body, recognizing these spikes as foreign, develops both antibodies and T cells that kill the virus. The coding in the vaccination can be changed quickly as the virus mutates in the human population. The technological advances in development of the mRNA vaccine will find many future applications.

SARS-CoV-2 has been found in other animals including minks, ferrets, and zoo animals that have had contact with humans infected with COVID-19 naturally; this also includes snow leopards, pumas, gorillas, and other wildlife after experimental transmission from humans. Antibodies against SARS-CoV-2 have been found in about 40% of samples of white-tailed deer taken in the United States in 2021. Given the documented spillover of COVID-19 from humans back to minks, and with frequent contact between humans and wildlife such as deer, a concern is that a large sylvatic cycle may become established with continued spillover to humans (Gao and Wang 2021).

Influenzas (flus) are grouped into types A, B, C, or D. Type A includes zoonotic flus and are the only type that have caused human pandemics. Both types A and B cause seasonal flus in humans, sometimes severe. Type C generally causes only mild infections, and type D is found mostly in animals like cattle, not in humans. Type A viruses are classified into subtypes based on variations in glycoproteins and hemagglutinins on their surfaces. Type B, found only in humans and seals, is classified as either B/Yamagata or B/Victoria, the only two known lineages for circulating type B viruses. Both type A and B flus are further classified into clades (groups) and subclades (subgroups) based on similarities of when and where genetic changes have occurred (WHO 1980). This approach allows construction of phylogenetic trees so that one can visualize the origins and mutational paths of the viruses.

The influenza A virus genome is made of eight segments of negative-stranded RNA. Two of the 11 proteins coded by RNA are used to classify influenza viruses in wild birds and poultry. Eighteen subtypes of the one glycoprotein, hemagglutinin (HA), and 11 subtypes of another, neuraminidase (NA), have been found. The name of a particular influenza is determined by the combination of HA and NA subtypes: H5N1 is composed of the fifth subtype of hemagglutinin combined with first the subtype of neuraminidase (fig. 9.4). Influenza A has been identified in many mammalian and avian taxa, but the natural reservoir is believed to be wildfowl and waterbirds. Found mostly in a low-pathogenic (LPAI) form, H5 and H7 subtypes may become highly pathogenic (HPAI, formerly "fowl plague"; Clark and Hall 2006; Olsen et al. 2006).

Olsen et al. (2006) describe a series of Eurasian outbreaks of H5N1. An HPAI outbreak of this virus occurred on poultry farms and markets in Hong Kong in 1997, resulting in the first documented case of human H5N1 influenza and associated fatality. The H5N1 outbreak recurred in 2002. It was found in waterfowl in two parks in Hong Kong and in other wild birds. By 2003 it had resurfaced again, this time wreaking havoc with the whole of the poultry industry in Southeast Asia. In 2005, the virus was found in migratory birds in Qinghai Lake, China, and subsequently affected wild birds broadly. That epidemic alone reduced the global numbers of bar-headed geese (*Anser indicus*) by about 10%.

More recently, the virus has been found in Asia, Europe, the Middle East, and some countries in Africa. In Europe, it has affected mute swans (*Cygnus olor*) and whooper swans (*C. cygnus*) as well as other wild waterfowl and some raptors, gulls, and herons. The disease

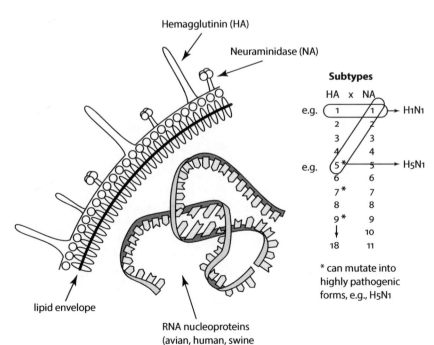

Figure 9.4 Avian influenzas are named according to subtypes of hemagglutinin and neuraminidase. Illustration by Lamar Henderson, Wildhaven Creative LLC.

has caused mortality in over 60 species of wild birds. It has been transmitted to at least 175 humans, causing 95 deaths, and has been found in pigs, cats, tigers, and leopards (Olsen et al. 2006). In the past decade, H5N1 outbreaks have occurred in Asia, Russia, the Middle East, Europe, and Africa; H5N2 in Italy and Texas; H7N1 in Australia, Pakistan, Chile, and Canada; H7N4 in Australia; and H7N7 in the Netherlands.

Beginning in March 2013, annual epidemics of human infections with H7N9 ("Asian H7N9") have been reported in China; in the largest, 766 people were infected from October 2016 to September 2017. About 39% of infected people have died, with cardiac injury sometimes occurring as well as acute respiratory problems (Gao et al. 2020).

Another zoonotic virus, Nipah virus is an emerging disease that causes encephalitis (inflammation of the brain) in infected people (Ang et al. 2018). Fruit bats (*Pteropodidae*) are a reservoir, henipaviruses (a category of virus that includes Nipah virus) having been found in fruit bats ranging from Australia (*Pteropus*) to Africa (*Eidolon*). The bats appear asymptomatic. The disease is transmitted from animals such as pigs to humans as well as from humans to humans. Since it was recognized in pig farmers in Malaysia in 1999, there have been repeated outbreaks, all in Southeast Asia, resulting in hundreds of human deaths (e.g., Sharma et al. 2019).

In Bangladesh, an outbreak of Nipah virus was related to the consumption of raw date palm juice that was contaminated with urine from fruit bats (Luby et al. 2006). Chua (2010) evaluated factors associated with the epidemic in Malaysia and concluded that the emergent disease was driven by climatic as well as anthropogenic factors. Deforestation for pulpwood and industry resulted in fewer native trees for pteropid bats. The problem was exacerbated by an El Niño–caused drought, which, along with slash and burn–induced haze, reduced the flowering and fruiting of forest trees. Fruit bats moved into cultivated fruit trees grown near confined pig operations. The bats then had access to pigsties where they hung to eat fruit, sometimes dropping fruit or urinating, spreading potential sources of the virus.

Rabid animals were described in Mesopotamia as early as the 23rd century BC (Fu 1997). Centuries later, Louis Pasteur described the disease as viral (before the term "virus" was clearly defined) and developed an attenuated preparation from rabbits that was used, with some risk, to treat patients bitten by rabid dogs.

Rabies is caused by rhabdoviruses (rod-shaped viruses) of the genus *Lyssavirus*. Rhabdoviruses infect plants and animals, including insects, fish, and mammals. These viruses are single-, negative-stranded RNA surrounded by a matrix protein, in turn surrounded by a bilayered envelope, with glycoproteins shaped into spikes on an outer surface. The RNA contains the codes for five proteins, enough information for replication of the virus in the cytoplasm of a cell. The viruses are transmitted by contact with the saliva of an infected animal (e.g., by bites, scratches, or licks on open skin or mucous membranes). In humans, the symptoms appear at first to be flulike, for weeks to years, and then they progress rapidly to paralysis, anxiety, confusion, paranoia, hallucinations, delirium, and death. Paralysis includes the throat and jaw, resulting in an inability to swallow to quench thirst and consequent hydrophobia (fear of water) and collection of saliva in the mouth until it overflows, hence "foaming."

Vaccination of pets, careful monitoring for presence of diseased animals, and safe and effective treatment immediately after exposure to rabid animals has greatly reduced the human incidence of rabies in Europe and North America. The cost for such efforts is high, over $300 million per annum in the United States, mostly paid by pet owners. In regions of the world where pets have been widely vaccinated, most human cases are now transmitted by wildlife; these wild cases are sometimes collectively called "sylvatic " rabies. In Europe, fox and raccoon dogs serve as reservoirs. In the United States, primary hosts include raccoons along the Eastern Seaboard, skunks and foxes in the south central states, skunks (*Mephitis mephitis*) in the north-central and southwestern states, foxes in Alaska, mongooses in Puerto Rico, and bats broadly (Slate et al. 2002). A recent outbreak in coyotes in Texas was controlled.

Often caused by dog bites, rabies is rampant in Africa and Asia. There, tens of thousands of people die annually (WHO 2021). WHO has declared September 28 Rabies Day to increase global awareness of the continued presence and importance of the disease.

West Nile virus is a *flavivirus* that was first isolated in the West Nile Province of Uganda in 1937 and that occurred in epidemics in parts of Europe during the 1950s, 1960s, 1990s, and 2000s. An outbreak occurred in New York City in 1999. West Nile virus infects many vertebrate species, but infected animals are asymptomatic in the normal range of the virus—Africa, the Middle East, western Asia, and Europe. This virus is transmitted by mosquitoes (*Culex* mainly, but also *Aedes* and *Anopheles*).

Most infected people never develop symptoms. Those who do experience fever, headache, fatigue, skin rash, and occasionally swollen lymph glands and eye pain. In the severe form of West Nile virus, the central nervous system is infected, and meningitis or encephalitis may occur. According to the CDC (2020), there have been over 51,800 cases of West Nile virus with 2,390 deaths reported in the United States between 1999 and 2019.

"Beaver fever," or giardiasis, is a mammalian zoonotic disease transmitted by a flagellated protozoan that causes mild to severe stomach distress in people. Giardiasis can be transmitted from wildlife such as beavers to people by fecal contamination of drinking water. For example, the clear water of mountain streams may appear pristine and safe and yet contain *Giardia* cysts if beavers upstream are infected. There were about 15,600 cases reported to the CDC in the United States during 2018, with numbers of cases increasing during the summer months.

Histoplasmosis is a fungal infection caused by *Histoplasma capsulatum*. The fungus grows in soil enriched with bird or bat guano and has been found in the litter of poultry barns, in bat roosts in caves and houses, and under bird roosts. Histoplasmosis occurs globally, but it is most common in North and Central America. In the United States, histoplasmosis mostly occurs in the Mississippi and Ohio River valleys. Spores can reside for years in guano. The spores become airborne when disturbed—e.g., with movement of soil for development of a subdivision—and can then cause infections upon inhalation. Histoplasmosis most often affects the lungs of infected people, who then think they have a cold. Particularly for those with emphysema or compromised immune systems such as those with human immunodeficiency virus (HIV), the disease can progress into a disseminated form that affects other organs (Bongomin et al. 2019). Reddy et al. (1970) found the disseminated disease in 25 of 530 patients who had been diagnosed with active histoplasmosis. In that study, all untreated cases of disseminated histoplasmosis were fatal. In an outbreak involving over 100,000 residents of Indianapolis, Indiana, in 1978–1979, 46 had progressive disseminated infection, and 15 died (Wheat et al. 1981).

Summary

- Wildlife diseases are abnormal conditions that impair functions of nondomesticated organisms; if they infect humans, they are also zoonoses.
- All wildlife diseases have sylvatic cycles wherein vectors move diseases between wildlife hosts; zoonoses also spillover to humans as, e.g., Lyme disease carried by ticks to humans, field mice, and deer.
- RNA- and DNA-based pathogens mutate and sometimes mix with other pathogens in hosts, thereby adapting to changing environmental conditions and sometimes allowing species jumps.
- Frequency of wildlife disease and zoonoses appears to be increasing, probably due to burgeoning human populations and increased contact with wildlife.
- Native pathogens serve important ecosystem functions and are of little or no concern; exotic pathogens have had no time for coevolution with affected species and can disrupt normal ecosystem function.
- Examples of wildlife diseases include chestnut blight, dogwood anthracnose, oak wilt, crayfish plague, frog chytrid fungus, largemouth bass virus, avian malaria, and chronic wasting disease.
- Examples of zoonoses include influenza A, Nipah virus, rabies, West Nile virus, giardiasis, toxoplasmosis, and histoplasmosis.

Review and Discussion Questions

1. If wildlife diseases are a normal part of ecosystems, why would they be of interest to people resolving human-wildlife conflicts? Please explain with specific examples.
2. What unique problems are posed by wildlife diseases of aquatic organisms? How, for example, does one get a pathogen out of a pond, a stream, or a lake?
3. Do you think it is just by chance that many wildlife viruses are made of RNA? How might RNA relate to the survivability of wildlife diseases? Are DNA-based diseases at a disadvantage? Share your thoughts along with some specific examples.
4. Might the notion of propagule pressure have an application in the study of wildlife diseases? Explain with examples. Conversely, might the notion of basic reproduction number, R_0, be applied to invasive species? Explain with examples.
5. Do wildlife diseases impact threatened or endangered species? How about, for example, the African wild dog? Or the black-footed ferret? Explain your answer.

PART IV • METHODS

In this part, I describe methods used to resolve human-wildlife conflicts, organizing them into physical, chemical, and biological techniques.

10

Physical Methods

Physical methods may be used to exclude, restrain, sample, or kill species; frighten, intermittently repel, or persistently repel a species; reduce natality of populations; or alter habitats or ecosystems to reduce presence of wildlife (table 10.1).

METHODS THAT RESTRAIN, SAMPLE, OR KILL

Statement
Wildlife may be physically restrained to prevent damage, to gather information, or to take other management actions; wildlife may be sampled to gather information; or wildlife may be killed using physical methods.

Explanation
Restraining devices include cage traps, foothold traps (open and enclosed), body-grip traps, cable devices (i.e., snares, powered and nonpowered), net traps, and net guns. Use of traps is guided by standards based on factors such as effectiveness, humaneness, economy, and safety (e.g., White et al. 2021).

Cage or box traps are designed so that animals can enter but not exit them. They may have triggers that release when animals move against them, closing one-way doors, or the doors may be controlled remotely. Or traps may have complex entrance designs that facilitate getting in but prevent animals from getting out. When closed, traps may send electronic signals to the trappers. Traps or cages may be baited with food, conspecifics or their effigies, or other attractants such as pheromones.

Cage traps are used on land or in water. They may be used for capture and release, relocation, or capture and euthanasia. Some traps resemble boxes with walls and drop-doors made of wire, solid wood, plastic, netting, or metal. Others, such as the Hancock or Bailey beaver traps, resemble suitcases. They can be mouse-sized or smaller, like minnow traps, or culvert-sized, like Stephenson box traps (McBeath 1941) or Clover (1954) traps made for animals like deer, bear, or feral pigs.

A foothold trap can be a long-spring, coil-spring, or jump-trap type with a spring underneath (fig. 10.1). Traps range in size from No. 0 for small rodents and some birds to No. 1 for a muskrat- or skunk-sized

TABLE 10.1   *Examples of physical methods used to manage human-wildlife conflicts*

Exclusionary Devices	Restraint Devices
Budcaps	Cable devices
Corrals	Cage traps
Crossings	Foothold traps
Fencing	Net guns
Grids	Net traps
Netting	
Spikes	SHOOTING DEVICES
Sticky substances	Blowpipes
Tubes	Handguns
	Rifles
HABITAT, ECOSYSTEMS ALTERATIONS	Tranquilizer guns
Pond levelers	
Roost pruning	STERILIZATION
	Addling
REPELLENT DEVICES	Chemical
Aircraft	Surgical
Bioacoustics	
Boats	
Drones	
Effigies	
Flags, tapes	
Lasers, strobe lights	
Pyrotechnics	
Scarecrows	

animal to No. 4.5 for a mountain lion–sized animal. Foothold traps have two opposing jaws attached to a baseplate. The jaws may be offset (i.e., have a small space between the gripping surfaces), double, laminated, or padded, or they may have additional springs (i.e., "four-coiling"). The baseplate has a pan and a dog (trigger) attached to it. The trigger is released when pressure on the pan exceeds a threshold, and the springs snap the jaws closed. The threshold can be adjusted, and the pan may have stops that limit the distance between it and the trap jaws.

Foothold traps may be modified for specific uses. For example, foothold traps are sometimes used to restrain raptors as part of a set called a "pole trap." The jaws may be wrapped in cloth and the coil springs heated to weaken them to further reduce chances of injury to the bird. The traps are set on top of posts but are designed to allow the restrained bird to drop to the ground where it may be more relaxed after capture. For example, Butchko (1990) used pole traps to protect endangered California least tern from predation by raptors.

Nets are also used to restrain and capture wildlife. Nets to capture fish include **seines** (e.g., beach seine nets, purse seine nets), dip nets, **trawl nets** (long bags or sock-shaped nets that are pulled through water), **gill nets** (designed so fish or reptiles of a particular range of sizes are caught when they try to pass through the net), **enmeshing** nets (not as selective as gill nets), and **trammel nets** (used for species not easily caught in gill nets, such as flatfish or sturgeon).

Net traps can be shaped in hoops and hand thrown; for example, hoop-nets designed by Ronconi et al. (2010) were used to capture sea birds including great and sooty shearwaters (*Puffinus gravis* and *P. griseus*) and red-necked and red phalaropes (*Phalaropus lobatus* and *P. fulicarius*). Specialized types such as **Fyke nets** have wings for use in lakes and internal cones that direct fish into a collecting area. Fyke nets can be further modified for turtles and other reptiles. Net traps can be fired from cannons or launched with rockets so that they quickly cover wildlife that has been lured to a baited site. Cannon or rocket nets have been used to capture and restrain many types of birds and mammals, including deer and bighorn sheep.

Net guns are handheld and are sometimes fired at wildlife from aircraft. O'Gara and Getz (1986), for example, used net guns to capture golden eagles.

A **cable device** (a **snare**) is a steel cable with a loop on one end. The loop draws closed when an animal attempts to pass through it, put a limb through it, or withdraw from it. Cables are made of wire strands, and the sizes vary. For example, a seven-by-seven cable is made of seven bundles, with each bundle containing seven wires. A nonpowered cable depends on the forward movement of the animals to close. A powered cable has a spring or other power source to place the cable on a limb of the animal or close the loop of the cable (see, e.g., Collarum Neck Restraint or the Wildlife Services' Turman snare; Shivik et al. 2005). Cables can have relaxing locks, breakaway hooks, loop-stop ferrules, in-line swivels, or anchor swivels. Relaxing locks allow closure of the loop when the animal pulls but no additional closure when the animal stops. Breakaway devices release animals that pull with greater strength than the targeted animal. A loop-stop ferrule prevents a loop from closing past a preset diameter. A maximum loop stop prevents too large an animal from entering the loop. A minimum loop stop prevents the loop from closing around the animal's foot. Cable devices can be set horizontally or vertically, and they may include a spring-activated throw arm to capture the animal's leg.

Restraining devices may also be especially designed for specific needs. An example is the "EGG Trap," a form of spring trap used in the United States to restrain raccoons.

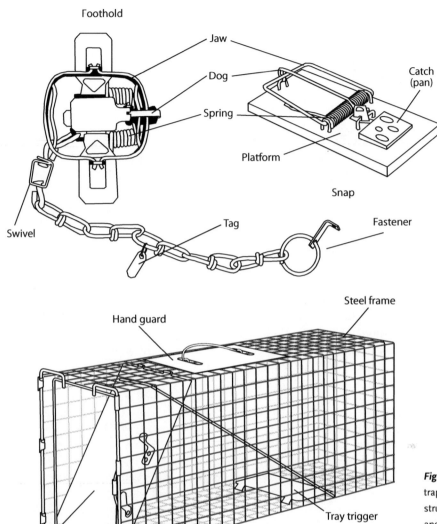

Foothold

Jaw

Dog

Spring

Platform

Catch
(pan)

Snap

Swivel

Tag

Fastener

Steel frame

Hand guard

Tray trigger

Door

Cage

Figure 10.1 Different looking traps often share common structures, such as jaws, dogs, and springs. Illustration by Lamar Henderson, Wildhaven Creative LLC.

Mechanically powered killing traps are designed to kill quickly. They are often powered by one or more springs. Examples are snap and body-grip (e.g., conibear) traps. Snap traps include "four-way" traps (so called because a rodent will be caught regardless from which direction it approaches the trigger) and "museum specials" designed to withstand the vigor of frequent use, both made for mice- or rat-sized animals. Most snap traps break the back or neck of the animal, which dies quickly. Conibear traps are often used for aquatic animals and are made in different sizes, such as 110 (muskrat size), 220, or 330 (beaver size). Other kill traps include gopher and mole traps that are set in burrow openings.

Restraining systems include foothold or cable devices, swivels, and anchors. Swivels allow movements that keep the anchor or cable from breaking and help to reduce injury. One swivel is usually placed on the base plate of a foothold trap, and another two or more are placed along the anchoring system. Anchors can be stakes or cable stakes, but they need to be strong enough to hold the largest animal that can be caught. In-line shock springs are sometimes also used to cushion lunges of the captured animal and reduce chance of injury or escape. A submersion trapping system might be used to kill an animal that lives in aquatic habitats or near water, such as a beaver or muskrat. The system is designed with a restraining device (e.g., a body-grip trap, suitcase, or cable device) and cables to hold the animal underwater.

The restraining system also includes the "set." The quality of the set depends largely on the experience of the practitioner. The set attracts intended animals and discourages others. The set may include baits such as food, scents, or lures as well as specific configurations. A myriad of baits and attractants are used successfully. Grains such as corn, milo, wheat, and oats, sometimes dyed in colors and soaked to swell and look like berries, have been used to attract granivorous and frugivorous birds. Live mice have served to visually attract predatory birds. Jesses (cables with loops, usually made of monofilament plastic) may be attached to cages to catch raptors. A mix of peanut butter and oatmeal is a common bait for rodents. Punctured cans of dog or cat food and predator urine, real or synthetic, with glycerin sometimes added as a preservative, are often used to attract predators. Rotted meat or eggs also effectively attract predators and scavengers. Synthetic fermented egg was designed as a predator attractant (and deer repellent; Bullard et al. 1978). Pheromones such as musks and beaver castor are sometimes used to attract predators and other mam-

mals. The set may also include sieved dirt, branches, or other vegetation to camouflage the trap or direct an animal toward the pan.

Pistols, rifles, or shotguns may be used to shoot and kill animals, from the ground or from aircraft. Shooting can also be part of a remote delivery system used to deliver agents such as anesthetics, immobilizing agents, contraceptives, or immunogenic compounds. Blowpipes (powered or not) and dart guns are also used, the agent delivered in a matrix such as a "biobullet" that breaks down once inside the animal.

Aerial gunning is performed with either light fixed-wing aircraft or helicopters. This type of gunning is sometimes used to manage predators such as coyotes in the western US states and wolves in interior Alaska on both public and private land.

Restraining and killing systems carry risks. One is injury to wildlife from devices such as leghold traps or cables. Another risk is capture-related **myopathy**, a muscle disease characterized by damage to muscle tissue caused by physiological changes due to extreme exertion, struggling, and stress (Breed et al. 2020). Some groups of animals may be particularly vulnerable and capture-related myopathy may underlie deaths when animals such as cetaceans are stranded (Herráez et al. 2013). Breed et al. (2020) suggest a broad range of factors be considered before attempting capture of an animal. Even then, they state that stress-related myopathy may occur. The researchers also suggest a broad range of treatments and management options including use of opioids, corticosteroids, dantrolene, sodium bicarbonate, fluid therapy, nutritional therapy, anxiolytics, and cooling with ice water immersion or water dousing, depending on the species and situation.

Noninvasive methods are increasingly being developed and used to replace obtrusive approaches when wildlife samples or population data are needed. Zemanova (2020) reviewed progress in this area. She described the use of camera traps to gather information on population dynamics of squirrels (*Sciurus vulgaris* and *S. carolinensis*), foxes (*Vulpes velox*), feral cats and wildcats (*Felis catus* and *F. silvestris*), dogs, tortoises (*Testudo hermanni*), northern flying squirrels (*Glaucomys sabrinus*), North American river otters (*Lontra canadensis*), and gray wolves. Zemanova (2020) points to the use of drones to collect exhaled breath to monitor the health of humpback whales (Apprill et al. 2017). She points to marking eggs, nests, or wildlife with dyes in short-term studies of lizards (*Anolis*) and rainbow trout, and to the identification of individual animals based on unique patterns or a wide range of species and studies. The reviewer describes technological improvements that allow identification of individuals or sex or age groups using footprints with greater than 90% accuracy or using vocalizations in great gray owls (*Strix nebulosa*), marmots (*Marmota olympus*), and Richardson's ground squirrels (*Spermophilus richardsonii*). She summarizes the use of hair traps, skin or buccal swabs (for frogs), feathers or eggshells, or fecal or saliva samples when environmental or DNA data are needed, again for a wide range of species from mussels (*Unionidae*) to harbor porpoises (*Phocoena phocoena*) to elephants.

Example

Alan Williams, a lock master, designed and tested a trap for the removal of carp from waterways in Australia. The "Williams' Carp Separation Cage" is set on the upstream side of vertical slots, bottlenecks through which migratory fish must pass. In prototype trials, about 83% of carp that passed through the slots and into the traps jumped into a containment area where they were subsequently harvested. Native fish passed through the trap. At Lock 1 on the lower Murray River, the trap was operated by a collaborative team for 11 years between 2007 and 2018. About 800 tons of adult carp (about 289,000 carp) were removed with as many as 6 tons per day. Sold on the market for AU$0.90 million (US$0.7 million), the trapped carp provided income that far exceeded the setup costs (Stuart and Conallin 2018). The traps may have applications in North America, New Zealand, and Europe for other Asian carp that tend to jump and in North America for separating sea lampreys from jumping fish such as rainbow trout.

METHODS THAT EXCLUDE

Statement

Exclusionary methods are often favored because they are nonlethal, but they must be maintained to retain their effectiveness.

Explanation

Exclusionary methods physically separate wildlife from a resource that humans want to protect. Exclusionary devices include barriers such as fencing and netting, tubing, wire grids, and bird spikes.

Barriers such as tubes and fencing have been designed in a myriad of shapes, sizes, and materials for exclusion of animals ranging from small fish to large elephants and for wildlife that swim, walk, climb, or fly. For example, "purpose-built barriers," i.e., those

that give a vertical rise of 2 to 4 feet and a have lips structured so that sea lamprey cannot attach themselves, help to prevent movement of sea lampreys upstream and into the Great Lakes while allowing jumping fish to pass. On land, barriers can protect individual plants or animals and they can protect herds, entire crops, and even entire natural ecosystems such as national parks.

Tubing is sometimes used to protect seedlings from herbivores such as gophers, mice, rabbits, mountain beavers (*Aplodontia rufa*), beavers, and deer (Marsh et al. 1990). For example, Durite and Tiller protection netting provide protection from pocket gophers, mountain beavers and lagomorphs for newly transplanted seedlings (e.g., Arjo 2003). Such tubing may be used in small gardens or in major reforestation programs, as in the Pacific Northwest of the United States. Tubing is placed over the transplant and held in place with bamboo or plastic stakes. The tubing is biodegradable and may be impregnated with a chemical repellent. Marada et al. (2019) found that an innovative type of tubing was effective in protecting orchards in the Czech Republic from damage by roe deer (*Capreolus capreolus*).

Bud caps can reduce browsing by deer, perhaps because the deer do not see the bud. Caps can be as simple as paper or plastic netting placed over the tops of the transplants and held in place with one or more staples. Caps are sometimes made by farmers using local materials, but commercial versions are available.

Exclusion fencing can be strictly physical or reinforced with repellent flavors, visual cues (such as flagging on fences used in **fladry** lines), or electrical charge (called **turbo-fladry** if used with fladry). Fences can be permanent or temporary and portable. Materials such as wood, concrete, plastic, or metal can be made into fencing as well as posts and corner braces. The material may be solid, woven, or braided. Fences may be placed vertically or at various angles. Fencing may be set partly underground to prevent the passage of burrowing animals and may include overhangs to prevent successful climbing. The overhangs may be solid or may give when an animal attempts to move over them. Fences can have aprons that extend outward on either side. A socket, a battery, or the sun can be used to charge fences with electricity. The charge may be lethal or repellent.

Fence designers need to consider where the fencing is supposed to function. Flat terrain is usually the easiest, whereas hilly terrain requires special consideration. For example, one side of a fence will be closer to the ground when traversing a hill, sometimes necessitating a taller fence. Rocks and other land-scape features need also to be considered. Special designs may be required for exclusion over streams or for fences that open onto bodies of water. Electric fences require moist soil to function most effectively. Design of gates and corners need to be given careful thought because these are often weak points exploited by wildlife. Severe storms and flooding can damage fences and reduce their effectiveness.

Fence designers should also consider the physical ability, behavior, and motivation of the targeted species as well as the purpose for excluding the wildlife. VerCauteren et al. (2006a) provide an illustration using fences intended to keep out white-tailed deer. The researchers found that a fence at least 9 feet tall was needed to exclude white-tailed deer in rough terrain. When motivated, deer can exhibit nontypical behaviors. For instance, a single-stranded electric fence may be sufficient to protect an orchard when other food deer eat is abundant, but a multiwired fence may be required when alternative food becomes unavailable. A fence that is 50% to 60% effective may be sufficient and economically advantageous in reducing deer damage to corn from 10% to 5%. However, if the intent is to prevent transmission of diseases to livestock or to prevent deer collisions with motor vehicles, no deer intrusions can be permitted.

High-tensile electric fences are increasingly used for effective exclusion of deer. These fences do not physically exclude deer but rather depend on avoidance reinforced with electric shock. The fences are usually 6 feet high, with parallel wires that vary from a foot apart at the top to 8 inches apart in the middle. The wires are either positively charged or alternate positive and negative. Slant designs are common in the eastern United States because they adapt to hilly terrain. These high-tensile fences have been used in New Zealand successfully for over 40 years (VerCauteren et al. 2006a).

Single-strand electric fences for deer are often used where portable temporary fencing or relatively inexpensive exclusion will provide satisfactory protection. Seventeen gauge steel wire can be used, but "poly-fences" that use polytape or polyrope with interwoven wires are more visible (e.g., Seamans and VerCauteren 2006). Peanut butter or other attractants can facilitate effectiveness of the fence designs by attracting deer to them (Hygnstromn and Craven 1988).

Sowka (2013) provides a variety of electric fence designs that can do everything from protecting garbage cans, birdfeeders, and apiaries from bears to excluding livestock from coyotes, wolves, and mountain lions. Their designs include permanent fencing as well as portable fences for campsites, coolers, or game carcasses. In one design, car tires are laid between two sections of cattle fencing (the lowest fencing is placed on the ground), and the garbage can is placed on top. The system is wired so that a bear receives a shock when it touches the can. Similar designs are used to shock bears that touch metal bird feeders. Breck et al. (2006) invented such a shocking device using batteries and an automobile vibrator coil/condenser that protected simulated bird feeders from black bears at test sites in rural Minnesota.

Barrier fences have also been designed to prevent movement of smaller animals such as rodents, rabbits, and raccoons. These are much shorter than deer fences and sometimes have a curtain or are placed partly underground to block digging. Designs may or may not be electrified. Ahmed and Fiedler (2002) found both lethal and nonlethal designs that kept rodents out of rice paddies in the Philippines. Reidinger et al. (1985) suggested that nonlethal electrical fencing might "train" rodents to establish territories along the fence; the trained rodents would then protect their territories, excluding conspecifics from the paddies.

Long and Robley (2004) evaluated the effectiveness of fence designs used in Australia to exclude certain animals (red foxes, feral cats, goats, pigs, and rabbits) from entire conservation areas. They found that effective feral pig and goat fences could be fabricated with one or two offset electric wires; the feral rabbit fences were 900-mm wire netting with an apron; the wild dog and dingo fences were made with aprons but varied considerably in overall design; and fox and feral cat designs also varied in design, but the most effective ones had floppy tops.

In Africa, beehive fences have been used to deter elephants (*Loxodonta africana*) from raiding local crops and gardens. (L. King et al. 2017). Elephants avoid even the sound of bees (fig. 10.2). In field trials in Kenya, fences were made with hives set at about 10 meters apart and connected with strong wire; others were made with a mix of actual and dummy hives. The fences were about 80% effective in repelling a total of 253 elephants from 10 farms between 2012 and 2015. The elephants were mixes of both individual and groups. The villagers were able to harvest and sell honey from the hives. Water et al. (2020) found beehive fences deterred 88.4% of individual Asian elephants and 64.3% of groups of elephants at pilot study sites in Thailand.

Barrier fences can also constrain the movements of reptiles, including snakes. Examples are the habu

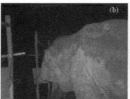

Figure 10.2 Elephants avoid bees or the sound of bees. Dr. Lucy King used this behavior to develop beehive fences that keep elephants from gardens and fields of people, reducing human-elephant conflicts. The approach, developed in Africa, has been extended to other regions of the world. These photos were taken in Thailand and provided by Antoinette van de Water/Bring the Elephant Home.

(*Trimeresurus flavoviridis*) and the brown tree snake. The habu is a poisonous sedentary snake in Japan. Electric fencing has been used to create snake-free zones and to keep snakes from entering villages on Tokunoshima and Amamioshima and electric substations in Okinawa Prefecture (Campbell 1999). For the brown tree snake, Campbell found a five-wire nylon netting fence to be most effective as a barrier, although a variety of designs and configurations can be effective (Clark et al. 2018).

Sometimes fences are used in combination with **wildlife crossings** to facilitate safe passage of wildlife over or under roads. The fences are designed to discourage wildlife crossing on the roadway and to encourage their movement toward an overpass or underpass that might be designed exclusively for wildlife or in combination with other uses

The crossings allow animals to get across human made barriers such as roadways. They include under-pass tunnels, viaducts, overpasses (e.g., green bridges and ecoducts), amphibian tunnels, fish ladders, tunnels, and culverts (fig. 10.3). Wildlife crossings and corridors already reduce vehicle collisions and allow safe passage for wildlife in parts of Europe (Bank et al. 2002) and Canada, and they are gaining more attention in the United States and other countries (Beckmann et al. 2010). For example, with about 25% of Dutch badgers (*Meles meles*) killed on highways each year, a special effort is being made in the Netherlands to use culvert systems that allow safe passage of the badgers. The US government's 2005 Safe, Accountable, Flexible and Efficient Transportation Act: A Legacy for the Future includes a directive to design projects and processes to reduce roadway impacts on wildlife habitat and driver safety, and wildlife crossing and corridors are now being designed in many states including Alaska, Arizona, Colorado, Florida, Montana, Vermont, and Washington.

Figure 10.3 Wildlife crossings over or under roads offer safe passage for wildlife while reducing collisions with vehicles. This bobcat (*Lynx rufus*) is about to leave an underpass. Photo courtesy of CSKT, MDT, and WTI-MSU.

One of the most extensive databases on wildlife crossings and fencing has been gathered from overpasses, underpasses, and fencing constructed across the Trans-Canada Highway for passage of wildlife in Banff National Park, Canada. Beckmann et al. (2010) collected information on more than 185,000 crossings in 12 years of monitoring, including crossings by grizzly bear, elk, red fox, striped skunk, and hoary marmot (*Marmota caligata*). Wolverines and lynxes have been observed using the passes. Given the number of crossings and composition of species, the findings suggest an enormous potential impact for wildlife crossings in reducing collisions with vehicles and thus human and wildlife injuries.

Crossings can serve as a landscape corridor, facilitating gene flow and genetic diversity within a metapopulation. This can be particularly important in situations where population sizes are small and where restricted gene flow might push a population below its critical minimum. An example is the Santa Monica mountain lion where there are only two breeding males (Kuykendall 2016). The mountain lions are isolated from a nearby population by a multilane highway (101 Freeway). Only one lion, a subadult male identified as P-12, has been known to have successfully crossed the freeway in recent years, and that event occurred in 2009. Plans are underway to build an overpass that will allow safe passage of wildlife over the highway, facilitating genetic exchange between the two nearby mountain lion populations.

Corrals have been used since ancient times to protect livestock from predators. Success depends largely on the worthiness of the structure in relation to the capabilities of the predator. Temporary and permanent corrals are used worldwide in Europe, the Americas, and in parts of Asia. In Asia, they protect livestock from wolves and snow leopards. Given the limited availability of timber in many higher elevation areas, traditional corrals have been built from stone, some maintained from generation to generation. The tops were often left open, however, leaving the goats or other livestock inside vulnerable to predation. Snow leopards would climb over the walls to enter the corral; once the leopards were inside, some goats would be killed by predation, and others by the turmoil. Retaliatory killings by humans would follow.

Recently, some international organizations such as Snow Leopard Conservancy have worked with local communities to completely close corrals. The roofing is expensive and will require maintenance; it is often made with a combination of wooden joists covered by screening. The results, however, are predator-proof exclosures (with enthusiastic owners; fig. 10.4). Although these roofs are expensive—costing from hundreds to thousands of US dollars each—the organizations see this as an opportunity to reduce conflicts and further coexistence between people and snow leopards (Palminteri 2015).

Netting is also used worldwide to exclude smaller mammals or birds from structures or crops. Cloth or

Figure 10.4 Herders with their sheep inside a predator-proofed corral. Photo by Milan Trykar. Courtesy of Snow Leopard Conservancy/Snow Leopard Conservancy-India Trust.

plastic netting may be treated to resist degradation by sunlight and weather. The netting might protect individual plants or trees, or orchard or vineyard rows; it may include suitable infrastructure to be pulled over entire high-cash-value crops such as cherries or other stone fruit (e.g., Dellamano 2006). Pochop et al. (2001) used plastic netting to drastically reduce nesting by ring-billed gulls (*Larus delawarensis*) on Upper Nelson Island in Washington State. The gulls feed on the fry of chinook and other salmon (*Oncorhynchus tshawytscha*) that pass through turbines on the Columbia and Snake Rivers, become disoriented, and float to the surface momentarily. Grid systems have also been used to protect the fry from the gulls (Steuber et al. 1995).

Netting has been used extensively in the Pacific Northwest to keep crabs and other predators from bivalves like the Manila clam (*Ruditapes philippinarum*) that are harvested commercially. In studies of the effectiveness of this netting, however, Bendell (2015) found the nets harbored predators rather than protected the clams, and they contributed to plastic pollution in the area. He suggested that the industry explore alternative methods for increasing bivalve productivity such as building "clam gardens" from walls of stones and boulders and set a low tide. Such gardens are an ancient approach for culturing bivalves that appears environmentally benign.

Grid systems of wires can be erected over bodies of water to exclude birds such as geese or ducks. Grid lines are often set 20 feet apart and suspended about 3 feet above the water. Monofilament and other types of string or wire can be used, but **Kevlar cord**, a para-aramid synthetic fiber, is a favorite because it is relatively easy to anchor and tighten and resists degradation by sunlight. Netting erected just 12 inches above the ground can be effective in preventing movement of birds such as geese, ducks, or gulls between the water and shorelines, encouraging the birds to find more suitable habitat. Blackwell et al. (2013) described the use of grid systems to reduce access of waterfowl to ponds at airports.

Although traps and cages are ordinarily thought of as devices to capture and retain wildlife, cages can also be used to exclude wildlife, such as keeping predators from ground-nesting birds. Major et al. (2015), for example, describe the use of nest cages to improve nesting success of endangered white-fronted chats (*Epthianura albifrons*). In the study's experimental tests, predation by corvids was reduced from 96% to 14%. Such success is not always the case. Sometimes parents abandon caged nests, and predators also can learn to associate the cages with nests.

Spikes have also found worldwide applications—excluding birds from perching, loafing, or roosting surfaces. Manufactured from either plastic or metal materials, spikes can be attached, for example, to overhangs of doorways or windowsills where the birds or their droppings are unwanted. Spikes have been used to keep birds from perching on antennae, signs, and ledges of structures at airports (Seamans et al. 2007).

Exclusionary devices have problems and carry some risks. Barriers must be maintained to retain their effectiveness. Metal parts corrode and need to be replaced. Storms and wind can damage barriers, either directly or from falling debris. Fire can destroy fencing. Vegetation can grow under electric fences and cause short-circuits. Vandalism is a common problem and can be expensive to monitor and repair (Long and Robley 2004). Fence lines sometimes exclude unintended species, can alter migratory routes, and can interrupt important wildlife behaviors such as reproduction. Further, fence-line contact between wildlife and livestock may facilitate transmission of diseases such as chronic wasting disease and bovine tuberculosis. Double rows of fences have been suggested as a means of eliminating the latter concern. Jakes et al. (2018) review the advantages and disadvantages of all types of fencing. The reviewers list exclusionary fencing as "impermeable to medium and large mammals" and "semi-permeable to small mammals, birds, and reptiles" (313).

Example

Predator fences have been used in Australia to keep dingoes and other feral dogs from livestock since the 1880s. The Dingo Fence extends about 5,320 km (3,306 miles) from Jimbour, Darling Downs, to Eyre Peninsula, Great Australian Bight. The fence is composed of the Great Barrier Fence (Wild Dog Fence, 1,553 miles) in Queensland, which has a full-time maintenance staff that patrols it weekly; the Queensland Border Fence; the South Australian Border Fence; and the Dog Fence in South Australia (about 1,383 miles long). Mostly wire mesh, the fence is about 6 feet high with an additional foot underground and metal posts about every 27 feet. Fitzwater (1972) recounted the situation of Beltana Pastoral Company, whose property was bisected by the Dingo Fence. The firm could run sheep only on the protected side of the fence; when sheep were placed on the unprotected side after a drought, the company suffered about a 3% loss due to predation.

Fisher et al. (2021) contrasted ecosystems on each side of the fence using methods that included analyses of nonphotosynthetic vegetation found in Landsat imagery. The researchers found observable influence of trophic cascades on both landscape and site scales and suggested that similar effects "may exist across the large areas of the planet over which apex predators have been extirpated" (1341).

SYSTEMS THAT REPEL

Statement

Physical methods can be used to repel wildlife. Some startle wildlife with sudden noise or lights. Others use visual or auditory cues to communicate to wildlife that an area is dangerous and should be avoided.

Explanation

Repellent methods include frightening devices (e.g., Avery and Werner 2017), **pyrotechnics** (self-contained reactions that can produce heat, fire, explosions), bright or flashing lights, and human activity. Liquid propane gas exploders have been used for years to frighten wildlife, including deer and birds. Some models discharge at random intervals, and some momentarily inflate pop-up scarecrows (such as Scarey Man) that add to the overall effect. Other pyrotechnic devices include ropes of firecrackers or individual firecrackers—e.g., 12-gauge exploding shell crackers and "bangers" and "screamers" that can be fired from specially designed 15-mm or 17-mm pistols. Pyrotechnics can be effective in dispersing birds from waterways or roosts, but habituation is a problem (Booth 1994; Avery and Werner 2017).

Effigies, or likenesses, can be effective for varied periods of time (Avery and Werner 2017). Effigies for repelling vultures have already been described. Stickley and King (1993) described the effectiveness of Scarey Man effigies in dispersing overwintering double-crested cormorants at a catfish fingerling complex in Mississippi. They found a "sudden and extreme drop in the number of cormorants flushed on the trial site in the first week of use" (91). Morrison and Allcorn (2006) found the Scarey Man effective as part of an integrated approach to disperse gulls (*Larus*) from a Coquet Island reserve in Northumberland, England. Gulls competed for nesting space with terns (*Sterna*) and were becoming an increasing problem on the island. In another case, Saul (1967) used effigies (gull carcasses preserved in formalin, wings outstretched) to repel gulls from the international airport in Auckland, New Zealand. Smith et al. (2020) tested the effectiveness of an inflatable human effigy they called "Fred-a-Scare" in scaring captive dingoes from accessing food. They found reduced access in the trials and suggested the

effigies had potential to reduce human-dingo conflicts in hotspots such as campgrounds and small livestock enterprises.

Okarma and Jędrzejewski (1997) reported that fladry—ropes with large rags—hung along forest paths were used to circle wolves for hunts by Polish kings in the fifteenth century, a practice picked up for hunting wolves in Russia in the 1800s. These researchers used fladry and netting to capture wolves for study in the Białowieża Primeval Forest in Poland. A concern is attenuation because avoidance can be a temporary neophobic response. Musiani et al. (2003), however, found fladry-based barriers protected ranches in Idaho and Alberta, Canada, from wolf depredation for up to 60 days. The researchers reported at least 23 approaches to barriers before four wolves crossed the barrier and killed cattle. Others (e.g., Bruns et al. 2020) have found that electrifying fladry fences can greatly increase their effective lives, sometimes to three years or more. A second concern, still largely unresolved and exacerbated by electrifying fladry fences, is the relatively high labor cost for installing and maintaining fladry systems.

Although designed for wolves, Young et al. (2019) modified fladry designs made for wolves by attaching the flags in a top-knot manner that reduced coiling (turning on itself and leaving a gap) and by reducing distances between the flags and tested the effectiveness of the modified design in discouraging captive coyotes from accessing food. They found the design improved overall effectiveness but that persistent coyotes crossed the fladry barrier, usually within the first four days.

Mylar tape, a metallic plastic tape, and holographic plastic tapes have been used to repel birds. These tapes are available in widths from less than an inch to several inches. They can be stretched from post to post to create a fencelike barrier or hung from string or posts at intervals. When in place, wind causes the tape to vibrate so that it hums and reflects light in patterns. Some practitioners believe that, from above, the red and silver color of some Mylar tapes give the appearance of fire. Bruggers et al. (1986) reported success with Mylar tape in reducing bird activity in the Philippines, India, and Bangladesh. Dolbeer et al. (1986) reported that the tape repelled blackbirds from millet, sunflowers, and sweet corn in the United States. However, the tape failed to protect blueberries from bird damage in field trials in New York State (Tobin et al. 1988). Tapes have subsequently been shown to be effective for keeping the Canada goose (*Branta canadensis*) from areas where they are not wanted, such as farmers' fields (e.g., Heinrich and Craven 1990), and herring gulls (*Larus argentatus*) and ring-billed gulls from certain areas such as loafing places. Tapes may be used as part of an integrated deterrent program. Marcus et al. (2007) used Mylar tape, along with other frightening devices, to redirect selection of nest sites by endangered interior least terns (*Sterna antillarum athalassos*) and threatened piping plovers (*Charadrius melodus*) away from gravel mines (where they would have had poor nesting success) in Nebraska.

Bioacoustical systems take advantage of distress or alarm calls of wildlife and have been used successfully in many situations and countries. Actual recordings of calls are used as are electronic simulations such as those generated by Av-Alarm or Bird-X systems. Holcomb (1976) tested Av-Alarm for effectiveness in protecting rice fields from red-billed quelea in Africa in 1975 and 1976. He found damage was lowest nearest the speakers and increased in a linear fashion as one moved away from the speakers, up to 450 feet. Gilsdorf et al. (2004) found that a bioacoustical system was unsuccessful in reducing deer damage to silk-stage corn in Nebraska. These researchers believed the large fields and height of the corn afforded safe areas for the deer during periods of activation; the deer would take corn after the calls ended.

Laser lights have been surprisingly effective in dispersing some bird roosts. Those tested have been class II (low power) or class III (moderate power). Baxter (2007) used two Lord-Ingerie Lem 50 lasers to completely remove 33,000 gulls nightly from ponds at a military airfield in England. The gulls moved back daily, but they were dispersed without habituation for at least 26 nights of use. Chipman et al. (2008) included lasers in 53% of their operations as part of an overall hazing program for dispersing American crow roosts in New York. Glahn et al. (2000) tested a Desman model FLR 005 class IIIB laser that directed red light with a 12-mm diameter at the source to see if it would disperse double-crested cormorants from night roosts. They also tested a Dissuader laser used for security; it was a class II diode-type laser with a 76-mm diameter at its source. Both failed to evoke avoidance in laboratory trials but effectively dispersed cormorants in field trials. The lasers offer a quiet, selective, and effective alternative to pyrotechnics for use in refuges or wetland habitat where cormorants tend to roost (Avery and Werner 2017).

Aircraft might be used to haze wildlife. Fixed-wing aircraft are often used over open, flat, or gently rolling terrain, whereas helicopters are used over bushy, timbered, broken, or mountainous areas. Use of aircraft involves maneuvering close to the ground at slow airspeeds, and the aircraft may be specially designed for that purpose and the pilots specially trained. Handegard (1988) reported use of fixed-wing aircraft to haze blackbirds damaging sunflowers in six districts and 7,000 to 10,000 square miles in North Dakota; these aircraft were used to harass depredating birds with low-level flying and shotguns. Although the overall efficacy has been questioned, the approach spreads damage among farmers and may allow plants to compensate for some damage by increasing the size of remaining sunflower heads. Remote-control model aircraft and boats have been used to haze birds at airports, in agricultural and aquaculture areas, and at landfill sites.

Unmanned aerial vehicles (drones) are also being tested for effectiveness in repelling birds from areas where they are unwanted, such as around croplands, in vineyards, over cereal crops, and at airports. In one study (Wang et al. 2019), the drones were equipped with a piezo tweeter and a crow effigy to imitate the sounds and look of a crow predator. The researchers found the drones effective in repelling Australian Raven (*Corvus coronoides*), common starling, sulphur-crested cockatoo (*Cacatua galerita*), and silvereye (*Zosterops lateralis*), the four main pest species in the four vineyards in New South Wales, Australia, where the drone was tested. The researchers concluded that the drone has potential as a highly effective bird repellent in these vineyards. Dayoub et al. (2021) have proposed the development of swarms of drones to repel the quelea weaver bird from millet crops in Africa.

Earthbound robots are also being developed to repel wildlife. One example is the "Goosinator," marketed as "a remote-controlled, space-aged design specifically made to chase geese off of your property" (https://goosinator.com/what-is-a-goosinator/). This is an orange-colored, amphibious robot that can move on land, ice, or water at up to 25 miles per hour and, with a better than 80-decibel whine, is used to chase geese. It cost about $3,775 at the time of this writing, and has been used in Denver, New York City, Boston, and Colorado Springs.

Efforts are underway to make robots autonomous. In a study aimed at developing drones to repel birds in vineyards, Grimm et al. (2013) designed a drone that looked and acted like a predatory bird. This unmanned aerial vehicle, however, could take off, land, and fly simulated attack maneuvers completely on its own. Referencing the limitations of the remote-controlled "Goosinator," Kim et al. (2019) have proposed the use of machine learning and artificial intelligence to develop an amphibious robot that will autonomously repel birds such as the Canada goose. The robot would patrol a defined area following predefined paths, detect birds within a 5-meter range, and chase the birds away. It would also gather locational and behavioral information for research.

Efforts are also being made to combine drones and robots with electronic sensors to develop repellent systems. In one such system, Dev et al. (2019) have proposed using modern technologies (Google's TensorFlow or the Internet of Things) integrated with a raspberry pi single-board computer used as a processor to detect the presence of wildlife and repel it. In the scenario described, the researchers use electronic sensors to detect presence of elephants invading a crop, the sensors trigger a video camera whose images are analyzed using TensorFlow technology to determine whether the animal is an elephant, and then select and broadcast a sound (12,000–16,000 Hz; imitating the sound of bees) to repel the elephants. Other animals would elicit frequencies irritating to them.

Example

Stevens et al. (2000) described a radar-activated system designed to keep waterfowl from large ponds contaminated with potentially lethal wastes from a power plant. The system was designed as "demand performance" in that detection of birds on radar initiated an integrated hazing program that included acoustic alarms, pyrotechnics, and chemical repellents. Because the hazing was a direct response to presence of waterfowl, habituation was not a problem. The system was effective but expensive to install, and it required skilled personnel for maintenance and operation. However, because failure of the system could result in substantial fines under the US Migratory Bird Treaty Act, the system was viewed as cost-effective. Commercial enterprises such as Accipiter Radar have continued to develop this technology and offer it as a solution to some environmental hazards for birds, including both tailings ponds and wind farms (Accipiter Radar Technologies 2017).

ADDLING AND PHYSICAL STERILIZATION

Statement

Physical methods can be used to reduce birth rates of populations or lower survival rates of young.

Explanation and Examples

Physical methods are sometimes used to reduce or eliminate fertility. Methods include physical addling

and surgical sterilization. Meaning "loss of development," **addling** in the strictest sense is destruction of eggs by shaking. Addling has come to mean destruction of eggs by any physical or chemical means—puncturing, freezing, or coating with vegetable oil. Addling is often used to manage goose populations in urban and suburban areas. It has also been used since 1978 to reduce the growth of invasive mute swan populations in Rhode Island. Nest destruction is a closely related activity and might be favored when eggs are not present (e.g., in controlling damage by barn swallows) or if the eggs are newly laid. Although we think of addling as a method for reducing overabundance of common bird species, Fernandez-Duque et al. (2019) have suggested that the method, along with the assistance of citizen scientists, might be effective in reducing impacts of invasive birds such as English sparrows (*Passer domesticus*) and European starlings in the United States.

Bromley and Gese (2001b) sterilized whole packs of coyotes and found reduced predation on sheep compared with coyote pairs having pups. The approach was cost effective and had no significant effects on territoriality or related behaviors (Bromley and Gese 2001a). Surgical sterilization has been shown as a viable technique with other wildlife species, e.g., African elephants (Marais et al. 2013) and eastern grey kangaroos (*Macropus giganteus*; Tribe et al. 2014). Reduced fertility, however, has been viewed as impractical for some wildlife situations such as management of large deer populations.

PHYSICAL ALTERATION OF HABITATS AND ECOSYSTEMS

Statement

Habitats or ecosystems can be physically altered, thereby reducing their carrying capacities and making them undesirable for offending wildlife.

Explanation

To make habitats unattractive to wildlife, roosts might be pruned, thinned, or removed; an ecosystem pushed to an earlier sere; dams removed physically or with pyrotechnics; water levels manipulated; or food, water, or shelter otherwise made unavailable to offending species. The desired consequence of such manipulations is reduction in human-wildlife conflicts, including wildlife diseases and zoonoses. However, such habitat manipulation can also negatively impact nontarget species, and such unintended consequences must be considered.

Pruning or removal of vegetation that serves as sites for bird roosts has been a common worldwide practice for management of bird damage, e.g., for management of red-billed quelea in Africa or the dispersal of nuisance house crows (*Corvus splendens*) in Singapore (Peh and Sodhi 2002). Manipulation of water levels can sometimes discourage the presence of wildlife such as muskrats or beavers. Raising winter water levels in ponds almost to flooding, then lowering levels in the summer can sometimes force muskrats to find more suitable locations. This action may also flood muskrat burrows in wintertime and make them more vulnerable to predators in the summer. Removing food sources—e.g., vegetation adjacent to beaver ponds—can sometimes also force beavers to abandon a pond.

Airports, landfills, and sewage treatment facilities are landscapes where presence of undesirable wildlife is reduced or eliminated by physical alteration. Here, the notion is to make unavailable a factor critical for wildlife, such as food, water, or shelter. Airports and landfills work from management plans that consider impacts and management of wildlife. Standard practices at landfills can be modified so that trash is covered regularly with soil or a repellent coating. Unnecessary ponds can be removed from airports, and necessary ponds can be covered with wire grids or chemical repellents. Vegetation that provides cover for undesired wildlife can be modified or removed. Fences can prevent the movement of wildlife into critical areas such as air operation areas of airports. All attractants, such as food or solid waste, can be cleared from operational areas, including those surrounding runways.

Example

Developed by Gene Wood at Clemson University, the "Clemson Beaver Pond Leveler" uses a PVC pipe placed within the beaver dam to drain a beaver pond; it works by having holes in the portion of the pipe inside the beaver pond and screening to prevent debris from plugging water intakes and access by beaver to the pipe on the pond side. An elbow and vertical standpipe riser at the exit end of the pipe allows regulation of the pond level. With this regulation, the pond can be maintained at a level where it is beneficial for people and wildlife and will not cause problems such as flooding standing timber, roads, or farmland (Miller and Yarrow 1994). The Leveler has also been modified to allow the passage of brook trout, allowing the device to be utilized as a bypass for these fish through beaver dams. The Leveler and similar designs have found use worldwide.

Summary

- Physical methods are used to restrain or kill animals, to frighten, sample, or repel them, to reduce populations of offending animals by reducing births, and to alter habitats to make them undesirable for undesired wildlife.
- Restraining devices include cage traps, foothold traps, net traps and guns, and cable devices. Systems include devices, swivels, and anchors to reduce injury to captured wildlife and to prevent breakage, and the "set" includes baits, attractants, and placement.
- Killing traps, such as rat or mouse snap traps, are designed to kill quickly.
- Shooting can be done with pistols, rifles, shotguns, blowpipes, or dart guns from the ground or from aircraft.
- Exclusionary methods separate wildlife from things humans want to protect and include fences, netting, tubing, wire grids, and bird spikes.
- Physical repellent methods include those that startle damaging wildlife with sudden sounds or lights, and those that use their senses to communicate that an area is dangerous and should be avoided.
- Physical methods that sample wildlife unobtrusively include camera and hair traps.
- Physical methods used to reduce the natality of undesired populations include addling and surgical sterilization.
- Physical methods that alter habitats to make them undesirable to wildlife include pruning, manipulating water levels, and limiting available food, water, or shelter.

Review and Discussion Questions

1. Are traps lethal or nonlethal devices? Explain your answer, using specific examples.
2. Explain the basis for the effectiveness of the Williams' Carp Separation Cage. How have these ideas already been incorporated in the design of some barriers for sea lampreys? How might they find other applications with damaging species in North America?
3. How important is the experience of the practitioner in using traps or snares to catch a coyote? Explain, please.
4. What techniques are used to ensure that a trap intended to catch a starling catches or holds only starlings? How about a snare set to catch a coyote, or a conibear trap set to catch a muskrat?

5. Do you think a pyrotechnic firecracker can be an effective tool for frightening geese away from a pond they might want to use as a spring nesting site? Will the firecracker be effective in removing a pair of geese that are already nesting? Explain.

6. What methods are available to gather information on individuals or wildlife populations unobtrusively? Give a specific example with references.

7. Using today's electronic and computer-based technologies, design a system that detects and repels coyotes (or wolves, or bears, or mountain lions), preventing them from attacking livestock on a high mountain pasture.

11

Chemical Methods

I review here chemical methods that are used to manage human-wildlife conflicts.

Statement

Chemicals have been discovered that sedate, repel, kill, or sterilize wildlife; these chemicals have been evaluated, and registered for use as pesticides following public policies and regulations particular to local and national governments worldwide.

Explanation

Most chemicals used to manage human-wildlife conflicts are considered pesticides. "**Pesticide**" means literally "to kill a pest," but that literal understanding can also be misleading. In a broad sense, the term includes any mix of chemical substances that affects behavior of pest plants or animals. The US Environmental Protection Agency (EPA) defines a pesticide (with certain minor exceptions) as "any substance or mixture of substances intended for preventing, destroying, repelling, or mitigating any pest" ("What Is a Pesticide?," https://www.epa.gov/minimum-risk -pesticides/what-pesticide). The USEPA also includes in this definition substances intended as plant regulators, defoliants, or desiccants, and nitrogen stabilizers. Pesticides may be chemicals such as **warfarin** (an anticoagulant poison), genetic parts of organisms, or whole living organisms such as viruses, bacteria, or protozoa. Pesticides made from living organisms or their genetic parts are called **biopesticides**.

Pesticides may also be described according to targeted organisms: e.g., antibiotics for bacteria, **avicides** for birds, **fungicides** for fungi, **herbicides** for plants, **molluscicides** for snails, **piscicides** for fish, **predacides** for predators, **reptilicides** for reptiles, and **rodenticides** for rodents and other small mammals.

Pesticides currently used in managing wildlife include chemical attractants and repellents; immobilizing (stupefying) agents; antifertility agents, including antibodies designed to immunize against fertility; biopesticides, including bacterial or viral agents used for biological control (see chapter 12); and toxicants including herbicides (table 11.1). Because of their biological or pharmacological activity, pesticides, as

TABLE 11.1   *Examples of chemicals used in human-wildlife conflicts*

Mode	Target	Active ingredient
Antifertility	Birds	Nicarbazin
	Mammals	Gonadotropin-releasing hormone (GnRH)
		Porcine zona pellucida (PZP)
Immobilizing	Birds	α-Chloralose (also toxicant, birds, mammals)
	Mammals	Propiomazine HCl
Repellent	Birds	4-Aminopyridine
		Anthraquinone
		Methiocarb
		Methyl (or dimethyl) anthranilate
		Polybutene
	Mammals	Capsaicin
		Egg acrylic
		Semiochemicals (e.g., dried blood, urine)
		Thiram
	Reptiles	Cinnamon, clove, and anise oil
Toxicant	Birds	3-Chloro-4-methylanine HCl (DRC-1339)
		Oil (addling)
		Sodium lauryl sulfate
	Fish	Antimycin A
		Niclosamide
		4-Nitro-3-(trifluoro methyl) phenol HCl
		Rotenone
		Saponins
	Mammals	Aluminum phosphide
		Anticoagulants
		Bromethalin
		Cholecalciferol
		Magnesium phosphide
		Para-aminopropiophenone (PAPP)
		Polybutene
		Sodium cyanide
		Sodium fluoroacetate (Compound 1080)
		Sodium nitrate, carbon (gas cartridge)
		Sodium nitrite
		Strychnine
		Zinc phosphide
	Plants	Glyphosate (Rodeo)
		Other herbicides
	Reptiles	Acetaminophen
		Methyl bromide

active ingredients, are the components of greatest importance for registration and legal use of pesticidal methods.

Most pesticides are formulated into baits designed to selectively attract targeted animals and encourage

them to feed. **Baits** may include mixes of liquids, solids, or powders formed into any of a variety of shapes and sizes. Baits may be complex and include whole organisms. For example, dead neonatal mice are used to deliver the toxicant acetaminophen (Tylenol) to brown tree snakes in Guam and Saipan (Shivik et al. 2002). Along with the active ingredients in pesticides, baits include **inactive ingredients** that serve other functions (box 11.1).

Delivery to targeted wildlife can be performed in various ways. Liquid baits may be put in a station, as with warehouses for managing rodents, or delivered with a sprayer, as for managing weeds. Powders may be injected, as with some zinc phosphide products designed to be placed within walls. Rodents get this powder onto their fur and ingest lethal amounts while grooming. Solid baits may be broadcast by hand or with specialized equipment, as with rodenticides applied in orchards, reforestation or island habitat, restoration programs, or pasture crops such as alfalfa (e.g., Witmer et al. 2007). Sometimes baits are delivered by aircraft over large areas, such as in baiting programs to immunize raccoons for rabies in the United States or the red fox and raccoon dog in Europe or to manage invading species such as rodents or predators on islands (e.g., Garden et al. 2019).

Baits may be put into specially designed containers at baiting stations. Locally available materials may serve as bait stations. The containers may be designed not just to hold the formulation but to keep it from the weather, to attract and allow selective access by the targeted wildlife, and to keep children or unintended wildlife out (fig. 11.1). Draw stations, such as carcasses from prior predator kills, may be used to attract predators and scavengers to a general area. Selective attractants may then be used to entice targeted species to other locations where there are baited devices. For example, an M-44 containing sodium cyanide and gauze soaked in a selective attractant may be placed under a rock near the draw station. The attractant may be designed to induce bite-and-pull behavior in canids, making the M-44 specific for targeted predators and less attractive to other species.

Immunizations and antifertility compounds have been delivered to members of small-sized populations of wildlife—e.g., feral horses at Assateague Island, Maryland, and elephants in small refugia in Africa—using rifles and blowguns. Bacteria and viruses may someday be used to carry active ingredients such as antifertility compounds to larger-sized populations.

In most countries, pesticides are registered at the national level by one or more agencies (see chapter 16).

BOX 11.1 Inactive Ingredients in Baits

Inactive ingredients can include a **carrier**, the matrix in which all of the ingredients are mixed; a **binder** or sticker, used to ensure the active ingredient adheres to the carrier; an **attractant**, which selectively attracts intended animals; an **enhancer**, which encourages consumption of sufficient amounts of the formulation; a **masking agent**, which overshadows or otherwise hides flavors in baits that might induce avoidance; and a **preservative**, which prevents degradation of the bait and protects it from the weather.

Carriers can be liquid or solid. Liquid matrices are sometimes used, such as in warehouses or arid environments where water might be particularly attractive to rodents or other wildlife. Carriers might be matrices that slow the release of volatile ingredients such as dimethyl anthranilate (a bird repellent). Often carriers are food familiar to and preferred by the targeted animals. Thus, ferns are a preferred food of the mountain beaver and also serve as carriers for pesticides. Cereals, starch, pastes, and meats are other common carriers. Waxes sometimes serve as weather-resistant carriers for baits, are well tolerated by some wildlife, and are included in some commercially available rodenticides.

Binders not only ensure active ingredients stick to carriers but can also weatherproof the formulations. For example, oils and fats are often used as binders for zinc phosphide (a rodenticide) baits in tropical regions. The oils keep ambient acidic moisture from the rodenticide until it is in the gut of the animal where the acids convert the rodenticide into phosphine gas.

Attractants tend to be aromatic and favored by targeted animals, such as apple or peanut butter flavorants for deer. Favorite attractants for rodents and predators include chicken, beef, or fish flavors. In aquatic systems, by-products of decomposition, such as the amino acid alanine, can attract scavengers such as catfish. Synthetic fermented egg, used in the United States for national surveys of coyote populations, also can be an attractant in predator baits. Highly attractive flavors have an important disadvantage: if they are unfamiliar, they tend to induce both neophobia and learned aversions, facilitating bait-shyness (Reidinger and Mason 1983).

Enhancers encourage consumption once the animal has found the formulation. For rodents and other herbivores, carbohydrates (putatively connote safe food) such as sugars often encourage optimal consumption. Fatty materials may be more effective for omnivores and predators.

Masking agents, or sometimes microencapsulation, may be used to hide flavors that the targeted animals find disgusting and avoid. For example, wildlife often perceive bitter flavors, such as those in some toxicants and antifertility agents, as potentially poisonous, so they avoid them (Reidinger and Mason 1983). Preservatives, such as EDTA, may also be added, or insect repellents to prevent consumption by unintended wildlife such as ants.

All inactive ingredients contribute to the final flavor of a bait and, in turn, its salience in inducing behavioral avoidance such as neophobia or bait-shyness.

In the United States, for example, pesticides are registered and regulated by the EPA. In November 2005, the US Food and Drug Administration (FDA) transferred to the EPA the authority and responsibility for registration of wildlife contraceptives (although the FDA retained registration of drugs used in managing wildlife). Examples of pesticide agencies in other countries include the Australian Pesticide and Veterinary Medicine Authority (APVMA) in Australia, the Environmental Risk Management Authority in New Zealand, and the European Commission (Environment) for the European Union. In many countries, use of pesticides also requires approval at lower governmental levels, such as states or municipalities.

Examples

Each registered pesticide has a label that provides detailed information on the product and terms and conditions for its use (e.g., acetaminophen; US EPA 2018) (fig. 11.2). It specifies that protective equipment and clothing must be worn when handling, mixing, or applying acetaminophen.

The label also specifies safety requirements and recommendations, environmental hazards (in this case, to avoid contaminating water), and storage and disposal requirements. It provides information on whom to consult, such as the Government of Guam Division of Aquatic and Wildlife Resources, and lists the specific threatened or endangered species that may be of concern in Guam and the Commonwealth of the Northern Mariana Islands. The label specifies

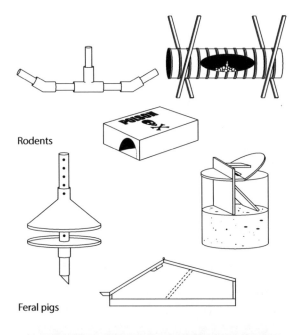

Rodents

Feral pigs

Figure 11.1 Variations in bait stations designed for rodents and feral pigs (top). A bear attempts to get bait from a BOS feeder (bottom). Illustration by Lamar Henderson, Wildhaven Creative LLC. Photo by Michael Avery, APHIS/WS/NWRC.

that actual use is restricted to employees of the US state and federal governments, the government of Guam, or the Commonwealth of the Northern Mariana Islands, and only employees trained in control of brown trees snakes or persons working directly under their supervision. The directions provide sufficient information so that a trained person can refer to it for guidance—e.g., "manually insert one 80-mg or two 40-mg acetaminophen tablets into the throat of a dead mouse pup (approximate age: 10–15 days)."

HERBICIDES

Statement

Herbicides are used to control damage caused by weeds such as exotic and invasive plant species, and to make habitat unsuitable for animals.

Explanation

Herbicides may be specific, in that they kill only targeted plants; selective, in that they kill only types of plants, such as broad-leafed ones; or general, in that they kill any plant. They can be pre-emergent, preventing germination or early growth, or postemergent, killing plants after they have emerged. These chemicals can depend on contact or can be systemic, wherein they are translocated throughout a plant and serve also as a soil sterilant, preventing plant emergence.

Whether to use herbicides and what types to use depend on the specific plants and the situation—i.e., managing exotic weeds versus altering habitat to reduce presence of wildlife or diseases. The reader is referred to county and state cooperative extension service guidelines or other local sources for recommendations because many herbicides are available.

Example

The herbicide glyphosate (Rodeo, 53.8%, CAS No. 38641-94-0) is a postemergent, nonselective herbicide formulated for use in multiple use sites, including aquatic environments. Rodeo is often delivered by aircraft and is used to control invasive cattails (*Typha*) and other vegetation in marshy areas where blackbirds roost, particularly in the prairie pothole regions of North America where sunflowers are grown (Linz and Homan 2011). Open potholes (ponds formed from glacier scours, important for wildlife and waterfowl production in North America) provide less favorable roost sites for blackbirds than potholes with thick cattails. Conversely, open potholes are preferred by desirable wildlife such as ducks and other waterfowl. The areas treated are those with the most severe damage from blackbirds, and they represent less than 1% of the estimated yearly total area of cattail stands in the state. From 1991 to 2011, about 1,400 hectares of cattails were treated annually in North Dakota as a demonstration by US Wildlife Services. The demonstration program ended in 2011 with the information made available for continued use by farmers.

{Container Label}

PRECAUTIONARY STATEMENTS HAZARDS TO HUMANS AND DOMESTIC ANIMALS

Si usted no entiende la etiqueta, busque alguien para que se la explique a usted en detalle. (If you do not understand the label, find someone to explain it to you in detail.)

READ THIS LABEL: Read the entire label and follow all applicable directions, restrictions and precautions. This label must be in the possession of the user at the time of pesticide application.

PERSONAL PROTECTIVE EQUIPMENT (PPE)
People who handle the product must wear:
- Long-sleeved shirt and long pants
- Socks and shoes
- Waterproof or rubber gloves

Applicators who handle bait must wear:
- Waterproof or rubber gloves

Follow manufacturer's instructions for cleaning/ maintaining PPE. If no such instructions for washables exist, use detergent and hot water. Keep and wash PPE separately from other laundry.

USER SAFETY RECOMMENDATIONS

Users should wash hands before eating, drinking, chewing gum, using tobacco, or using the toilet. Users should remove PPE immediately after handling this product. As soon as possible, wash thoroughly and change into clean clothing. Users should remove clothing immediately if pesticide gets inside. Then wash thoroughly and put on clean clothing.

ENVIRONMENTAL HAZARDS

Do not apply directly to water, or to areas where surface water is present or to intertidal areas below the mean high water mark. Do not contaminate water when cleaning equipment or disposing of equipment washwaters or rinsate.

Label ID 56228-34-Oct-5-2020
Container Label: Page 1 of 1

ACETAMINOPHEN FOR BROWN TREESNAKE CONTROL

ACTIVE INGREDIENT
Acetaminophen 72.73%
OTHER INGREDIENTS 27.27%
TOTAL...100.00%

This product may be used only to control brown treesnakes (Boiga irregularis) in non-crop areas in and around military bases, airports, shipping ports, urban and residential areas, and other areas where brown treesnakes may be present. For use only by employees of the U.S. State and Federal governments, the Government of Guam or the Commonwealth of the Northern Mariana Islands trained in brown treesnake control, or persons under their direct supervision.

KEEP OUT OF REACH OF CHILDREN
CAUTION

NOTE: This is the container label. See the enclosed full product label for Acetaminophen for Brown Treesnake Control (EPA Reg. No. 56228-34) for complete **Directions For Use** and all other label and use information, including Endangered Species Considerations. **It is a violation of Federal law to use this product in a manner inconsistent with its labeling.**

STORAGE AND DISPOSAL

Do not contaminate water, food, or feed by storage and disposal.

PESTICIDE STORAGE: Store only in a closed container in a freezer at or below 0 °F (-18 °C) that is inaccessible to children, pets and domestic animals.

PESTICIDE DISPOSAL: If the product cannot be disposed of through use in accordance with product label directions, contact your Territorial Pesticide or Environmental Control agency, or the Hazardous Waste representative at the nearest EPA Regional Office for guidance.

CONTAINER DISPOSAL: Nonrefillable container. Do not reuse or refill this container. Triple rinse container (or equivalent) promptly after emptying. Offer for recycling, if available. Otherwise, puncture and dispose of in a sanitary landfill or incinerator.

UNITED STATES DEPARTMENT OF AGRICULTURE
ANIMAL AND PLANT HEALTH INSPECTION SERVICE
4700 RIVER ROAD, UNIT 149
RIVERDALE, MD 20737

EPA Reg. No. 56228-34
EPA Est. No. 56228-CO-01
Net Contents: __________
Batch Code No.: __________

Figure 11.2 A pesticide container label issued by the US Environmental Protection Agency. Public domain.

DRUGS INCLUDING IMMOBILIZING AGENTS AND TRANQUILIZERS

Statement

Chemical drugs sedate wildlife so they can be captured and removed from an area or calmed to reduce self-inflicted injuries while being retained.

Explanation

Immobilizing agents, or stupefying agents, render wildlife temporarily unresponsive to environmental stimuli, so they can be readily captured, or to reduce resistance and injury when captured. Two of these, alpha-chloralose (CAS No. 15879-93-3) and propionylpromazine hydrochloride (CAS No. 7681-67-6) are registered for use in the United States under Investigational New Animal Drug (INAD) permits.

Alpha-chloralose is used as an anesthetic, hypnotic, immobilizing agent, sedative, and toxicant. It is used by trained and certified "applicators" of the USDA's Wildlife Services to capture and remove waterfowl and other birds—pigeons (*Columbidae*), ravens (*Corvus*), and sandhill cranes (*Grus canadensis*)—often from urban or suburban settings. It can be used in recreational and residential areas and near swimming pools, shoreline residential areas, golf courses, and resorts. This chemical is mixed in a single bread or corn bait for waterfowl and in corn baits for pigeons. From October 2004 through September 2005, 443 grams of alpha-chloralose were used to capture mostly Canada geese, with a mortality rate under 5% (O'Hare et al. 2007). In some European countries, alpha-chloralose is used to immobilize and kill birds and rodents. Its use as a rodenticide is being considered in the United States (Witmer et al. 2017). The chemical is registered in Australia and New Zealand to manage some bird species. As in the United States, use is restricted to certified applicators.

Propionylpromazine hydrochloride is being considered and tested in the United States and Australia as the active ingredient in a tranquilizer tab (rubber nipple) attached where caught wildlife will instinctively bite at traps and other restraints. The compound sedates—without loss of consciousness—coyotes, feral dogs, wolves, and other wildlife (Savarie et al. 2004; Fagerstone and Keirn 2012).

Example

The International Crane Foundation began capturing Greater Sandhill Cranes (*Grus canadensis tabida*) for a long-term ecological study in 1990. The findings from these studies have contributed to the effective management and survival of the cranes by unfolding details of their population dynamics and behaviors. Capture using alpha-chloralose was chosen because the method had lower morbidity and mortality rates than alternatives used for small groups of cranes, such as leg nooses or rocket nets. The use was authorized under the drug permit held by Wildlife Services. A concern, however, was capture myopathy. Some studies indicated the disease occurred more readily under lighter sedation during August or October than under the heavier sedation used in September. The foundation changed its research schedule in 2002, limiting captures to September and early October. Based on a comparative review of records of 317 cranes captured between 1990 and 2011 by Hartup et al. (2014), the new schedule was associated with modest improvements in overall efficacy and lowered morbidity and mortality. The improvements appeared mostly due to other changes in methodology, however, such as consistency of fluid administration rather than the sedation caused by alpha-chloralose.

CHEMICAL REPELLENTS

Statement

Chemical repellents use the sensory systems of wildlife to irritate or frighten them, thereby averting offending behavior. Learned effects need to be reinforced or they attenuate quickly, whereas unlearned effects can be long-lasting.

Explanation

I list common repellents (see table 11.1) and describe their uses here.

Anthraquinone (CAS No. 84-65-1) is a naturally occurring substance that causes gastric illness in learned flavor aversions (Avery et al. 1997; Werner and Provenza 2011; DeLiberto and Werner 2016). It was first evaluated as a bird repellent in the 1940s and was tested as a means of repelling blackbirds from rice seed in the 1950s. Anthraquinone is the active ingredient in products used for bird repellency in the United States (e.g., Flight Control, Avipel, and Airepel) and New Zealand (Avex); (Spurr and Coleman 2005a) and is now used as a bird repellent for crops including corn, rice, sunflower, wheat, sorghum, forest trees, turf, sugar beets, sweet corn, and fruit and nut crops (DeLiberto and Werner 2016). A critical step for effective use on corn was developing a formulation that encouraged sufficient consumption to stimulate learned aversions by birds such as Canada geese, red-winged blackbirds, and ring-necked pheasants (*Phasianus colchicus*; Werner et al. 2009). Anthraquinone has also been tested for repellency in some mammals including feral pig, thirteen-lined ground squirrels (*Ictidomys tridecemlineatus*), black-tailed prairie dogs (*Cynomys ludovicianus*), common voles (*Microtus arvalis*), house mice, Tristram's jirds (*Meriones tristrami*), and black rats (DeLiberto and Werner 2016).

4-aminopyridine (CAS No. 504-24-5) induces an alarm response in birds that eat it. These birds move erratically and emit calls that frighten other birds before dying. Surviving birds leave or avoid the area. The chemical is the active ingredient in Avitrol grain baits (Avitrol Corporation) in the United States and Australia for use with birds such as pigeons, blackbirds, house sparrows, and European starlings that can be legally killed (Werner and Avery 2017).

Capsaicin (e.g., capsicum oleoresin, CAS 404-86-4), the main spicy chemical in hot capsicum peppers, is the sole active ingredient in some repellents (e.g., Bonide Hot Pepper Wax Animal Repellent, Hot Sauce) and one of several active ingredients in others (e.g., Deer Off, Havahart). The products are sold as repellents for wildlife including deer, predators, and birds. Red pepper, which contains mostly capsaicin and dihydrocapsaicin, is the active ingredient in some sprays that are used to protect humans from attacks by dogs and predators such as bears.

A variety of capsaicin-based formulations have been developed or proposed to repel elephants in Africa and Asia. These methods also differ in effectiveness (Von Hagen et al. 2020). In one commonly used formulation, hot peppers are mixed with motor oil to generate pungent odors. Cloths or ropes soaked in the mixture are then used to build fences. Capsaicin is also used in aerosol sprays, in smoke canister bombs shot from cannons, and in pepper dung (a mixture in which hot peppers and dung are burned, creating a pungent smoke). Capsaicin serves as a trigeminal irritant, and its effectiveness is related to its concentration (Andelt et al. 1994). Von Hagen et al. (2020) have used liquid chromatography–mass spectrometry to quantify the amounts of capsaicinoids in the formulations and in relating the amounts to differences in Scoville Heat Units (a measure of chili pepper "hotness" to humans) for several fence designs. The researchers have proposed that this analytical approach be used more generally. Knowing the potency levels, Von Hagen and Schulte say, help to determine whether "elephants or other wildlife overcome a particular chili fence because of a lack of potency or other factors."

To repel snakes, Wildlife Services (USDA) has used a mix of cinnamon, clove, or eugenol oil as aerosol sprays and as vapor fumigants. One or more of the ingredients are included in some commercial snake repellents such as Ortho Snake B Gon or Bonide Snake Stopper. Aerosols of one or more of these oils at 2% of the total formulation, sprayed directly at the head of a snake or as a fumigant, will repel brown tree snakes and other snakes (Clark and Shivik 2002). These oil mixtures are exempted from US federal registration because they pose minimal risk to users and the environment. Wildlife Services provides a Technical Note describing general information on use of these repellents (see National Wildlife Research Center, Research Gateway). Practical use of these mixtures for repelling brown tree snakes from cargo areas has remained elusive, however (Clark et al. 2018). Many of the formulations have plasticizing properties and cannot be used in cargo areas. Technical problems with vaporizing methods have prevented the aerosols from inducing reliable escape behaviors in simulated cargo areas (Kraus et al. 2015).

DMA (synthesized dimethyl anthranilate, CAS No. 85-91-6) and MA (natural methyl anthranilate, CAS No. 134-20-3) irritate the trigeminal nerve (which innervates the common chemical sense) of most birds and repel them (Werner and Avery 2017). Birds subsequently avoid this unlearned irritation. DMA and MA are grape flavorants used in candy, drinks, chewing gum, and other products. The compounds are listed on the USFDA's GRAS (Generally Recognized as Safe) list. DMA and MA, formulated for controlled release, are used to repel birds from areas such as airports, golf courses, landfills, turf-farms, and lawns. The natural form, MA, is less expensive than the synthesized form, DMA, and is the active ingredient in products such as Avex, Avian Control, Bird Shield, Bird Stop, and EcoBird 4.0. Effectiveness varies, and some applications need to be repeated after rainfall. DMA and MA break down quickly under ultraviolet light, often retaining effectiveness for less than a week. Economic use therefore depends on damage carrying high risks and costs (e.g., Spurr and Coleman 2005b).

Methiocarb (CAS No. 2032-65-7), or Mesurol, is registered in more than 60 countries as an insecticide and molluscicide. It is registered in New Zealand, South Africa, and some European countries as a bird repellent for use on emerging seeds and nonfood crops (e.g., Spurr and Coleman 2005a). It serves as an illness-inducing agent in learned flavor aversions (Reidinger and Mason 1983). The chemical has been extensively tested and shown to be effective as a bird repellent in the United States. However, the product can be hazardous to aquatic wildlife, is on the European Union's second Watch List for water contamination, and is restricted in the United States because of concerns surrounding human health and safety (Werner and Avery 2017). One use of methiocarb, conditioning ravens to avoid the eggs of endangered birds and turtles, is maintained by US Wildlife Services.

Polybutene (CAS No.9003-29-6) is transparent and sticky and can be applied in a liquid or solid formulation to discourage birds from roosting on perches or ledges. Products are sold under such trade names as Bird-X Bird Repellent Liquid and Bird-B-Gone Transparent Bird Gel. Some formulations may also have other active ingredients, such as capsaicin (to give birds a "hot foot" if they alight on the surface). Glue boards using polybutene are also sold to capture snakes and commensal rodents and rats under trade names such as J. T. Eaton's Stick-Em, but these are intended as lethal rather than repellent devices.

There are many other mixes of compounds designed to repel wildlife. These include combinations of dried blood, lipids, putrid compounds including real and synthetic fermented egg, and predator urine. Some include human hair and soaps. Effectiveness varies broadly, but in general most have at least temporary neophobic effects, and frequent movement and rotation of compounds can sometimes extend protection.

Examples

Smith et al. (2008) reviewed data from 83 encounters in which humans used sprays to stop undesired behaviors of bears (brown, black, and polar bears). All encounters were in Alaska between 1985 and 2006. In 72 cases, sprays were used to defend the people from the bear, whereas in the other 11 cases, persons sprayed an object or an area to protect it. Each spray had capsaicin as its active ingredient. Most (69%) of the encounters were with brown bears, and only 3% were with polar bears. The researchers sorted negative behavior into aggressive, defensive, or nuisance, and they separated food searching from other activities such as curiosity.

Regardless of brand or canister size, effective spraying distance (often advertised at 15–20 feet), or nature of behavior or activity, spraying stopped undesired behaviors in 92% of the encounters, and 98% of the people involved were uninjured. Of those injured, one required stitches for minor lacerations, and none were hospitalized. Although wind influenced accuracy, sprays always reached bears. In two cases, the spray nearly incapacitated the person as well as the bear. Overall, Smith et al. (2008) concluded that the capsaicin-based repellents served as effective alternatives to shooting.

Guerisoli and Pereira (2020) conducted a systematic literature review on the effectiveness of deer repellents. The researchers extracted from the scientific literature 246 "essays" testing 10 repellents and 236 testing four application methods. Included in the search were all types of vegetation. Of the 10 repellents, all but one (sound, lasers) were chemicals or mixtures of chemicals: denatonium benzoate; sulfurous compounds including eggs, garlic derivate; casein, other milk proteins, or feathers; meat or blood; citric oils; urine, hair, or feces of predators; hot pepper or capsaicin; and a mix of repellents such as garlic plus oil. Application methods were "sprayed/dusted" where spray or powder application or resin adhesive was tested; "packaging" where a sponge, capsule, sachet, or hang balls was tested; "in" where the products tested were in seeds, in pregrow treatments, or in altered diets; and "other" where a physical, painting, or sound method was tested. The researchers used a Beta regression with percentage of vegetation "unbrowsed" as the response variable. They found that unbrowsed vegetation was most strongly associated with "meat and blood," which they interpreted as due to their fear-inducing qualities. They also found a strong association with urine, hair, and feces of predators, with samples taken mostly from bobcat, wolf, and puma.

TOXICANTS

Statement

Toxicants can be effective and selective tools for managing wildlife damage.

Explanation

Uses of toxicants are carefully regulated by governmental agencies. Many can be utilized only by specially trained persons, sometimes also of designated agencies. Common toxicants (see table 11.1) and their uses are described here.

AVICIDES. The only avicide currently registered in the United States is compound DRC-1339 (CAS No. 7745-89-3). DRC-1339, chemically known as 3-chloro-4-methylanin was the 1,339th compound tested by the Denver Research Center (now the National Wildlife Research Center) as a potential avicide. It is the active ingredient in products sold as Starlicide. DRC-1339 is slow acting and lethal in a single feeding. This chemical deposits uric acid in kidneys and blood vessels, causing necrosis and impaired circulation. Death is from uremic poisoning and congestion of major organs.

DRC-1339 formulations include a 98% active ingredient concentrate to mix in grain, french fries, eggs, or meat baits. Its primary uses in the United States are to reduce consumption and contamination of livestock feed by starlings and blackbirds; to control starlings, pigeons, crows, and grackles in structures; and to control starling, pigeons, and grackles in staging areas. DRC-1339 is also used with bread-cube baits as a gull toxicant along coastal breeding areas and in egg or meat baits for the control of ravens, crows, and magpies and for livestock protection. A premixed pelleted bait called Starlicide Complete is available for restricted use by certified applicators to control starlings and blackbirds at feeding stations near livestock and poultry (Linz et al. 2018). DRC-1339 is registered for use in New Zealand but not in Australia (Lapidge et al. 2005).

Sodium lauryl sulfate (CAS No. 151-21-3), a wetting agent exempted from federal registration in the United States, is also a lethal control agent for managing blackbird roosts. The product is intended for use on upland roosts that are located away from bodies of water. Roosts are sprayed with water containing the wetting agent when weather and conditions are appropriate. The compound allows water to saturate feathers so birds die of hypothermia at ambient temperatures under 41° Fahrenheit. During 2004–2007 field trials in Missouri, birds died quickly, as soon as 30 minutes, after exposure (USDA 2021).

RODENTICIDES. Rodenticides can be **chronic** (require repeated feedings, mostly anticoagulants) or **acute** (effective in single doses). Warfarin (CAS No. 81-81-2), the active ingredient in baits such as COV-R-TOX, Co-Rax, Mouse pak, Rat-O-cide, RAX, and Waran, was the first anticoagulant rodenticide. Discovered (with research funded by the Wisconsin Alumni Research Foundation) in the 1920s as the sweet-smelling compound with anticoagulant properties in decomposing sweet clover hay, the name warfarin is an amalgam of the funding organization and the active ingredient, coumarin. The compound found heavy use worldwide, mostly for control of commensal rodents in urban areas and on farms. Its use as a blood-thinning medication began in the 1950s, and today warfarin remains the most widely used anticoagulant medication in the world. Other coumarin-based compounds are brodifacoum and bromadiolone (with these second-generation compounds, single feedings are lethal). Chlorophacinone (registered for mountain beaver control in Oregon and Washington), valone, and pindone are also first-generation (multiple feedings required) compounds, but these are based on indandione rather than coumarin. Diphacinone is a second-generation indandione-based anticoagulant.

Genetic resistance to warfarin by Norway rats was first discovered on a dairy farm in Scotland (Boyle 1960). Subsequently, genetic resistance was documented in rural areas of Europe, in major cities around the globe, and with both first-generation rodenticides such as pival and diphacinone and second-generation anticoagulants such as difenacoum and bromadiolone (Jackson and Ashton 1986). Blood samples are sometimes taken and monitored for clotting time to detect genetic resistance. The geographic focus of resistance can then be contained or eradicated using a different chemical or control method.

Whereas anticoagulant poisons have been mostly used to manage commensal and agricultural rodents, the compounds have also been utilized to manage other mammals, including muskrats, monkeys, and feral pigs. Uses now include eradications of invasive rodents on islands for conservation purposes. US Wildlife Services maintains three such registrations: two for brodifacoum (see below) and one for diphacinone. Because brodifacoum has secondary hazards, diphacinone is increasingly the compound of choice for these conservation applications.

Brodifacoum (CAS No. 56073-10-0) is a second-generation anticoagulant rodenticide that has seen broad application worldwide as the active ingredient in rodenticide baits such as Final, Havoc, Jaguar, Klerat, Ratak, Ropax, and Talon. It is effective against warfarin-resistant rodents. Several formulations include a wax base that helps resist the effects of weather. These compounds must be used outside in ways that limit access to nontargeted wildlife because of secondary hazards to raptors and other predators and scavengers.

Cholecalciferol (CAS No. 67-97-0) is vitamin D_3, a human health supplement when taken in low doses. In higher doses, however, cholecalciferol mobilizes calcium. The calcium then mineralizes organs, having a lethal effect within three to five days. Cholecalciferol is the active ingredient in products such as Agrid$_3$, d-Con, and Quintox, rodenticides that have found particular use against anticoagulant-resistant rodents. The compound can be hazardous to non-

target animals such as pet dogs, which limits its use in many countries.

Acute rodenticides are fast acting but can cause bait-shyness. These types of rodenticides might be formulated as fumigants, powders, or food baits. **Prebaiting** (using the bait without the active ingredient) is sometimes done to encourage consumption. Examples of acute rodenticides follow.

Aluminum phosphide (CAS No. 20859-73-8) is formulated as a tablet that converts to phosphine gas in a wet acidic environment. Trade names include Al-Phos, Detia, Farmoz, Phostoxin, Phosphume, and Phostex. The tablet, used for burrowing rodents, is placed in a burrow to fumigate it. Although secondary hazards are rare because the gas quickly disperses, primary hazards to unintended wildlife that may also be in the burrow, such as pigmy owls, are a concern.

Bromethalin (CAS No. 63333-35-7) is a single-dose neurotoxic rodenticide that is marketed under trade names such as Fastrac Soft Bait, Gladiator All Weather Bait, Rampage, T1 Mouse, and Top Gun. The chemical has been used effectively against anticoagulant-resistant rodents. The toxicant works by uncoupling mitochondrial oxidative phosphorylation, thereby decreasing adenosine triphosphate (ATP) synthesis, in turn reducing amounts of an ATPase enzyme that causes a buildup of cerebral spinal fluid. The consequence is damage to the central nervous system and death. Because of the general nature of its mode of action, bromethalin is potentially toxic to all living organisms including pets and children. Tamperproof bait stations are a key to its safe use; no major effects or deaths occurred among 2,227 pediatric exposures reported to US poison centers between 2008 and 2017 (Feldman et al. 2019). Bromethalin toxicosis was implicated in the death of four feral conures (parakeet-type birds) taken near San Francisco in 2019 (van Sant et al. 2019).

Gas cartridges, with such trade names as Giant Destroyer, Smoke'Em, Amdro Gopher Gasser, Revenge Mole, Revenge Rodent Smoke Bomb, Gopher Smoke Bombs, and Dexol Gasser, are also fumigants. Sodium nitrate (CAS No. 7631-99-4) and charcoal are the active ingredients. US Wildlife Services maintains registrations for large and small cartridges. Small cartridges are used for ground squirrels, pocket gophers, woodchucks (marmots), and prairie dogs. Larger cartridges are for coyotes, red foxes, and striped skunks in rangeland, crop, and noncrop areas. The cartridges are lit with a fuse and placed into a burrow, and they burn very rapidly and inefficiently. The consequence is generation of lethal carbon monoxide.

Zinc phosphide (CAS No. 1314-84-7) is one of the most widely used rodenticides in the world (Eisemann et al. 2003). Trade names in the United States include Arrex, Denkarin Grains, Gopha-Rid, Phosvin, Pollux, Ridall, Ratol, Rodenticide AG, and ZP. It acts by converting to phosphine gas on ingestion, activated by the acidic nature of the stomach. Because it acts quickly, often killing the rodent within a few feet of a bait station, it is a farmer favorite. Zinc phosphide is used to reduce damage by rodents such as mice, voles, rats, prairie dogs, ground squirrels, muskrats, and nutria to cereal crops, pastures, and forage crops, orchards, rangeland, and field borders. This chemical is effective in a single dose but is believed to have a strong taste to wildlife, thus requiring prebaiting to minimize bait-shyness.

Strychnine (CAS Nos. 57-24-9, 60-41-3) is used as a rodenticide in the United States and as an avicide in other countries. Its use is restricted to burrows in some countries, including the United States, because of concerns for nontarget hazards. Strychnine remains undigested in affected animals, which then become hazards to scavengers and other wildlife. By restricting use to below ground, these hazards can be minimized. US Wildlife Services maintains four registrations, all for managing pocket gophers.

In February 2021, the EPA announced additional restrictions on rodenticide products available in the United States for consumer use. For the consumer market, these products need to be ready-to-use bait stations. Available rodenticides in these types of stations are bromethalin, chlorophacinone, or diphacinone in block or past form. Pelleted rodenticide baits are available for use by professional applicators in or near buildings and other structures, and for use in or near agricultural buildings and man-made agricultural structures ("Restriction on Rodenticide Products," 2021, https://www.epa.gov/rodenticides/restrictions -rodenticide-products).

PISCICIDES. Piscicides are chemicals poisonous to fish and include rotenone (CAS No. 83-79-4), antimycin A (CAS No. 642-15-9, marketed as Fintrol), saponins (amphipathetic glycosides, a group of chemicals abundant in some plants that have soaplike qualities); sea lamprey larvicide [4-nitro-3-(trifluoromethyl)phenol, TFM, CAS No. 88-30-2]; niclosamide (CAS No. 50-65-7); and ethanolamine salt of niclosamide (Bayluscide). Piscicides are used worldwide in hatcheries, ponds, lakes, and streams to eradicate or reduce the presence of undesired dominant, parasitic, diseased, or invasive fish. Entire ponds or lakes are sometimes treated with nonselective piscicides—e.g., rotenone, to eliminate a community of fishes—and then stocked with more desirable species.

Rotenone is sold as Chem Fish, Fish Tox, Liquid Derris, Nox-Fish, Nusyn NoxFish, Prentox, and Parderil. It is both a piscicide and an insecticide, used worldwide to control some insect pests and remove unwanted, often invasive, fish species from ponds and streams (e.g., Rayner and Creese 2006). It interferes with the electron transport chain in mitochondria. This compound is absorbed through gills of fish, making them particularly sensitive to it. Indigenous people in some Southeast Asian countries found that if they crushed the roots of certain plants and placed them in water, fish would become paralyzed, float to the surface, and be easily harvested. The plants contained rotenone. Of the four piscicides registered for use in the United States, rotenone is by far the most used compound (Finlayson et al. 2000).

Derris Dust and Liquid Derris, and other rotenone-based pesticides that have been used as general insecticides, are no longer registered for use on food products by the EPA effective March 23, 2011. Nonfish uses of rotenone have also been banned by the European Union, the Canadian Pest Management Regulatory Agency, and many other regulatory agencies worldwide. The concern is a link between rotenone and Parkinson's disease, a progressive neurological disorder (e.g., Lawana and Cannon 2020).

Sea lamprey larvicide (Lamprecide, TFM, 4-nitro-3-trifluoromethylphenol) is registered in the United States for control of the sea lamprey. It is applied as part of an integrated program of control in the tributaries of the Great Lakes (Marsden and Siefkes 2019).

PREDACIDES. Predacides are used to kill predators including coyotes, skunk, raccoons, and foxes. Two, sodium cyanide and Compound 1080, are used in the United States and Australia, and three others—para-aminopropiophenone (PAPP), sodium nitrite, and theobromine with caffeine—are under development and evaluation.

Sodium cyanide (CAS No. 143-33-9) produces the nerve gas hydrogen cyanide when it enters the mouth or stomach of an animal, causing death within minutes. This compound is the active ingredient in a capsule placed in the holder on top of the M-44 mechanical ejector. The M-44 ejector is tube-shaped, about 7 inches long by 1 inch in diameter. It is made of a base containing a loaded spring that is anchored in the ground, the capsule and holder, and an ejector mechanism including a spring-driven plunger. Gauze containing an attractant for coyotes is placed over the capsule above the ground. A chemical may be added that stimulates bite-and-pull behavior in canids. When pulled, tension on the spring is released, and the plunger pushes up into the capsule, ejecting the sodium cyanide into the mouth of the coyote. Death ensues quickly. Because cyanide gas dissipates rapidly, there is also little concern for secondary hazards. US Wildlife Services maintains two federal restricted-use registrations for M-44 cyanide capsules: one for managing coyotes, foxes, and wild dogs nationwide where state conditions and registrations allow; and one for managing the Arctic fox (*Vulpes lagopus*) in the Aleutian Islands.

Cyanide is registered for control of possum (*Trichosurus vulpecula*) and wallaby (*Macropus*) in New Zealand and registered experimentally for control of foxes in Australia (Eason et al. 2017). It is considered the most humane compound for control of possum when delivered in an optimized system (Gregory et al. 1998).

Sodium fluoroacetate (Compound 1080, CAS No. 62-74-8) is a competitive inhibitor of acetate in the citric acid cycle and so has potential toxicity to kill any living organism. It affects energy production in cells and organs, and death is usually due to ventricular fibrillation. Compound 1080 is particularly toxic to canids. It is less toxic to herbivores and some omnivores such as rats, but it has been used as a rodenticide.

In the United States, the only currently used form of Compound 1080 is in the Livestock Protection Collar. This collar is attached to the neck (which is usually the site of attack) of a lamb or goat that is tethered for accessibility to an attacking coyote. On attack, the rubber bladder containing the toxicant is punctured, and the liquid is squirted into the mouth of the attacking animal. Death usually ensues within five hours. The collar is expensive and is viewed as a last resort in removing an offending coyote that has eluded such methods as trapping and M-44s.

Compound 1080 has been used successfully to eradicate rodents or predators from islands. For example, it has been utilized with the Arctic fox on Kiska Island to protect the endangered Aleutian blue goose (Palmateer 1987) and has found recent interest as a compound of choice to eliminate invasive species on other islands (Veitch 2001; Burbidge and Morris 2002; Nogales et al. 2004). Compound 1080 has a relatively low toxicity to species of particular concern in Australia and New Zealand (D. R. King et al. 1989), including wombats (*Vombatidae*), quolls, and Tasmanian devils. It is therefore used widely in both countries for the control of feral animals, including pigs, dogs, foxes, and rabbits (Eason et al. 2017). About 80% of the world's use of Compound 1080 is in New Zealand, but this chemical has also served as a rodenticide and a predacide in many other countries in Africa, Asia, and the Americas.

REPTILICIDE. Snakes, including brown tree snakes, are particularly sensitive to acetaminophen (CAS No. 103-90-2), the active ingredient in pain relievers such as Tylenol. Acetaminophen was registered in 2003 as a restricted-use compound for lethal control of brown tree snakes in Guam and the Commonwealth of the Northern Mariana Islands. A tablet is placed in the throat of a dead mouse pup as bait, and this, in turn, is placed in a bait station or broadcast by hand or airplane. Baits broadcast by helicopter have small parachutes attached to them and many get caught in trees, where they are taken by brown tree snakes (fig. 11.3). This approach is used as part of an integrated pest management strategy for control of the snake on these islands (Engeman et al. 2018). Acetaminophen has also been found an effective toxicant for juvenile Nile monitors (*Varanus niloticus*) and Burmese pythons, and it may play a role in managing these invasive species in the United States (Mauldin and Savarie 2010).

Examples

Downtown Omaha, Nebraska, was selected by starlings as the site for a major winter roosting area (Thiele et al. 2012). Several parking garages and high-rise office complexes became roosting areas, as did a new mall complex with groves of older trees. The birds benefited from warm air associated with vents and ducts, and the roost size expanded to the tens of thousands. Droppings accumulated on parked cars under roosting areas and rained onto people. Business at the

Figure 11.3 A parachute-equipped bait being test-dropped from a helicopter. The parachutes entangle in the forest canopy in Guam where the mouse baits containing acetaminophen are ingested by brown tree snakes (*Boiga irregularis*). Photo by Kenneth L. Tope, USDA/APHIS/WS-National Wildlife Research Center Archives.

new mall suffered because of the mess and concerns for disease. The City of Omaha asked for assistance from federal agencies, and US Wildlife Services took a lead role in developing a management plan.

In the first phase of the plan, the damage practitioners captured, marked, and released blackbirds. About 6,000 blackbirds, mostly starlings, were tagged with streamers that could be readily identified from a distance. The practitioners sought tagged blackbirds at feeding sites, beginning just outside of Omaha. They first found tagged blackbirds at a cattle feedlot about 20 miles from Omaha. Over time, they found six cattle feedlots with blackbirds, at distances up to 60 miles from the city, and were able to show some movement of individual tagged birds between roosts over years. The feedlot operators were also concerned about the birds.

To address the damage, practitioners treated several feedlots with Starlicide bait, after meeting regulatory legal requirements and informing and involving the public. The overall population and size of the winter roost in Omaha were greatly reduced. Building owners added exclusionary devices and screening to make their buildings less suitable for roosting. The city removed larger trees from the mall area and replaced them with new landscapes less usable as roosts. Pyrotechnics and other repellent devices were utilized to disperse roosts when birds attempted to form them. The program continues at the time of this writing but has already been considered a success.

ANTIFERTILITY COMPOUNDS

Statement
Antifertility compounds offer the potential of reducing growth of populations by reducing birth rate; finding antifertility agents with the characteristics needed for resolving human-wildlife conflicts, particularly multiyear effectiveness with free-ranging wildlife, has been difficult, but products are now available for commercial and restricted use.

Explanation
Antifertility agents have been sought and used in zoos since the mid-1970s, with Asa and Moresco (2019) reporting almost 45,000 records of their use in zoo-maintained databases. In zoos, contraception plays an important role in managing overpopulation of wildlife such as great apes while maintaining fertility for release back into the wild.

Ideal antifertility agents for free-ranging wildlife would have certain characteristics, including single-dose effectiveness that lasts one or more seasons or continued acceptability of delivery compounds; reversibility, wherein wildlife could return to full reproductive function after the period of treatment; species specificity, in that only targeted wildlife would be contracepted; deliverability to free-ranging wildlife; and economic benefits that clearly outweighed management costs. Such an ideal antifertility agent has been elusive and may not be necessary for some uses. For example, delivery to individuals rather than large, free-roaming populations may be sufficient for managing confined wildlife populations in refuges and zoos (Rutberg 2005) or populations of wildlife such as some horse populations that are rounded up annually.

Antifertility agents for free-ranging wildlife have also been sought for many years, with limited success (e.g., Fagerstone et al. 2010). Most extensively tested are immunologically based contraceptives, including those targeting the zona pellucida (membrane covering the egg; ZonaStat-H, ZonaStat-D, PZP-22, SpayVac, Recombinant ZP) and gonadotropin-releasing hormone (GnRH): GonaCon, Improvac, Improvest). ZonaStat-H and PZP-22 are zona pellucida–based contraceptives registered as restricted-use pesticides for contraception of wildlife burros and horses. ZonaStat-H (Humane Society of the United States, Washington DC) is prepared by the Science and Conservation Center in Billings, Montana; PZP-22 by the University of Iowa School of Pharmacy; and SpayVac by ImmunoVaccine Technologies in Halifax, Nova Scotia, Canada. PZP formulations have been evaluated for effectiveness in African elephants, water buffalo, wild horses and burros, moose, bison, deer, elk, rats, yak, feral pigs, and other species (Naz and Saver 2016).

ZonaStat-H is approved by EPA for use in wild horses and burros. PZP vaccines, delivered remotely using biobullets and dart guns, successfully reduced population growth of wild horses on Assateague Island National Seashore within two years. ZonaStat-D is a zona pellucida–based immune-contraceptive registered to manage deer. The registration allows remote delivery but requires a prime dose followed with a second booster dose delivered within 2 weeks, and an annual dose thereafter, making it impractical for use in many free-ranging deer populations (Curtis 2020). Improvac was developed in the late 1990s in Australia as an anti-GnRH approach to castrating boars. It has been marketed as Improva, Improvest, and other brand names and is approved as safe to use in many countries globally.

GonaCon (and a variant GonaCon-Blue) is also registered in the United States for use as an immunologically based antifertility agent. It blocks GnRH production and has been tested for efficacy in wildlife including deer, horses, squirrels, prairie dogs, elk, elephants, brushtail possums, and feral pigs (Naz and Saver 2016). GonaCon is approved for inhibiting deer (GonaCon) and horse reproduction (GonaCon-Equine), but the agent is available only as a restricted-use compound. The label also requires that deer be captured and the vaccine hand-injected, making it impractical for wide-scale use in managing free-ranging deer populations (Curtis 2020).

At the time I write this chapter, OvoControl P is the only antifertility agent registered and commercially available in the United States to control reproduction of free-ranging wildlife. The product is for control of pigeons and has nicarbazin as the active ingredient, which affects the membrane separating the egg white from the yolk, making conditions unsuitable for growth of the embryo. OvoControl G, for geese, was removed from the US market because of insufficient sales.

Egg oil, specifically corn oil, is used as an addling agent, sprayed on gull and goose eggs to prevent them from hatching. It is also available for use in the United States under restricted conditions determined by federal and individual state laws. Corn oil is exempt from registration requirements of the EPA.

Example

Assateague Island National Seashore (ASIS) is home to two herds of feral horses, one in Maryland and one in Virginia. The ancestors of today's horses were most likely put on the island in the late 17th century by mainland horse owners to avoid fencing laws and taxation of livestock. To maintain a healthy ecosystem, the US Fish and Wildlife Service allows a grazing permit for a total population of about 150 horses. The herd in Virginia is managed by Chincoteague Volunteer Fire Company. The company holds annual pony pennings where excess ponies are sold, proceeds supporting the fire company.

The National Park Service is responsible for the Maryland horses and, to maintain their cultural and historical significance, manages the herd at a stable size. This population (fig. 11.4) has been the focus of the world's longest running feral horse

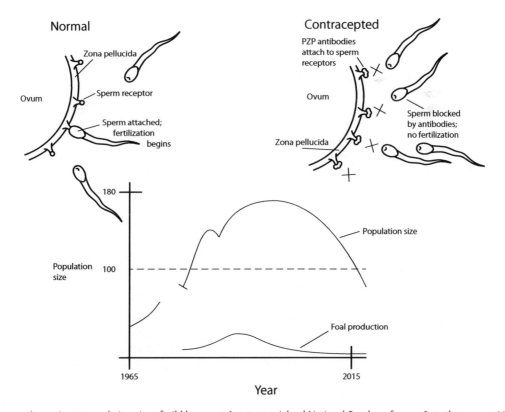

Figure 11.4 Approximate population size of wild horses at Assateague Island National Seashore from 1965 to the present. Horses were contracepted with porcine zona pellucida (PZP) to manage their numbers until 2015. PZP generates antibodies that block sperm receptors on the surface of ova. No contraception is being used now because production of foals is being encouraged. Illustration by Lamar Henderson, Wildhaven Creative LLC based on public information provided by the National Park Service.

immunocontraception program. At the onset of the program, there were about 166 horses, enough to negatively impact the Assateague Island ecosystem. In 1995, Kirkpatrick and Turner (2008) began administering PZP contraceptives to all mares 2-years-old or older. The contraceptives were given using dart guns and biobullets. Subsequent treatments were based on the Park Service goals of maintaining genetic diversity of the herd (by allowing young mares to foal) and moving toward a reduced but stable population size. Conditions of the mares were monitored bimonthly by ASIS employees. Population growth was reduced to zero within 2 years, and the population size began declining in 8 years. The delay was due to improved longevity of vaccinated mares from 6.5 years to about 20 years. The increased longevity was due to improved health of the mares by reduced energy stress of pregnancy and lactation. The extent of treatment has varied depending on the population status. For example, the National Park Service suspended treatment temporarily in 2017 to facilitate an increase in herd size.

As pointed out by Hobbs and Hinds (2018), the Assateague Island situation is unique in that all mares were treated, the population was manageable, individuals were easily identifiable, the area was small, the population was isolated, and the animals were approachable by foot. Nonetheless, the program serves as an example of the potential for contraceptive technology as a management tool for free-ranging wildlife.

Summary

- Pesticides are substances or devices, including antibiotics, avicides, fungicides, herbicides, molluscicides, piscicides, predacides, reptilicides, and rodenticides, that affect the behavior of pest plants or animals.
- Pesticides are active ingredients in formulations, mostly baits, that also include inactive ingredients, such as carriers, binders, attractants, enhancers, masking agents, and preservatives.
- Pesticides are delivered by broadcasting, by spraying, by placement in bait stations, with specialized devices such as M-44s, or with rifles or blowguns.
- Pesticides are registered and have labels that describe allowable and restricted uses as well as conditions for use.

- Many herbicides are available to manage weeds or to make habitat unsuitable for animals.
- Alpha-chloralose is an immobilizing agent used to capture wildlife such as geese and pigeons; propionylpromazine is being tested as a sedative for canids caught in restraining devices.
- Repellents that use sensory systems to irritate or frighten wildlife include anthraquinone, 4-aminopyridine, capsaicin, mixtures of oil scents, DMA and MA, methiocarb, and polybutene.
- Toxicants include avicides such as DRC-1339; chronic rodenticides such as warfarin, brodifacoum, and cholecalciferol; acute rodenticides such as aluminum and zinc phosphide, gas cartridges, and strychnine; piscicides such as rotenone and sea lamprey larvicide; predacides such as sodium cyanide and sodium fluoroacetate (Compound 1080); reptilicides such as acetaminophen; and antifertility compounds such as Ovotrol P, GonaCon, ZonaStat-H, and egg oil.

Review and Discussion Questions

1. If you extracted the odors from human hair, distilled and concentrated them, and then set a vial of the extract in a field of alfalfa to keep out deer, is the extract considered a pesticide? Would it require registration by the EPA?
2. How might herbicides be used to break the sylvatic cycle of a zoonosis? Give a specific example.
3. How important is rotenone as a toxicant in wildlife damage management? Are there suitable alternative piscicides to, for example, clear a pond of undesired fish before stocking it with desired ones? Would it be better to just drain the pond?
4. Is bait-shyness learned from the flavor of an active ingredient (e.g., warfarin or zinc phosphide), from the composite flavor of a formulation, or both? How might baits be designed to minimize bait-shyness? Give an example.
5. Some argue that wildlife contraception as a management tool takes too long to be practical for farmers and ranchers who need quick solutions. Deer continue feeding and coyotes continue killing during the time it takes contraception to reduce populations. Do you agree or disagree? Explain your view, including specific references.

12

Biological Methods

In this chapter I review biological methods used to manage human-wildlife conflicts. I include immunologically based disease control—e.g., rabies control—here because the antigens are live or attenuated viruses or bacteria and depend on biological responses to foreign materials in blood. Bacterial- or viral-based delivery systems for chemical pesticides are simultaneously chemical and biological methods. None are developed enough for routine use, so I describe those along with genetically based biotechnologies (genetic biocontrol) in the chapter on future directions (chapter 17).

Statement

With biocontrol, management is accomplished by introducing a living agent such as an alternative plant or prey, pathogen, parasite, herbivore, predator, or an interspecific aggressor. Success with biocontrol offers alternatives to chemical pesticides or reduced amounts of chemicals.

Explanation

Biological control (**biocontrol**) uses interactions among organisms to limit damage or the presence of damaging species. Biological methods include predator and prey interactions, parasite and host interactions, disease and host interactions, and interspecific aggression for territorial or social defenses, such as guard animals attacking predators of livestock. Biocontrol also includes **lure** (or **decoy**) crops as alternate foods and **diversionary feeding**.

Biocontrol has had particular success in managing damage caused by weeds and insects. The use of the insecticide-producing bacterium *Bacillus thuringiensis* (Bt) and Bt-engineered plants (plants that have Bt genes incorporated into their genetic makeup to provide protection from insects) are an example of biocontrol familiar to people who garden or farm. Parasitic wasps and praying mantises are also familiar biocontrol agents. When seeking a biological agent, it is better to enlist an organism already available in the ecosystem, such as Bt, than to import one from a distant land. An organism already adapted to the ecosystem and vice versa probably reduce opportunities for **ecological backlash**.

Biocontrol includes classical, inundative (augmentative), and inoculative (conservation) methods for introducing biocontrol agents into ecosystems, classical methods being most used. With **classical methods**, one or more agents are found and, after the appropriate guidelines are followed to allow its/their importation, the agent(s) is/are released one or more times in an exotic location. Over subsequent generations, the agent becomes established and gradually manages the pest. An example is the release of several parasitoid wasps in the United States for management of the emerald ash borer. With **inundative** (sometimes also called augmentative) **methods**, large numbers of agents (sometimes using sterile ones as a tactic) are released periodically throughout a single season. The sheer numbers overwhelm the problem population. Inundative methods usually require mass-producing the agents. **Conservation** methods are similar to classical methods, but the natural enemies are already present in the environment, and plans are made to supply them through a period of re-establishment.

Biological control of weeds has been applied in aquatic (e.g., using common grass carp) as well as terrestrial ecosystems. Winston et al. (2014) provided the fifth edition of a world catalog of biocontrol agents used for controlling weeds through 2012. According to an analysis of the catalog by Schwarzländer et al. (2018), there were at least 1,555 intentional releases of 468 biocontrol agents used to manage 175 target weeds in 48 plant families in 90 countries. Common lantana (*Lantana camara*), a native to Central and South America but considered invasive or naturalized in much of the tropical world, was the most targeted species. About 80% of the agents were from three insect orders: Coleoptera, Lepidoptera, and Diptera. About half of the releases caused medium or heavier damage to the targeted weeds, and about 66% of the weeds experienced some level of control. The analysis informs that biocontrol has value in weed management, either alone or as part of a broader integrated management program (see chapter 16).

Biocontrol of weeds or other wildlife may be successful in one location but not in another (McFadyen 2000). For example, the moth *Cactoblastis cactorum* was released in Australia in 1926 to limit the spread of prickly pear cactus (*Opuntia inermis* and *O. stricta*). The cacti were themselves introduced around 1839 as horticultural curiosities, and they quickly moved into arid areas as problematic feral plants. Cochineal insects (*Dactylopius*), including four species that came along with the first European settlers, were also used. The moth and other insects controlled the cacti, a success story in biocontrol (Zimmermann et al. 2000). The cactoblastis moth was also introduced into the Caribbean for biocontrol. It is believed to have been subsequently moved into Florida on at least three occasions, where it was first discovered in 1989, causing concerns for unintended impacts on the cacti of Florida, the southwestern United States, and Mexico (D. Johnson and Stiling 1998). The cactoblastis spread northward to Charleston, South Carolina, and westward to New Orleans, Louisiana, but it has not moved farther north or west during the last 15 years, suggesting that these concerns may be unfounded, at least in the short term. Although it is speculation at this time, the cactoblastis larvae may be limited by macronutrients availability in the more western cacti. Studies by Schartel and Brooks (2018), for example, have suggested a negative relationship between the presence of crude fiber content in the pachyderma of the host cactus, the presence of the larvae, and an increasing fiber content in the western types of cacti. Studies exploring this and other possibilities are continuing.

The viruses *Myxoma* and rabbit hemorrhagic disease (calicivirus) have been used to control the European rabbit in Australia and New Zealand, the first time biocontrol was used successfully to manage a mammal. Here, however, the effectiveness of *Myxoma* in Australia was limited to about 40 years (1950–1990), after which resistant rabbits began repopulating the country. Rabbit hemorrhagic disease effectively reduced rabbit populations after the disease was illegally (New Zealand) or erroneously (Australia) introduced in 1995 (see the Examples section).

Biocontrol has been used successfully to break the sylvatic cycle of rabies in wildlife, first in Europe and later in North America. Reservoirs of rabies, such as foxes, dogs, raccoon dogs, and raccoons, are provided with antigens that result in the production of antibodies and immunity from the disease (see the Examples section).

Guard animals have been used successfully worldwide to protect domestic livestock from predators such as coyotes and wolves (e.g., van Eeden et al. 2018) (fig. 12.1). Guard animals are often dogs (**livestock protection dogs**, or **LPDs**) but can also be cattle, goats, llamas (*Lama glama*; e.g., Franklin and Powell 1994), mules, or donkeys (e.g., Walton and Feild 1989). Rigg (2001) lists over 44 varieties of LPDs and at least 28 countries where LPDs are used, although Kinka and Young (2018) found breed made little difference in effectiveness of guard dogs. Rigg (2001) points to mongrels as effective guard dogs—for example, mongrels are used by the Navajo. LPDs may see additional uses,

Figure 12.1 A Great Pyrenees guard dog protecting sheep. Photo by Don DeBold from San Jose, California.

such as protecting livestock from diseases, conserving wildlife (Rigg 2001; Gehring et al. 2010), and protecting high-cash-value crops from herbivory (see the Examples section).

Depending on the feeding preferences of species, lure crops or diversionary feeding can be utilized. These are, therefore, forms of biocontrol and have many uses in managing human-wildlife conflicts. In one scenario, a crop or feed that is particularly attractive to pests is used to lure them away from a more expensive crop or from a situation where they can harm humans or their property (e.g., black bears at a campsite), thereby reducing human injury or economic losses. The approach is sometimes used by wildlife refuge managers to keep waterfowl in refuges and away from neighbors' fields. Crops sought by waterfowl are planted in the refuge, and the waterfowl feed in the lure fields unmolested. Sometimes, neighboring fields where waterfowl feeding patterns are already established are purchased for refuge purposes. Scare devices are sometimes used in surrounding fields to further encourage use of the lure crop. Cummings et al. (1987) suggested decoy plantings as a potentially effective means to reduce blackbird damage to commercial sunflower fields, and Linz et al. (2004) resurrected the idea in the form of "Wildlife Conservation Sunflower Plots" that would provide habitat for some 49 species of fall migrating birds, reduce damage to commercial sunflowers, and garner support from both agriculturists and conservationists.

In other scenarios, wildlife are attracted to lure crops where they may be captured and removed. For example, the trap barrier system uses rice planted earlier than the main crop as a lure crop. In some designs, rats wanting to access the rice need to swim across a moat, where they are trapped and removed (e.g., Singleton et al. 1998). Sullivan and Klenner (1993) used food to divert red squirrels from damaging crop trees in lodgepole pine (*Pinus contorta*) forests, and Witmer and VerCauteren (2001) reported use of cracked corn or soybean to divert voles from drilled seeds on no-till cropland.

Lure crops and diversionary feeding raise concerns, especially when they are used to manage the damage caused by game species such as bears (e.g., Ziegltrum 2008). Lure crops can be viewed as not only providing a food choice to wildlife but also as supplemental feeding. The Wildlife Society (2020) has expressed concern that unregulated supplemental feeding by the public "may convey the erroneous concept that such practices are suitable replacements for adequate habitat and scientific management of wildlife." Supplemental feed could have other impacts, such as reducing home ranges and territories or increasing contact among wildlife species, wildlife habituation to humans, or stress among wildlife populations. "A fed bear results in a dead bear" is a scenario sometimes suggested by wildlife agencies because of bear habituation to artificial food sources, e.g., garbage and bird feeders.

The Wildlife Society encourages federal, state, and academic institutions to monitor the full spectrum of impacts on wildlife affected by baiting and supplemental feeding. As an example, Rogers (2011) studied over eight years of diversionary feeding of bears at US Forest

Service campgrounds and residences. Only one bear, a transient subadult male, was removed. This researcher looked at other locations and found fewer house break-ins, attacks, and bear removals where diversionary feeding was used. He also found that food-conditioned bears did not jeopardize public safety, and he encouraged reevaluation of policies regarding use of diversionary feeding.

Garshelis et al. (2017) noted that diversionary feeding is used to manage bears more commonly in Europe than in the United States. The researchers reviewed the literature and summarized case studies from both regions. They concluded that there is still "a dearth of information" and recommended that if diversionary feeding is used it should be overseen by professionals, rather than an "ad hoc effort of a local organization or community," and conducted in the context of an experimental adaptive management program.

Some attempts at biocontrol have failed spectacularly. The introductions of monitor lizards, giant cane toads, and the Asian mongoose on Pacific islands to manage rats are examples already described (see chapter 4). Primarily because of concerns about ecological backlash, biocontrol through species introductions remains a controversial method (e.g., Louda and Stiling 2004). This is true not only for the introduction of exotic agents but also for live bacteria or RNA-based viruses that can mutate quickly (see part V). Correctly or not, people tend to see agents of biocontrol as easily adapting to new environments in ways not anticipated by experts, then becoming pests. Given such controversy, why do intensive and expensive searches for new agents continue? At least partly, says McFadyen (2000), because new biocontrol agents offer the potential for long-term solutions to wildlife problems. Despite this promise, people are often unaware that biocontrol agents have been part of successful solutions, so they tend to distrust these agents or remember failures rather than successes.

Governments, international organizations (e.g., United Nations, Food and Agriculture Organization 2017; US Department of Agriculture 2020a), and individual experts (e.g., Balciunas 2000) have developed guidelines and regulations intended to minimize opportunities for ecological backlash when using biocontrol. In general, these guidelines ensure that potential benefits exceed potential risks, that the public is informed throughout the project, that all regulations and policies are followed, and that results, and impacts are monitored so adaptive practices can be used. As described by Ward (2016) and Clark et al.

(2020), the regulatory landscape for biological control can quickly become complex.

Transfer of successful methods between countries can also be complex because exotic agents may be involved and assistance may be needed for technical aspects of the program, e.g., implementation of releases, program monitoring, and rearing of agents. Bilateral agreements are often involved, and sometimes more than two countries need to cooperate. Despite these obstacles, there have been many successful transfers (see the water hyacinth example below).

Examples

ZOONOSES. Efforts to manage Zika virus and dengue fever have focused on controlling *Aedes aegypti* and *A. albopictus*, mosquitoes that either act as vectors for the diseases or can be potential vectors in cooler ambient temperatures. Two companies, one called MosquitoMate and the other the World Mosquito Program, have developed approaches using bacteria. Both take advantage of *Wolbachia*, a naturally occurring bacteria found in 60% of all insect species and considered safe for humans and the environment. Both companies produce male mosquitoes that carry the bacteria. (Recall that only female mosquitoes bite humans and transmit diseases to people, so introducing large numbers of male mosquitoes into a population will not affect the number of mosquito bites.)

Here is how repeated introduction of infected male mosquitoes works. Infected male and infected female mosquitoes produce viable young, but they are also all infected. Uninfected males mating with infected females produce viable young that are also all infected. However, infected males mating with uninfected females produce offspring that die in the embryonic stage. With repeated additions of infected males into a population, mating that yields infected offspring greatly outnumbers the mating by uninfected males and uninfected females producing uninfected offspring. The portion of infected mosquitoes increases rapidly from generation to generation until all members of the population are infected (fig. 12.2). Because infected mosquitoes cannot transmit dengue fever or other diseases, the disease diminishes and eventually disappears in people.

The World Mosquito Program has begun releases in programs in Africa and elsewhere aimed at reducing dengue and other mosquito-vectored diseases. The United States tested *Wolbachia*-infected mosquitoes provided by MosquitoMate in the Florida Keys beginning in 2017.

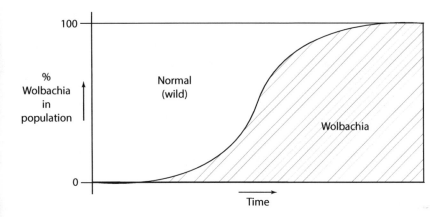

	Wolbachia female	Normal (wild) female
Wolbachia male	All Wolbachia offspring	None— eggs do not hatch
Normal (wild) male	All Wolbachia offspring	All normal (wild) offspring

Figure 12.2 Mosquitoes infected with the bacteria *Wolbachia* cannot transmit diseases such as dengue fever. By repeated releases of infected males and with some crosses resulting in sterile offspring, the population eventually becomes completely infected. Illustrations by Lamar Henderson, Wildhaven Creative LLC.

The near future may include even more advanced methodologies, such as engineered mosquitoes and gene drives, such as **CRISPR** (clustered regularly interspaced short palindromic repeats)/Cas9 (CRISPR-associated protein 9 enabled) genes (Berube 2020; see chapter 17). Tests are underway in the Florida Keys as I write using mosquitoes genetically modified with an OX5034 self-limiting gene provided by Ozitec, a British company. These mosquitoes are designed to produce male offspring that survive to mix with the wild population but females that do not survive long enough to reproduce. The mosquitoes are being tested in Brazil and elsewhere as a means of preventing the spread of vector-borne diseases rather than as a method of controlling disease outbreaks.

VOLES. Sullivan et al. (2001) explored using diversionary food to protect lodgepole pine seedlings from voles (three *Microtus* and the red-backed vole (*Myodes gapperi*) in old-field habitats near Summerland, British Columbia, Canada, and in forested areas). The intent was to provide food that was more attractive than seedlings but not so nutritious as to trigger re-production in the voles. The researchers made "logs" of alfalfa pellets, wood pellets, or bark mulch mixed with wax and sunflower oil. These scientists found that bark mulch, followed by alfalfa pellet logs, reduced damage to seedlings in both old-field and forested habitats, with the voles feeding 2.6 to 2.8 times greater on seedlings in untreated than in the treated sites (biologically, not statistically, significant, $P=0.09$). No change in reproduction of the voles was noted. The results supported supplemental feeding as a potentially effective method for reducing vole damage.

WATER HYACINTH. In the 1970s, three species (*Neochetina bruchi*, *N. eichhorniae*, and *Niphograpta albiguttalis*) that feed on water hyacinth (*Eichornia crassipes*) were introduced by US Department of Agriculture (USDA) researchers to manage this weed in North America. The results, as part of an integrated program that includes use of herbicides, have been successful. Australia, partly based on experience in the United States, imported one weevil (*N. eichhorniae*) and a moth (*Niphograpta albiguttalis*) and released them in the 1970s. Later, in the 1990s the Australians released *N. bruchi*.

The releases were also successful, and the Australians then aided other countries having problems with water hyacinth. Weevils released in the Sepik River lagoons of Papua New Guinea cleared formerly choked waterways in less than five years. Choked waterways in Lake Victoria across Kenya, Uganda, and Tanzania were cleared in fewer than three years. Australians also helped Benin, South Africa (Cilliers 1991), Nigeria (Farri and Boroffice 1999), and Thailand to control water hyacinth. In Rwanda, weevils from Uganda were introduced with help from the United States (Agaba et al. 2009). The British assisted weevil introductions on the Shire River in Malawi.

In spite of at least 22 introductions in countries since the first releases in the United States, biocontrol is not a panacea for managing water hyacinth. The weevils have slow life cycles compared with the rapid growth of water hyacinth. Herbicides and insects or pathogens are also needed to manage the hyacinths. The method is best viewed as an effective part of a broader integrated pest management program. Tipping et al. (2017), for example, found that the presence of biological control agents more than quadrupled the effectiveness of the herbicide 2,4-D against *E. crassipes*.

LPDS (LIVESTOCK PROTECTION DOGS). VerCauteren et al. (2008) provided experimental evidence that LPDs might be used to protect livestock from wildlife diseases. They evaluated whether four LPDs would keep potentially infected white-tailed deer at least 15 feet (in radii) from cattle, thereby protecting them from bovine tuberculosis. For this study they chose enclosures with high deer densities, about 35 deer/mi^2 and 93 deer/mi^2, respectively, and Great Pyrenees dogs. The dogs were bonded with the cattle and trained for about one year. Fenced paddocks were set up within the larger enclosures, and deer became accustomed to the areas. Shelters were built within the paddocks, and electric fencing was provided to contain calves. An invisible electric fence (invisible so that it would not influence fence-line contact between deer and calves) was used to contain the LPDs, which were trained to stay within the fence. VerCauteren et al. (2005) found the dogs to be highly effective in keeping deer from hay bales within the paddocks and also keeping deer from approaching the calves.

MYXOMATOSIS. **Myxomatosis** was first seen in rabbits in Uruguay in the late 1800s. In the European rabbit, the disease causes skin lumps and puffiness, sometimes conjunctivitis and blindness, and reduces resistance to secondary bacterial infections such as pneumonia. Infected rabbits become listless, lose appetite, and die within 2 weeks. The disease has less severe effects on New World rabbits (*Sylvilagus*). The idea of introducing the myxoma virus into Australia's rabbit population came from a Brazilian scientist in 1919.

The Australian government initially considered the introduction too dangerous. However, a prominent Melbourne pediatrician urged reconsideration; after 14 years of tests, the disease was introduced in 1950. Rabbit fleas were introduced to facilitate spread of the disease. The rabbit population declined 99% within 2 years, and the livestock industry received enormous economic benefits. By the end of the 1950s, partly through a natural selective process acting on beneficial interferon proteins already in the rabbit population and that offered gene immunity (Alves et al. 2019), resistance to the disease emerged, myxomatosis became less effective, and the rabbit population gradually began to increase.

RHDV. **Rabbit hemorrhagic disease virus** (RHDV, or rabbit calicivirus) was considered as a second biological control agent in Australia. The virus causes rabbit hemorrhagic disease, a highly contagious illness specific to rabbits both domestic and wild. The disease was first discovered in China in the 1980s, spread to Europe by 1988, and subsequently appeared in Australia, Cuba, Mexico, New Zealand, and the United States. RHDV was evaluated for efficacy and nontarget hazards on Wardang Island, off the coast of Australia, and the pros and cons were heavily debated. In 1995 the disease escaped the island and made it to the mainland on its own whereupon the pros and cons became moot points. The rabbit population has shown some resistance to the RHDV virus, although the immunity might involve complex carbohydrate structures on the surfaces of rabbit epithelial cells and multiple pathways. Efforts are underway to discover the genes supporting the immunity (Elfekih et al. 2021).

The ecosystem effects of landscape-level rabbit control in arid parts of Australia are confounded by livestock grazing and other environmental factors, but introduction of myxomatosis and RHDV each appeared to have been associated with pulses of increased recruitment of plant species, particularly highly palatable ones (Mutze 2016). Two native rodents—the dusky hopping mouse (*Notomys fuscus*) and the plains mouse (*Pseudomys australis*)—have markedly increased their occurrence, as has the crest-tailed mulgara (*Dasycercus cristicauda*), a threatened marsupial. All three species now qualify for threat-category downgrading on the IUCN Red List (Pedler et al. 2016).

In 1997, the government of New Zealand determined that it would not allow introduction of RHDV.

However, the disease was deliberately introduced on the South Island by individuals who opposed the decision. The New Zealand government attempted to contain the disease while farmers, using diseased rabbits and blenders, furthered the disease's spread. Unfortunately, the releases were not only illegal but also poorly timed, allowing immune response by rabbits under two weeks old and the rapid growth of a resistant population. The disease has, nonetheless, heavily impacted rabbit populations in New Zealand, particularly in humid areas. The natural resistance of young rabbits and the presence of a benign form of RHDV that conferred some natural immunity to rabbit populations (Nicholson et al. 2017) has lessened the overall effectiveness of the method.

ORAL RABIES VACCINATION. An epidemic of rabies spread from Poland to the eastern part of Europe at 12 to 36 miles per annum beginning in 1939. By the 1950s, most cases of rabies in Europe were sylvatic rather than urban, mainly due to red foxes rather than pets (pets were mostly immunized). Raccoon dogs, introduced from Asia into western Russia for hunting around 1920, subsequently moved into other parts of Europe and became a wild species important for transmitting rabies The epidemic reached Switzerland in 1967 and parts of central Europe by the 1970s. At this time, American scientists at the US Centers for Disease Control showed that red foxes could be immunized with attenuated virus vaccines delivered orally. The European and Canadian communities applied these concepts as a broadscale biological method to reduce or eliminate rabies in wild red fox and raccoon dogs (Vitasek 2004).

In 1978 Switzerland became the first country to attempt mass vaccination, and in doing so freed much of the country from sylvatic rabies within four years. Other countries followed, including Austria, Belgium, France, Luxembourg, Finland, the Netherlands, and then central and eastern European countries (Vitasek 2004). During the first years, baits made of chicken heads with capsules containing the vaccination (mostly SAD Bern strain) were placed by hand. Subsequently, manufactured baits were developed, and alternative vaccinations became available. The Bavarian method, in which hunters were enlisted to help place baits, was developed and used in some countries; in another technique, baits were broadcast by helicopter and fixed-wing aircraft.

European countries found that by coordinating their efforts and timing the treatments by season (e.g., raccoon dogs hibernate, making treatment less fruitful during that time), optimal effectiveness for cost of treatment could be achieved. These countries also learned to expect an increase in the size of fox populations as one consequence of effective rabies control, and they adjusted baiting densities to maintain effectiveness. The campaigns were intensified in the late 1980s after a resurgence of sylvatic rabies in some countries. Sylvatic rabies then declined precipitously, pushing the eastern boundary of the epizootic to the eastern border of Poland and western edge of Italy. In 2010, there were still over 1,500 cases of rabies in nine member states of the European Union. By 2018, with continued oral vaccination of foxes and other wildlife, there were only eight (six in wild animals and two in domestic); zero cases was seen as a realistic short-term goal (Robardet et al. 2019).

Over 40,000 cases of sylvatic rabies were diagnosed in Ontario, Canada, between 1958 and 1988 and over 3,500 cases in 1986. In 1989, the province of Ontario began a program of oral vaccination of foxes, the main sylvatic reservoir. The plan was to continue vaccination for at least one complete rabies cycle, about four years. At its peak, over 500,000 baits were distributed by aircraft in one season. By 2012, more than 12 million baits were distributed in southern Ontario. Baits were made of small plastic packets that contained a mix of fat and wax—i.e., oleo stock, a strengthening wax, and mineral or fish-liver oil—and chicken flavorant. Tetracycline served as a marker. The vaccine was an attenuated but live strain of rabies embedded in a blister packet of bait. Over seven years, an average of 357 cases per annum of sylvatic rabies was reduced to none, and the program was declared a success (MacInnes et al. 2001). In addition, the incidence of rabies in skunks paralleled the reduction in foxes, a consequence of treatment predicted from the onset.

An outbreak of raccoon rabies began in Hamilton, Ontario, in December 2015, with 449 animal cases by December 2018. Originally thought to be caused by a cross-border spread from the Niagara region where raccoon rabies has persisted, phylogenetic analyses of whole genome sequences identified the cause as likely due to the long-distance translocation of a diseased animal from southeastern New York (Nadin-Davis et al. 2020).

Oral rabies vaccination was begun in the United States in 1994, with field trials in Virginia, Pennsylvania, and New Jersey. A national management program (Slate et al. 2005) gradually emerged, in which oral vaccinations were distributed by aircraft in strategic locations to prevent movement of rabies from states where the virus already existed (Slate et al. 2002). In the United States, only Raboral V-RG, is registered for use.

Raboral V-RG is a vaccinia-based virus; vaccinia is the large, DNA-based poxvirus that was used to eradicate smallpox. Raboral V-RG was made using recombinant techniques in which the only rabies genes incorporated into the vector virus were those for the glycoprotein outer surface; this virus cannot cause rabies and is less sensitive to heat than other oral vaccines.

Vaccinia is effective in producing antibodies against several variants of the rabies virus. For raccoons, the vaccine is dyed pink, placed in a plastic sachet resembling a fast-food packet of ketchup, and coated with a mix of fishmeal and oil. The baits are distributed by ground and aircraft. In 2020, APHIS Wildlife Services and their cooperators planned to deliver over 4 million packets by hand and aircraft to strategic locations in at least nine states (US Department of Agriculture 2020b). Achieving herd immunity has been difficult, partly because of habitat fragmentation and competition for baits by nontargeted species, leading to reduced bait encounters by raccoons, lower than desired seroconversion rates, and patchy bait patterns. New technologies are being employed along with additional research on raccoon, skunk, and opossum ecology to help overcome these obstacles (Stengel et al. 2019).

Summary

- Biocontrol uses interactions among organisms to resolve human-wildlife conflicts or the presence of damaging species. It uses both classical and inundative methods.
- At least 41 weed species have been managed successfully using biocontrol as the only or the preferred method.
- Biocontrol using oral vaccination has been used successfully to break the sylvatic cycle of rabies in Europe and North America.
- Over 44 varieties of livestock protection dogs are used in at least 28 countries to protect livestock from predators. Other livestock protection animals include cattle, goats, llamas, mules, and donkeys.

- Myxoma virus and rabbit hemorrhagic disease have been used to manage feral rabbits in Australia and New Zealand.
- Lure crops or diversionary feeding are sometimes used to entice damaging species away from more expensive crops or from a situation where they can injure humans or their property. The Wildlife Society has expressed concerns about the use of artificial or supplemental feeding by the public because of potential adverse impacts on game and nontarget wildlife species.
- Spectacular failures using biocontrol have included introductions of monitor lizards, the mongoose, and cane toads on islands to control rats or other species. Such failures are often remembered longer than successes.

Review and Discussion Questions

1. Based on our previous discussions regarding exotic invasive species, why might it be preferable to find an endemic rather than a foreign biological control agent?
2. Based on our previous discussions regarding ecosystems, why might it be necessary to import a foreign biological control agent to manage an invasive exotic pest?
3. Why do you think The Wildlife Society has expressed concerns about public use of supplemental feeding, such as diversionary food or lure crops, in managing wildlife? Explain why the Society is especially concerned about game and nontarget wildlife species.
4. What has been the impact of oral rabies vaccination on fox and raccoon dog populations in Europe? Can the same impacts be expected when oral disease vaccinations are applied to other species and diseases? Explain.
5. Do you think an oral rabies vaccination program can be developed for management of bat rabies in North America? Why or why not? How would such a program differ from the one currently in place for control of raccoon rabies?

PART V • **HUMAN DIMENSIONS**

In this part, I explore the human aspects of an effective management action, including its economic, political, legal, religious, cultural, and social dimensions.

13

Economic Dimensions

Natural resource managers live in a world of limited resources and competing needs, where opportunities to accomplish an important action are sometimes forgone to accomplish another even more important one. How the manager decides to allocate resources is often subject to public scrutiny and debate. Given the range of public views on decisions involving natural resources, universal acceptance of any decision is unlikely. One can therefore anticipate vocal and sometimes legal challenges on any management action. Thus, it behooves the practitioner to base any action on sound judgment, using the best available information and analytical techniques. Fortunately, economics offers concepts and methodologies that apply broadly to human-wildlife conflicts, helping the practitioner make informed trade-offs among available management options.

DAMAGE ASSESSMENT

Statement
Most meaningful economic assessments of human-wildlife conflicts that involve wildlife damage require statistically based sampling and analytical methods.

Explanation
With sampling methods, assessments of damage (or some variable related to damage) are made in portions of the total area (or population), and damage found in the portions is used to interpolate or extrapolate an estimate of damage for the entire area (or population). For example, the practitioner might be able to assess only 1% of a forest damaged by North American porcupines because it would be too costly to examine every tree. This might mean the practitioner measures damage in about 100 samples of 50 trees each. From the samples, he or she then extrapolates damage for the forest. Quadrat sampling and minimum distance are two methods used to assess wildlife damage (box 13.1).

With sampling, the potential for errors—systematic (e.g., miscalibrated mechanical damage counters that consistently add 10 to the total

In **quadrat sampling**, a grid of specific dimension— for instance, a 10×10 grid of corn stalks—is used to sample damage, starting with a randomly determined coordinate. In this example, all ears damaged by birds might be counted. Depending on need, the surface area of damage on each ear might also be recorded. A partially damaged ear of field corn might have economic value in that it can still be fed to cattle. However, for sweet corn a partially damaged ear would probably be counted as a total loss because people will not buy damaged ears. Quadrat sampling is based on assumptions usually accommodated in the patterns of wildlife damage, such as random clumped distributions. The method tends to be labor intensive and expensive, however (Engeman 2002).

With **minimum distance sampling**, a point is randomly selected within a field and the distance measured between that point and the nearest damage. Then the distance is measured between that damage and the nearest second point of damage, then between the second and third points, and so on. The procedure is repeated a predetermined number of times, and an estimate of damage is garnered from these data. The method is also based on assumptions of symmetry, but these have been modified so that minimum distance sampling offers quality as well as reduced labor and cost (Engeman 2002).

pling. Costs of sampling vary with layer, and variability at one layer contributes to the next so that biases and errors are carried through all the higher layers. Layers contributing the greatest variability should be sampled most heavily (Engeman 2002).

When it is not practical to measure damage itself, another variable must be found that can be quantitatively related to the damage. Thus, missing lambs might be used to indicate predation on sheep. Often, population sizes of damaging species are assumed to be related to levels of damage. For example, population measures of muskrats in constructed wetlands of sewage treatment facilities might be used as a measure of the level of damage that can be expected to the wetland vegetation as well as a measure of the effectiveness of any control program that has been attempted. In these cases, a clear relationship must be established between populations and damage. At that point populations, or some index of the populations, can be used to estimate damage.

A similar approach is used with human survey methods, but here respondents, in sampling the affected population, are asked to judge aspects of damage. For instance, selected people from the affected population might be asked to describe their impressions of damage. Because sampling is involved, variability and errors associated with all sampling techniques need to be considered. In addition, biases associated with human perceptions of damage and dishonest responses should also be considered. For example, a farmer who believes any federal activity is a waste of money and who wants to avoid a possible federal response may deny having damage, even though he has experienced a substantial amount. Conversely, a farmer might bias upward her responses because she believes this will get more attention for a problem. Or a farmer might simply see damage differently from how it would be interpreted by researchers.

Statistical analysis of the sample results is used to provide information on damage for the area (or population) sampled. The results are often presented as **mean** (average) or **median** (the middle value when all are ranked from lowest to highest) levels of damage, along with some measure of variation (e.g., standard deviation or standard error of the mean). For instance, results of the analysis of porcupine damage may have shown that mean damage to the forest was $8.2 \pm 0.5\%$ of the trees at the 95% confidence level (i.e., at least 95 of 100 field assessments would yield means within 0.5% of 8.2).

count) or random errors—becomes a factor. Even with appropriate methods and no errors, variability is part of sampling. For example, suppose the practitioner examines all 100 tree plots for porcupine damage. It is unlikely that all plots will yield exactly the same number—say, eight damaged trees—because porcupines are not uniform in either distribution among trees or inclination to damage them.

Depending on the overall distribution of damage, stratification may be needed. For instance, if a field of corn is being assessed for bird damage, and most of the damage occurs along fence rows and edges, the areas of more intense damage may need to be sampled separately from the rest of the fields. Sometimes sampling occurs at different levels; e.g., a survey of coyote damage to livestock might be done at county, state, and national levels. Each level represents a layer for sam-

COSTS OF DAMAGE AND ITS MANAGEMENT

Statement

Cost of damage provides one measure of its importance and can be used in determining the level of management that is warranted. Assessments of cost are available for major areas of human-wildlife conflict, such as animal-vehicle collisions and invasive species.

Explanation

Costs of damage are gathered by tallying the numbers of goods damaged and assessing their value. Goods include agricultural commodities such as crops or livestock, properties, human health, and other wildlife. For example, the US National Agricultural Statistical Services surveys ranchers annually to derive a national estimate of sheep and goat deaths, and these data are periodically analyzed to separate deaths due to nonpredation and due to predation (US Department of Agriculture 2015).

Costs can be direct or indirect. The direct value of a good is not necessarily what it sells for in the local market—i.e., its **market value**—but what a person is willing to pay for the good (e.g., Schuhmann and Schwabe 2002). Nonetheless, market value is often easy to obtain, and it serves as a minimum value. For agricultural crops including timber, livestock, and fish, price at market reflects its actual consumptive value.

The direct value of damaged property, of human health, health threats, and health impacts, and of livestock or pets can sometimes be estimated from insurance payments or medical expenses. If the event involved injuries or fatalities to humans, livestock, or pets, then medical or veterinary expenses and potentially funeral and burial expenses need to be included. If the event involved livestock, loss of market value should be considered in addition to veterinary or other medical expenses. Zinsstag et al. (2007) found that cost-effectiveness for management of some zoonoses (e.g., mass vaccination for brucellosis in Mongolia and vaccination of dogs for rabies in Chad) was realized only when total societal costs were considered.

Indirect costs must also be considered, and these can be greater than direct costs. For example, wild rodents gnaw between the internodes of canes of sugar, often causing only minor amounts of direct damage and sugar loss. However, the wounds open the cane to secondary bacterial infections such as red rot, which sours the entire cane. For humans infected with a zoonosis or injured in an animal-vehicle collision, indirect costs may include a shortened productive life span, loss of productive work, health care, and disability compensation (e.g., Havelaar 2007). If a crop or livestock is insured against damage by wildlife (including compensation programs), the cost of insurance must be included. Sometimes crops or livestock are purposely not grown or raised in an area because of concerns for wildlife damage, known to economists as a **forgone opportunity cost**. To assess this cost, one would have to estimate the income that might have been generated had the area been cropped or had livestock been raised there, or, in the case of conservation, cropland retirement or logging bans (Yang et al. 2020).

Some plants can compensate for damage, an indirect capability that can reduce costs of damage. For example, Williams (1974) found rat damage to coconuts was compensated for by increased size of the remaining nuts in groves in Fiji, a factor that reduced costs of damage to coconuts. Rice plants sometimes compensate for rat damage to panicles early in the growing season by producing more panicles. If panicle compensation occurs early in the crop cycle, the new grains can contribute to overall yield. Often, however, the new grains are immature at harvest, and farmers are reluctant to conduct a second harvest unless substantial (over 40%) rat damage has occurred (e.g., Islam and Hossain 2003).

Values of wildlife include those considered consumptive (e.g., catching fish) and nonconsumptive (e.g., watching birds). Consider, for example, the cost of an endangered least tern egg taken by a raven. The cost goes beyond the egg's consumptive value (e.g., estimated as the market cost of a chicken egg) to include the potential contribution of the egg to the overall survival of the species.

Market values can be obtained in countries where wildlife is legally harvested and brought to market for sale or for which there is little concern or no legal regulation, such as commensal rats or some common birds. Wildlife regulations have also been used to estimate value. Here, costs in fines for illegal take (Engeman et al. 2002b), importation, or sales give an indication of value (Sterner 2009). Engeman et al. (2002a) use $100 per marine turtle, the Florida state minimum fine, to calculate cost-effectiveness of methods for reducing predation by raccoons and armadillos.

Estimates of the value of wildlife might also be based on **willingness to pay**—for a hunting, fishing, or hiking trip, for example. Humans ascribe some value to wildlife simply because it is there, known as its **existence value**. Sometimes people have neither the time nor the opportunity to use wildlife now but want it there for use at some other time—i.e., an **option**

value. For some people, it is important to know that wildlife is present for use by others now (**altruistic value**) or for future generations (**bequest value**).

Willingness to pay has been used to assess value of a broad range of wildlife, including eagles, grizzly bears, deer, and waterfowl (Schuhmann and Schwabe 2002). This concept has also been utilized to assess the willingness of villagers in the eastern Caprivi Region of Namibia to pay for technologies, e.g., fencing, to reduce damage and attacks by wildlife such as elephants (Sutton et al. 2004) and to model economic considerations of community-based wildlife damage management in Zimbabwe (Mhuriro-Mashapa et al. 2017). With this method, people are asked either how much they are willing to pay for a specific change in access to the wildlife or whether they are willing to pay a specific amount for the change. Using the method, Bhattarai et al. (2021) found that visitors to Nepal Park were willing to pay over US$20 above the current entrance fees to experience tigers and other wildlife in the park, and, using a variation of the method, Schutgens et al. (2019) found visitors willing to pay US$59 above the entrance fee to observe wildlife including snow leopards (*Panthera uncia*) in the Annapurna Conservation Area in Nepal.

In addition to costs from damage, costs to manage damage need to be assessed. Costs can be substantial, sometimes precluding management. Economic analyses comparing relative costs of management options may help the manager decide which, if any, action to take. For example, Hygnstrom and VerCauteren (2000) compared the costs and effectiveness of five different burrow fumigants for managing black-tailed prairie dogs (*Cynomys ludovicianus*) in colonies in central Nebraska. The researchers found that the products were about equally efficacious but that application costs were higher for pressurized fumigants.

As with damage costs, costs of management are indirect as well as direct. Direct costs include those for personnel, equipment, travel, materials and supplies, permits or licenses, insurance or compensation, and overhead and administration. Management costs can be actual or anticipated and can be collected and tabulated from fiscal budgets and accounts. Costs accrued help to evaluate economic benefits of completed management actions.

Indirect costs might encompass such items as loss of wildlife or delays in traffic due to management activities at an urban bird roost. Overall ecosystem impacts, often indirect, need to be considered and can be difficult to value. Oppel et al. (2011) listed the following indirect costs to an island community during an eradication campaign: biosecurity and monitoring costs after the eradication; loss of income from commercial activities during the eradication; costs of volunteering; and inconvenience from new practices (such as garbage disposal and animal husbandry) and regulations. In the United States and many other countries, if the proposed management involves a major federal action, an environmental assessment is required by law (see chapter 15). Indirect costs of management can sometimes be calculated from these government or other documents.

Compensation and abatement programs are a cost of management. In the United States, most states and related federal programs have **abatement programs** (i.e., consultation services, direct support from agency experts, or subsidies for abatement technologies), and about 25 states have compensation programs (Yoder 2002). For example, a farmer or rancher might be paid for verified wildlife damage to her or his crops, livestock, or property. Or he or she might receive educational and technical support from state or federal extension specialists or agents or financial assistance from state agencies to install deer fences or other forms of damage abatement. In Germany, landowners may be entitled to reimbursement for damage to forest species such as Norway spruce (*Picea abies*) by ungulates such as red deer, roe deer, European beavers (*Castor fiber*), hares (*Lepus europaeus*), and voles (*Muridae, Arvicolidae*). The compensation is usually paid by the shooting tenant (the individual or group who contracts with the landowner for hunting rights) of the land (Schaller 2002). In South Africa, farmers are compensated to relocate problem animals, e.g., cheetahs (*Acinonyx jubatus*), to approved areas (Cilliers 2003).

Increasingly, conservation organizations have supported compensation programs to encourage conservation of wildlife by local people. Whereas the costs of damage lie with the ranchers or farmers surrounding the refugia, the benefits of retaining wildlife are spread globally across the human community, which supports the presence of wildlife for existence, option, or bequest values. Compensation can be used to facilitate a more equitable distribution of costs and benefits.

Haney (2007) listed 48 wildlife species around the globe for which compensation has been paid, including many **charismatic megafauna** such as African elephants, African lions, American black bear, Eurasian lynx, gray wolves, and snow leopards. Compensation was paid by national, state, or regional governments and by nongovernmental organizations (NGOs) such as Mbirikani Predator Compensation Fund, Defendcrs of Wildlife, World Wildlife Fund, and Project Snow Leopard. Of interest is the fact that the presence of wilderness offered alternative prey for predators, tending to reduce the costs of compensation for livestock depredations, a form of **wilderness discount** (Haney et al. 2007). Compensation for losses was key to gaining support of ranchers for the reintroduction of the wolf into Yellowstone National Park in the United States. Funds for the program were initially provided by Defenders of Wildlife.

Despite their popularity, compensation programs remain controversial. Some experts (e.g., Bodenchuk et al. 2002) argue that compensation programs are an efficient means of protecting damaging wildlife, whereas others (e.g., Hoare 2003) argue that with an open market economy and inefficiencies in distribution of funds, compensation diminishes the value of the wildlife. Others have stated that compensation does not make economic sense because the practice reinforces inappropriate husbandry practices, encourages economic dishonesty in both claims of damage and distribution of funds for compensation, and may even increase resentment of damaging wildlife and subsequent retaliation. For example, Fernandez et al. (2009) reviewed community-based conservation programs in Zambia that compensated villagers for wildlife damage. As wildlife populations increased, so did damage. Program policies, however, did not allow adjustment of compensation for increasing damage. Villagers became frustrated with uncompensated damage and looked to their own solutions, leading to additional human-wildlife conflicts. Montag (2003) argued that compensation programs were fraught with additional economic problems. She posited that the issues might be framed differently by ranchers or farmers who may see compensation programs as federal management of private rights, affecting equity, public grazing, and public or private land management.

Shwiff et al. (2020) presented a general framework for assessing the economic impacts of vertebrate invasive species in the United States. They separated types of damage into destruction (i.e., damage to things such as statues, golf courses, ecosystems, vehicle collision, nonconsumptive crop damage), depredation (i.e., consumption of crops, livestock, companion animals, livestock), and disease (i.e., morbidity and mortality in humans, companion animals, livestock, or wildlife) and examined how the framework fit into published literature. Based on the proposed framework, the scientists reviewed the existing literature on costs of damage and its management in the United States for rats, brown tree snakes, Burmese pythons,

starlings, nutria, and feral swine. The researchers concluded that available data are variable, often with different focuses within a category and using different methodologies, making comparisons difficult. They suggested that data on diseases are particularly difficult to measure and scanty. Shwiff et al. (2020) found that the most complete information was available on feral swine, a species they suggested is "poised to become the most significant contributor to damage" among all vertebrate invasive species in the United States (3112). Shwiff et al. (2020) suggested that the framework be used for future research on economics of vertebrate invasive species, suggesting further that doing this would allow useful comparisons of data and use of the data in regional and national models.

Examples

COSTS OF WILDLIFE DAMAGE TO CROPS USING CROP INSURANCE. McKee et al. (2021) used federal crop insurance data to assess wildlife damage in the United States over the periods 2015 to 2019 to corn, soybean, wheat, and cotton. They found that the eastern and southern regions of the United States were most susceptible to wildlife damage, estimating a combined total loss at about US$593 million. Soybeans incurred the greatest losses (US$324 million) followed by corn (US$194 million). The researchers described the method as a reliable way to not only obtain estimates of wildlife damage to the crops but also evaluate geographic and temporal heterogeneity and hotspots.

COSTS OF WILDLIFE-AIRCRAFT COLLISIONS. Wildlife collisions with aircraft occur worldwide, with airlines incurring costs for losses and airports incurring costs for control. Mammals on runways or bird species weighing over 2 kg (e.g., geese in North America, lapwings *Vanellus vanellus* in the United Kingdom) tend to cause the most damage (Allan et al. 1999). Dolbeer et al. (2021) summarized wildlife strikes to civil aircraft in the United States from 1990 to 2019. The number of reported strikes increased from 1,850 in 1990 to 17,228 in 2019. Of 30,498 reported incidents during those 30 years, 4,610 provided estimates of repair costs that totaled US$748.5 million, a mean of about US$162,000 per incident. Assuming this was about the mean cost for repairs of all collisions, the total repair costs (mean times 30,498 incidents) would have been about US$4.9 billion dollars. Total costs were estimated at about US$5.9 billion. The estimates did not include indirect costs such as putting passenger in hotels, rescheduling aircraft, and flight cancellations. Seventy-three aircraft were damaged beyond repair with one such loss occurring in 2019. Terrestrial mammals were involved in 42% of these total losses, and geese and vultures in 41% of the 27 incidents involving birds.

Allan (2002) projected that an average bird strike worldwide costs US$39,705 (about 11% for damage repair and about 89% for delays and cancellations). Based on air transport movement worldwide, Allan estimated costs of damage at of US$1 to $1.5 billion per year, or about US$64.50 per flight. The estimates did not include helicopter traffic or costs to military aviation. This researcher estimated costs from bird damage to the US Air Force at about US$33 million per year and to the Royal Air Force (RAF) of the United Kingdom at about US$23.3 million per year.

Thorpe (2003) reported the loss of 231 lives in civil aviation between 1912 and 2002, and Richardson and West (2000) estimated about 141 deaths in collisions with military aircraft between 1959 and 1999 in Western nations. Dolbeer et al. (2021) reported 36 human fatalities in US civil wildlife strikes between 1990 and 2018. Specific costs associated with this loss of lives were not provided.

COSTS TO MANAGE DAMAGE AT AIRPORTS. Allan (2002) considered the costs of minimizing habitat in and around an airport at about US$75,000 per annum in Western Europe. The estimate was based on costs for making and maintaining grass swards that discourage the presence of species such as gulls, lapwings, golden plovers, and starlings on airfields in the United Kingdom, an approach that might also favor the presence of small mammals that attract predatory birds (Barras et al. 2000). The costs were estimated at about US$65,000 to $130,000 per airport per annum as contracted by the RAF for patrolling, bird dispersal, and wildlife depredation, either during daylight or during all times, respectively. Similar harassment programs in the United States were estimated to range between US$25,000 and $150,000 per airport. Thus, Allan (2002) estimated a cost of about US$200,000 per year to implement a full control program in the United Kingdom; less expensive were programs that might be cost-effective at airports with minimal needs for bird control.

Dolbeer and Chipman (1999) estimated that airstrikes were reduced from about 170 to 50 per year during the period when shooting teams were employed to prevent birds from entering airspace at John F. Kennedy International Airport in New York. He estimated the airlines using the airport saved about US$4,764,600 per year, at an annual cost to the airport of about US$120,000 per year.

COSTS OF ANIMAL-VEHICLE COLLISIONS. Hardy et al. (2007) estimated the costs of annual wildlife-vehicle collisions in the United States to be US$8.4 billion. This estimate is probably conservative. Many accidents involving damage to cars under $1,000 go unreported. And although Hardy et al. based their analyses on collisions with wildlife, actual collisions do not always occur. Sometimes, for example, drivers lose control of a vehicle while attempting to avoid wildlife, causing human injury or death. And sometimes drivers stop along or on roadways to remove wildlife from a dangerous situation and accidents ensue. Further, the report noted that travel delays as well as accidents to subsequent motorists if the vehicle or animal lies in the right of way were not considered, nor were the costs related to emotional trauma of the motorists.

Williams and Wells (2005), in a US study, found that when human injuries or fatalities occurred, they were costly; medical expenses averaged $2,702 per collision. Other costs averaged at $125 for towing and law enforcement services; $2,000 for the value of the deer; and $50 for carcass removal and disposal (Hardy et al. 2007). If a larger animal was involved, the costs were greater—e.g., $3,000 for an elk and $4,000 for a moose. If the involved species were federally or state listed as threatened or endangered, the costs were higher. Conover (2019) estimated 58,622 human injuries and 440 fatalities in collisions between deer and vehicles from June 2017 to July 2018, and 640 human injuries and six fatalities in vehicular collisions with moose, and 10 human injuries and one fatality in vehicular collisions with birds during the same period.

Langbein et al. (2010) reviewed the literature involving collisions between deer or other ungulates in Europe. They provided sources and summarized total costs for collisions with deer in Slovenia at about US$18 million per year (assuming a US dollar is about 1.21 Euros), and with ungulates in Finland at about US$197 million per year, in Sweden and France at around US$121 million per year, and in Germany at about US$541 million per year, cautioning that "it is difficult to interpret such global estimates." Assuming an average repair cost of about US$1,500 and excluding compensation for human injury or fatalities, the investigators estimated annual repair costs in Europe alone "to lie by now well in excess of US$1.2 billion."

Sáenz-de-Santa-María and Tellería (2015) conducted a comprehensive review of animal-vehicle collisions that occurred in Spain from 2006 to 2012. They reported a total of 74,600 collisions with 2,911 causing human injuries. The researchers found feral swine and red deer were most costly, with the former costing US$55.2 million per year and causing 3.6 human injuries and 0.049 fatalities per 100 collisions, and the latter costing US$8.0 million per year and causing 4.4 human injuries and 0.074 fatalities per 100 collisions. They reported an average yearly total cost of US$127 million.

AAMI, a company that insures motorists in Australia, reported 7,992 claims for collisions with kangaroos and 392 with wallabies between March 2018 and February 2019 (AAMI 2019). AAMI reported 21,000 claims for collisions with animals between February 1, 2019, and January 31, 2020. AAMI's data found motorists are most likely to experience a major collision with a kangaroo (84%), wallaby (5%), wombat (2%), deer (2%), or bird (1%) (AAMI 2020). Based on 2016–2017 claims, the average costs of repairs were about US$3,080 (assuming US$1 is about AUD$0.77); including a surcharge for policy excesses, the costs of collisions with kangaroos cost about US$26 million (Huddle Insurance 2019).

Abra et al. (2019) found that the Military Highway Police of São Paulo State, Brazil, reported 2,611 animal-vehicle collisions per year with 18.5% of these resulting in human injuries or fatalities. The average cost was US$9,629 regardless of whether human injuries or fatalities occurred. Animals included the lowland tapir (*Tapirus terrestris*), capybara, and large domesticated species and large domesticated species including cattle (*Bos taurus*) and horses.

Collisions with animals, including camels, are also a problem in Asia and the Middle East. DeNicola et al. (2016) reported over 600 collisions with camels each year in Saudi Arabia. Most occur on straight sections of roads at night, and many collisions result in human injuries or fatalities.

COSTS OF INVASIVE SPECIES. Pimentel et al. (2001) estimated that over 120,000 species of wildlife, both plants and animals, have invaded the United States, the United Kingdom, Australia, Africa, India, and Brazil. A few adapted to domestication—crops and animals that became livestock provide over 98% of the human food supply globally. Others cause economic damage to human interests, including the environment. Diagne et al. (2021) used records included in the InvaCost database (Diagne et al. 2020) to estimate costs for damage due to invasive species globally between 1970 and 2017. The researchers found minimum accumulated losses of US$1.288 trillion, with mean costs of US$26.8 billion. They found that mosquitoes (*Aedes*) contributed the most to the cumulative costs, almost US$150 billion, followed by rats and mice, cats, and the Formosan subterranean termite (*Coptotermes formosanus*; fig. 13.1). Of further concern, the scientists noted that mean costs were trebling every 10 years,

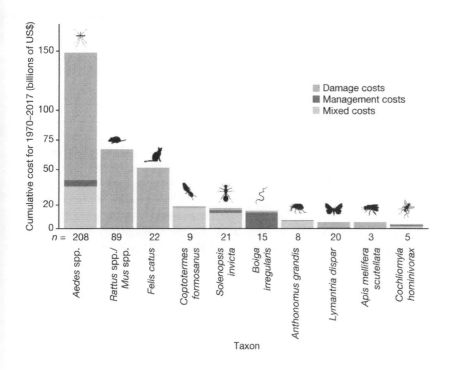

estimating an annual mean cost of about US$162.7 billion in 2017, and noted additionally that "these costs remain strongly underestimated and do not show any sign of slowing down" (571).

COSTS OF COVID-19. Cutler and Summers (2020) estimated that the direct and indirect costs associated with the COVID-19 zoonotic pandemic, in the United States alone, is about US$16 trillion or about 90% of the gross domestic product (GDP). This includes about $4.4 trillion due to premature deaths, US$2.6 trillion due to long-term health impairment, US$1.6 trillion due to mental health impairment, or about US$200,000 for an average family of four. We are all aware of the incredibly costly human impact of the COVID-19 pandemic, but the pandemic also serves as a stark reminder of the economic costs sometimes associated with zoonoses and other human-wildlife conflicts.

THE BENEFIT/COST ANALYSIS AND BENEFITS OF MANAGEMENT

Statement

Assessment of potential benefits in relation to management costs help practitioners decide how to judge trade-offs—e.g., whether management action is justified—and to choose among management options. Assessment of actual benefits helps practitioners determine potential values of adaptive improvements and whether a program should continue.

Explanation

Cost-effectiveness is the economic bottom line of wildlife damage management. To assess net benefits, one must know the full range of benefits in relation to costs of damage and costs of management (see the previous section in this chapter). In wildlife damage management, benefits are sometimes viewed as avoided costs. Sometimes benefits can be calculated either by measuring the reduction in damage and associated losses or by measuring a concomitant increase in production. This direct approach applies most readily to wildlife damage to livestock, crops, forestry, and aquaculture production. If the damage is to property such as houses, automobiles, or aircraft, a reduced rate of loss over some reasonable period might allow an estimate of benefits of management (e.g., 10% fewer wildlife-vehicle collisions) and a measure of the effectiveness of the control action. If there is a strong statistical relationship between damage and its actual cost or between activity of the population and cost of damage, then reduced damage or activity can be used to estimate benefits of the management.

Also important is to consider indirect or "spillover" benefits to secondary entities—e.g., increased livestock production from management of coyote predation to improve antelope production in Wyoming (Shwiff and Merrell 2004). If human health is improved because of management, as with malaria control programs or reduced incidences of ehrlichiosis or Lyme disease, indirect benefits accrue as fewer sick days and increased productivity as well as reduced costs for medical treatment and insurance payouts. There might be other induced or intangible benefits—hard to assess but real nevertheless (Shwiff 2004). For example, jobs might be created or additional monies flow into local economies from improved production. Distribution of benefits, the environment, and resources may also be positively affected.

The tools available for economic assessments include the following: studies of economic feasibility (i.e., is cost of damage equal to or greater than cost of management?); economic efficiency (i.e., which of the available management options is least expensive?); cost-effectiveness analysis (CEA, i.e., which of the available management options offers the best return for investment?); and **benefit-cost analysis** (**BCA**, i.e., what is the actual benefit to be gained in relation to cost of management?).

McCoy (2002) provides a generic example of an economic analysis, using a hypothetical multiyear goose management program. The analysis begins by assessing management costs, including initial investments, total (life of program) operating costs, and secondary costs such as unintended loss of other wildlife. The total benefits—i.e., value of reduced goose damage—need to be assessed. Future expenditures and benefits can then be incorporated into present values using a future discount rate (i.e., roughly annual inflation), resulting in **net present values** (NPV).

Expected value (EV) analyses can be used if there is uncertainty surrounding benefits. In his example, McCoy (2002) spends U$100,000 for goose management that includes habitat changes around an airport. There is a 10% chance that management action to reduce goose strikes to aircraft at an airport will produce no beneficial results, a 60% chance that the actions will reduce losses from US$1 million to US$650,000, a 25% chance that the actions will reduce losses to US$500,000, and a 5% chance that losses will be reduced to $0. An EV of US$315,000 is calculated from the probabilities, after subtracting the US$100,000 for cost of management (see McCoy 2002 for calculations). This information allows several economic analyses. For example, economic feasibility can be calculated. Here, any effort that yields an NPV greater than $0—i.e., a cost of US$315,000 or less—is considered economically feasible. Economic efficiency might be calculated by comparing potential benefits of habitat changes with other management strategies, such as hazing, repelling, or shooting geese, an approach McCoy (2002, 121) suggests can help the practitioner get the "biggest bang for the buck."

If only costs of management are known and benefits can be quantified on some scale (e.g., percent effectiveness of the method), a CEA can still be completed (McCoy 2002). With CEA, costs are compared to relative benefit. In the goose management example, McCoy provides the following management options: hazing with US$10,000 for 10% population reduction; repellents with US$20,000 for 10% population reduction; habitat modification with US$20,000 for 20% population reduction; habitat modification and hazing with US$25,000 for 25% population reduction; and habitat modification and repellent with US$35,000 for 30% population reduction. By comparing these figures, one can see that use of repellents alone costs relatively more than the other methods and that hazing alone can achieve the same results at lower cost.

To complete a full-fledged BCA, all benefits and costs, direct and indirect, present and future (discounted to present value), must first be estimated. Total costs are then subtracted from total benefits to derive the **net benefit**. By comparing net benefits of different approaches, the practitioner can compare economic differences among the options (Portney 2020). Shwiff and Sterner (2002) presented a general model that applies BCA to hypothetical proposed approaches for reducing wildlife hazards to aircraft. Four approaches were considered: bird harassment, habitat management, fencing, and doing nothing. This model is helpful in explaining the use of dependent variables, spreadsheets, decision trees, calculation of NPV, and sensitivity analyses (to assess impacts on outcomes when variables change)—all important components of BCAs, which can be applied to other wildlife damage situations.

BCA models are increasingly available to help make informed decisions for resolving human-wildlife conflicts. The practicality of BCAs and their applications to such issues have risen dramatically, along with the rising capacities and iterative calculating capabilities of computers. Many BCAs have specific applications; for example, Caudell et al. (2010) provide a BCA model to assess when OvoControl G is more cost-effective for controlling Canada geese than other methods, such as egg oiling or other forms of addling.

Sterner and Tope (2002) provide models to determine when it is cost-effective to use a commercial turf repellent (Rejex-It) to repel Canada geese from golf fairways and a commercial shrub/plant repellent (Deer I Repellent) to deter deer from eating landscape shrubbery. VerCauteren et al. (2002) have developed models that generate BCA analyses for various levels of rodent control at confined swine facilities. Choquenot and Hone (2002) provide models to be used to evaluate BCA of helicopter shooting and Compound 1080 poisoning to reduce feral pig predation on newborn lambs in Australia.

BCAs have also assisted conservation efforts. For example, Emerton (1999) provided a BCA for conservation efforts at Lake Mburo National Park, Uganda. Her evaluations included costs of damage by wildlife to crops, from livestock kills, and from transmission of diseases to domestic herds. She used the analyses to suggest cost-effective approaches for managing damage.

Examples

BCA COMPARING METHODS FOR MANAGING COYOTE ATTACKS ON LIVESTOCK. Brewster et al. (2019) analyzed the benefits to costs of 88 scenarios using aerial gunning, M-44 devices, snares, livestock guard animals, calling and shooting, foothold traps, and combinations of the methods. The researchers applied the scenarios once or twice yearly or continuously in a model for an average-sized (1,000-hectare) cattle operation over a conceptual 10-year period in Texas. The scientists used data available from the scientific literature, subject matter experts, and anecdotal literature from 1960 to 2019. They found the most cost-effective method for reducing calf depredation was to use snares in combination with livestock guarding animals beginning one month prior to the calving season. When applied once during the season, the approach provided an 81% decrease in overall costs of calf depredation and predator management (reducing losses during the 10-year period from US$79,852 to US$15,456 on an average-sized ranch). The researchers point out that the study did not consider potentially important intangible benefits that ranchers should consider, such as grazing benefits of coyotes (e.g., reduced numbers of rabbits competing with cattle) and ecological benefits (e.g., mitigation of mesopredator release).

BCA USING ORCHARD NEST BOXES FOR FALCONS TO REDUCE DAMAGE TO FRUIT. As part of an integrated pest management program in an area of Michigan where sweet cherries are grown, Shave et al. (2018) evaluated the economic benefits of encouraging the presence of American kestrels to the costs of installing nest boxes and artificial perches. The researchers found that fruit-eating bird counts were greatly reduced at orchards with active kestrel boxes and that from US$84 to US$357 worth of sweet cherries would be saved for every dollar spent on the nest boxes. On a regional basis, the scientists projected that increased sweet cherry production would create 46 to 50 new jobs and about US$2.3 million additional revenue for Michigan over a 5-year period. The scientists concluded that adding kestrel nest boxes to the pest management repertoire was highly cost-effective.

BCA COMPARING DEER FENCES. VerCauteren et al. (2006b) developed a model that provides BCAs for different fence designs to protect crops from deer damage. The model uses simulation software that can be adjusted for: acreage (0–200 acres), field perimeter (254–11,390 meters), percent of damage (0.00–1.00), crop value per acre ($0.00–$4500), fence efficacy (0.00–0.99), material costs per meter of fencing ($10–$15), labor costs per meter of fencing ($0–$15), and discount rate (0.00–0.15) for the life of the fence.

The user provides information specific to his or her situation. The model then calculates the NPV. By using iterative calculations, practitioners can compare impacts of factors such as different fence designs and labor costs on net benefits. In their simulations, for example, fences designed to protect high-cash-value crops such as apple orchards tended to benefit from more expensive, long-term fences made from woven or welded wire. For lower cash-value crops, such as soybeans and alfalfa, fences were only marginally cost beneficial, even with less expensive designs.

ANIMAL-VEHICLE COLLISIONS. Huge costs are incurred by governments in attempts to prevent or reduce collisions. Hardy et al. (2007) identified 34 methods of alleviating this problem and compared BCAs for these techniques. Fencing, in combination with underpasses and overpasses, had the best economic return for investment, resulting in reduction in deer-vehicle collisions of over 80% and bringing net benefits of about $24,000 to $33,000 per km of a hypothetical highway per year. Standard signs and deer reflectors and mirrors were ineffective, costing more to install and use than was gained in benefits.

Summary

- Most assessments of wildlife damage are based on statistical sampling and surveys, using methods such as quadrat sampling or minimum distance sampling. When feasible, the best measure is damage itself. When this is impractical, a variable related to damage must be found. Damage can be direct or indirect.
- Cost of damage is based on numbers of goods damaged and the value of the goods. Goods can be property, crops, livestock, human health, or other wildlife. Value can be based on market value or economic concepts such as willingness to pay. Costs can be direct or indirect.
- Costs of management can be direct or indirect. Direct costs can be obtained from fiscal budgets and records. Indirect costs may be harder to value. Costs are available for major areas of wildlife damage such as animal-vehicle collisions or invasive species.
- Compensation and abatement programs are a cost of management. Abatement programs reduce damage by providing technical and material assistance for both preventing and managing damage. Over 48 wildlife species worldwide have been part of compensation programs, often underwritten by conservation organizations as a means of facilitating a more equitable distribution between benefits and costs to communities bearing the brunt of damage.
- To assess net benefits from managing wildlife, one must know all benefits in relation to all costs of damage and its management. Tools available for economic assessments include economic feasibility, economic efficiency, cost-effectiveness analysis, and benefit/cost analysis (BCA).
- Economic tools provide a means for assessing potential benefits in relation to management costs, to decide whether management action is justified and to choose management options.
- Many BCA models are focused on specific issues, such as various levels of rodent control at confined swine facilities, comparisons of economic differences among deer fence designs in relation to situations, and economics of use of rodenticides in alfalfa fields to manage vole damage.
- Other BCA analyses focus on a particular area of wildlife damage, such as predator management or relative merits of animal-vehicle mitigation measures.

Review and Discussion Questions

1. Choose a primary publication that provides statistical information on wildlife damage. What information is provided on variability of the results around the means or medians? At what level of confidence? Choose a test comparing two or more means. Relate variability to significant difference (if there is one) or no significance difference (if

there is not one). A diagram or two using normal bell curves (i.e., normal distribution) might help.

2. Compensation programs are sometimes used by conservation organizations to affect a more equitable economic distribution between those bearing the costs of damage and those benefiting from effective damage management. Does this approach improve the value of wildlife or not? Choose a specific damage situation, choose a specific position, and defend your position with references to specific literature.

3. We have seen an increase in economic computer models that allow the practitioner to assess relative economic merits of control options. What is the role of iterative calculations in these models? What is the importance of calculating a threshold value?

4. Federal predator management programs in the United States have been criticized as being noneconomical subsidy programs for the livestock industry, but Bodenchuk et al. (2002) calculate a 12–3 benefit-to-cost for all livestock protection programs. Who is correct? Defend your view.

5. Are signs and deer reflectors an economically effective means of reducing collisions with vehicles in North America? Explain your answer using the literature.

14

Human Perceptions and Responses

In this chapter I explore societal, cultural, religious, and personal factors influencing human-wildlife conflicts and their resolutions.

Statement

How one responds to a human-wildlife conflict is often based on one's perceptions of its seriousness, especially the sense of personal risk and whether one is personally suffering damage. Responses fall within the limits of one's values, beliefs, attitudes (VBAs), and mental state, such as fear or anger.

Explanation

VALUES. **Social values** are conceptions of what is good or bad, desirable or undesirable. Values form slowly during youth, change little later in life, and strongly influence attitudes and norms (group-held rules for acceptable behavior within social settings, or things one "ought" to do; Manfredo and Teel 2008). Social values therefore provide a broad frame within which we think and behave, and they serve to transmit social culture across generations, ensuring that individual behavior is in line with societal norms. Social values underlie how people view human-wildlife conflicts and their thoughts about which benefits are most desirable (Allen et al. 2009).

Wilson (1984) proposed an innately emotional connection between humans and other living things driven by a long coevolutionary history. He argued that natural selection favored individuals well-tuned to nature and life, leading to a "cognitive archeology" with detailed attention to plants, animals, time, space, weather, water, and wayfinding (Kellert 2007). The result was what he called **biophilia** (love for living things), genetically based in modern humans. Biophilia serves as a form of paradigm, filtering what humans pay attention to and what responses they learn or resist learning (Wilson 1993).

Kellert (1993) theorized that biophilia manifests in different people as different typologies (types of values) or biophilic values (table 14.1). He found typologies particularly useful in relating peoples' values to their behaviors. For example, Kellert and Berry (1982a) found that hunters scored high on dominionistic values (other researchers use the

TABLE 14.1   *A typology of biophilia values*

Term	Definition	Function
Utilitarian	Practical and material exploration of nature	Physical sustenance/security
Naturalistic	Satisfaction from direct experience/contact with nature	Curiosity, outdoor skills
Ecological-Scientific	Systematic study of structure, function, and relationship in nature	Knowledge, understanding, observational skills
Aesthetic	Physical appeal and beauty of nature	Inspiration, harmony, peace, security
Symbolic	Use of nature for metaphorical expression, language, expressive thought	Communication, mental development
Humanistic	Strong affection, emotional attachment, "love" for nature	Group bonding, sharing, cooperation, companionship
Moralistic	Strong affinity, spiritual reverence, ethical concern for nature	Order and meaning in life, kinship and affiliational ties
Dominionistic	Mastery, physical control, dominance of nature	Mechanical skills, physical prowess, ability to subdue
Negativistic	Fear, aversion, alienation from nature	Security, protection, safety

Source: From *Biophilia Hypothesis* by Stephen R. Kellert and Edward O. Wilson. Copyright © 1993 by Island Press. Reproduced with permission of Island Press, Washington, DC.

term "doministic," and these terms are often used interchangeably although they are not identical), whereas anti-hunters scored low on dominionistic values but high on humanistic and moralistic values.

Manfredo et al. (2009; see also Kluckhohn 1951) refer to values that characterize a culture as **value orientations**. Value orientations are a form of ideology. For example, the authors suggested that Navajo are oriented toward harmony with nature whereas Mormons are oriented toward mastery over nature. Two value orientations are particularly important for understanding human relations with wildlife in North America. One is a **doministic value orientation**, in which human well-being is seen as a higher priority than the well-being of wildlife. Doministic value orientation leads to acceptance of lethal or intrusive methods of control and utilitarian evaluation of methods. The other is a **mutualistic value orientation**, in which wildlife are seen in trusting relationships with humans and are perceived as having rights like those of humans as part of an extended family. Individuals with mutualistic orientations oppose lethal or intrusive control methods and tend to see wildlife from an anthropomorphizing view. Manfredo et al. (2009) thought doministic orientations were important during colonial times in North America, during the period of western expansion, and during the industrial revolution. The researchers believe that the mutualistic orientation is now, during the postindustrial era, regaining prominence in North America.

BELIEFS. **Beliefs** are "judgments about what is true or false" (Allen et al. 2009, 5). "Knowing" what is true and false can come from many sources, such as scientific information, feelings, intuition, or cultural norms (Allen et al. 2009). Beliefs can be found in individuals or groups. For practitioners resolving human-wildlife conflicts, "knowing" can be based on experience as well as peer-reviewed scientific information. An experienced feral swine trapper, for example, has an intimate personal knowledge and understanding of the habits and behaviors of these swine. This is known as "local knowledge" or "traditional ecological knowledge" if it passes from generation to generation.

Everyday beliefs can go wrong in at least three ways (Babbie 2001). First, the strength of one's commitment to a belief may be so strong that contrary evidence is ignored. Strong religious or political ideologies or family loyalties are examples. Second, one's analyses may be too casual, often relying on anecdotal evidence. Third, there may be no clear link between observations and explanations.

Beliefs, though they may not be correct, can nonetheless be culturally defined (Romney et al. 1986). For example, 62.3% of respondents from villages in the Tunduru, Liwale, Nachingwea, and Masasi districts of Tanzania believed that humans in their villages were attacked and killed by ghost lions, not real lions (Nyahongo and Røskaft 2011). The villagers believed the ghost lions were made by witch doctors for revenge. Casual observations about the lions and the circumstances surrounding their appearances supported their being ghosts.

Many societal values and beliefs are rooted in religion. All major religions view animals as sentient and therefore foster humane treatment. Each religion teaches beliefs on how humans relate to other animals, communities, and ecosystems, and these in turn affect how human-wildlife conflicts are perceived and contribute to or constrain allowable management practices. Intertwined with religious beliefs are those values and beliefs provided by society as norms, ethics, and culture. **Ethics** are sets of moral values defining good and evil, right and wrong, by which people behave. **Culture** is the set of shared values, beliefs, and behaviors that characterize an institution, organization, or group. Understanding such collective values and beliefs can be important for practitioners resolving human-wildlife conflicts. For example, in pantheistic cultures in which humans believe the divine exists in all living creatures and some nonliving things, a direct attack on wildlife can also be a direct affront on God (box 14.1). If animals are believed to possess souls that survive death, these animals might take revenge directly or through a living animal relative. Persons holding either belief are likely to choose nonlethal methods of management.

In the 1980s, Cornell University's Human Dimensions Research Unit developed a scale for assessing human attitudes toward wildlife (i.e., a wildlife attitudes and values scale, or WAVS). Applying the WAVS to a variety of human and wildlife situations in New York State and synthesizing the information, the research unit (Decker et al. 2002) related attitudes to basic beliefs. Although the beliefs fell into four basic categories, those relating to problem tolerance (i.e., **human tolerance**, beliefs surrounding acceptable risks associated with wildlife) seem particularly important to human-wildlife conflicts. For example, Decker et al. (2002) pointed to a trend analysis by Butler et al. (2001) involving WAVS analyses of New York residents between 1984 and 1996. The analysis shows a declining tolerance for problems related to wildlife among both rural and nonrural residents. Mcgovern and Kretser (2015) used WAVS to predict support for the recolonization of mountain lions in Adirondack Park, New York.

ATTITUDES. **Attitudes** are "learned tendencies to react favorably or unfavorably to a situation, individual, object or concept" (Allen et al. 2009, 5). Predicting support or opposition to actions surrounding human-wildlife conflicts, attitudes can be modified more rapidly than values or beliefs by such factors as education and experience.

In a classic survey presented to the US Fish and Wildlife Service, Kellert (1982b) showed many insights into the attitudes of the public regarding wildlife damage and its management. Only 27% of the public were aware of issues surrounding the killing of live-

BOX 14.1 Karni Mata Temple

Karni Mata Temple is located in Deshnoke, India, about 30 miles south of Bikaner. The following summarizes information on the temple from several internet sources. Karni Mata lived in the 1400s and was believed to be the incarnate of Durga, a highly revered Hindu goddess. According to believers, the following is true. Karni Mata lived for 151 years and performed many miracles throughout her life, even as an infant. She disappeared into a divine light, altered the hand of an aunt, saved the lives of her father and others by rendering harmless otherwise toxic snake venoms, fed many soldiers from a single small pot, and provided water for multitudes on several occasions.

When a son-in-law drowned and the body was brought before her, Karni Mata remained in a closed cave with the body for three days. On the fourth day, the resurrected son-in-law left the cave. "Kabas" (ratlike beings, a reincarnate step between humans) began emerging from the cave, and subsequently they were born and lived in the Karni Mata Temple. When Karni Mata facilitated rebirth of her son-in-law, she also changed the natural laws. All humans had to be reborn as Kabas before they can be born again as humans; the Kabas may comingle with their human relatives under the auspices of the temple. The Kabas never leave the temple, complete their life cycles within it, and the temple itself provides their sustenance.

For the wildlife conflict practitioner charged with managing rat problems in this part of India, it matters little whether the story of Karni Mata is true. The belief matters, and it must be respected and considered if any strategy to control rats (not the Kabas) in the area is to succeed. Should you find yourself as a professional wildlife specialist in just that situation, and should you accidentally kill a Kaba, your transgression can be rectified. You need only provide a likeness in size and shape of the Kaba that you killed . . . and, oh, yes, it needs to be made of pure gold!

stock by coyotes, and 52% were unaware. Regardless of awareness, the public opposed trapping and poisoning for coyote management but favored hunting offending individual coyotes. Livestock producers favored trapping and poisoning over hunting. The public and livestock producers opposed the use of public funds to compensate producers. Respondents from concentrated populations over 1 million opposed (64%) shooting or trapping coyotes, whereas respondents from populations under 500 favored (56%) the methods. The general public disagreed with shooting (61%) or poisoning (about 90%) golden eagles if they were killing sheep, whereas livestock producers (81% for sheep producers and 72% for cattlemen) favored shooting the eagles. The public (about 65%) also opposed poisoning blackbirds that were affecting crops, assuming that some unintended (but not endangered) animals would also be poisoned. However, the general public supported (71% informed and 77% uninformed) killing rats to protect agriculture even if some unintended animals were killed.

In general, Kellert found that persons holding utilitarian attitudes (later, *typologies*) supported more obtrusive management methods, including lethal control options, than persons holding humanistic attitudes. Persons with utilitarian attitudes were predominantly rural, had limited mobility and education, and were older males with lower incomes who had direct experience dealing with wildlife damage. Although these findings were based on studies from North America, Treves (2009) pointed to similar findings from studies in other countries.

The society in which one lives and its culture—and through it the ethical standards and norms of the society—can strongly influence one's attitudes about management of human-wildlife conflicts. Individual factors such as sex, age, income, and education can also affect attitudes. People having experienced a conflict tend to take a stronger interest in resolving it than people more removed from the problem. Thus, farmers adjacent to wildlife parks in Africa are often less tolerant of damage caused by elephants than those individuals less affected by the damage, just as persons experiencing damage from prairie dogs or geese are less tolerant of the animals than those who have not experienced it.

Attitudes are closely related to subsequent behavior (Manfredo and Bright 2008). Attitudes can predict behavior, but general attitudes do not predict specific behaviors. In wildlife damage management, for example, a generally positive attitude about the use of traps would not necessarily predict support for trapping feral cats. Salient beliefs—those that come to mind without prompting—are more likely to influence attitudes than those that are nonsalient. Strong attitudes—stable ones that remain relatively unchanged over time—are more influential than weak

ones in guiding behavior. Attitudes also may be extreme, ambivalent, or carry some other level of conviction. For example, Manfredo and Bright (2008) found that two people may agree that shooting from a helicopter is good for controlling coyotes. One, however, may see the method as excellent whereas the other may find it only acceptable. Here, the more predictable behavior would be by the person seeing the method as excellent (the one with the extreme attitude). Attitudes may be centrally embedded (or not) within a broader base of beliefs. Each of these factors influences the strength of the attitude, its likelihood of being changed, and its likelihood of affecting one's behavior.

Attitudes underlie behaviors important to human-wildlife conflicts, such as taking a management action or voicing support or opposition to such an action. Often the individual evaluates her or his own attitude in relation to appropriate group norms before taking any action (Allen et al. 2009). Attitudes also can be influenced and changed by factors such as education and experience. Therefore, educational programs, including those provided by extension and natural resource agencies, play important roles in ensuring that accurate and objective information is available when individuals interested in human-wildlife conflicts are receptive to the information, with subsequent changes in attitude. Values may also be influenced by education during early formative years. Project Wild (United States), Project Wet Foundation, Project Learning Tree, 4-H Youth Development and Mentoring Organization, the World Organization of the Scout Movement, and World Association of Girl Guides and Girl Scouts are examples of international environmentally oriented youth educational programs that already have ties with natural resource agencies and extension programs and could serve as conduits for furthering education of youth in human-wildlife conflicts and their management.

RISKS. Beliefs involving **risks** can be particularly important in forming human attitudes about human-wildlife conflicts. Here, the probabilities of occurrence of damage and severities of the consequences (measured as level of the emotion dread) are important (Decker et al. 2002). A cougar attack, for example, may be unlikely but carries a high level of dread. Deer damage to rural gardens may be likely but carries low levels of dread. Other insights relating risk to attitudes include the following: tolerance for risk decreases as probability of the occurrence increases, perceptions of risk rather than objective risk assessments often drive peoples' actions, people accept greater risk if they assume it voluntarily, perceptions of risk are elevated when consequences can be severe or are not distributed equitably, risks to children are less tolerated than those to adults, and perceptions of risk decrease when associated benefits are understood (Decker et al. 2002). As summarized by Madden (2004), public outcry is often not proportional to actual damage but rather to the perceived risk and lack of control of managing a solution.

Perception of high risk may not always drive one's attitude. For example, Zinn and Pierce (2002), in a survey of wildlife value orientations of Colorado metropolitan residents toward mountain lions, found that women perceived more risk than men from a mountain lion but were less willing than men to accept destroying the mountain lion. Storm et al. (2007) evaluated attitudes of exurbanites (people who live between suburbia and rural areas) around Carbondale, Illinois, toward white-tailed deer. Concerns for deer-vehicle collisions (low probability but high level of dread) were common among respondents (84%), but damage to ornamentals or fruit producing plants (high probability but low level of dread), not fear of collisions, determined tolerance for deer.

EMOTIONS. As already indicated, understanding **emotions** is also necessary to comprehend how one might respond to situations involving human-wildlife conflicts and their resolutions. Emotions, along with moods, constitute "affect," a general class of feelings that humans experience (Manfredo 2008, 51). Emotions relate to specific events and are short-lived and conscious, whereas moods are longer lasting and in the background of consciousness. Primary emotions include happiness, sadness, anger, surprise, disgust, frustration, and fear. Secondary emotions are combinations of primary ones. Emotions are probably rooted in the long evolutionary history of humans and are modified by culture.

In human-wildlife conflicts, emotions are sometimes used in attempts to effect changes in attitudes and to raise funds. Most of us, for example, have seen photographs of animals caught in traps on fundraising mailers and brochures of organizations that oppose trapping. Do these appeals work? According to Manfredo (2008), research suggests such appeals can be effective, but it depends on the context. For example, Zinn and Manfredo (2000) found that emotional appeals, including pictures of animals in traps, were more memorable but no more effective than rational appeals in swaying people on banning trapping in Colorado.

Dread, or extreme fear, is another emotion important to human-wildlife conflicts. Decker et al. (2010) recommended that "practitioners in the area of human

dimensions of large herbivore restoration and management should assess levels of fear of the animal in question to gain a deeper understanding of public attitudes regarding management efforts" (49). In one study, for example, respondents for the Hochsuerlandkreis region had significantly more negative attitudes and less knowledge of bison than respondents in the Siegen-Wittgenstein region of Germany. The researchers used logistic regression analyses to show the role of fear in forming the public attitudes. The differences led to a decision to restore bison in the region in which residents had more favorable attitudes.

Personal involvement with a wildlife problem tends also to increase its importance. For example, ranchers near Yellowstone National Park expressed stronger concern about the release of wolves than people more remote from the introduction. Likewise, people living around wildlife refugia in Africa, who are likely to suffer crop losses or injury, tend to have strong emotional involvement with issues surrounding human-wildlife conflicts.

Personal moral decisions involve direct (i.e., I am doing it) harm to another person or wildlife. Impersonal moral decisions involve someone else doing the harm. Whereas personal moral decisions stimulate areas of the brain tied to cognition and emotion, impersonal moral decisions stimulate areas tied to in-depth cognitive processing. The difference may help to explain, for example, why some individuals will not trap and euthanize a squirrel from an attic but have no problem paying a management specialist to do the job. Here, Manfredo (2008) sees great opportunity for examining human-wildlife relationships: "Are groups that differ on wildlife values or on wildlife issues more similar in their personal moral judgments than impersonal judgments? Are impersonal judgments more susceptible to cultural difference and culture shift? . . . Can we explore apparent inconsistencies in people's attitudes or wildlife value orientations based on this distinction?" (61).

Memories are more easily made as well as recalled when tied to a strong emotional state. For example, decision-making is facilitated by a positive emotional state wherein people tend toward more creativity and belief in success; decision-making is opposed by a negative affect where people tend to be more vigilant and analytic. In the context of wildlife management, Manfredo (2008) asks, "What forms of conflict resolution and stakeholder-engagement might be devised that explore the utility of creating positive affect as a base for more effective and lasting compromise?" (60).

Examples

Not convinced that VBAs are important to specialists of human-wildlife conflicts worldwide? Here are some excerpts from a 2003 newspaper article in *China Daily* entitled "Monkeys Terrorize India Workers, Tourists":

In a capital city (New Delhi) where cows roam the streets and elephants plod along in the bus lanes, it's no surprise to find government building overrun with monkeys . . .

But the officials who work there [have] . . . been bitten, robbed and otherwise tormented by monkeys that ransack files, bring down power lines, screech at visitors and bang on office windows . . .

A past initiative to scare off the army of Rhesus macaques with ultrahigh frequency loudspeakers didn't work. A plan to deport them to distant regions has stalled because local governments refused to have them . . .

There's an ape patrol of fierce-looking primates called langurs, led about on leashes by keepers. But whenever a langur looms, the pink-faced, two-foot-tall hooligans simply move elsewhere on government grounds . . .

"Please do not feed the monkeys," implores a sign at Raisina Hill, the complex of colonnaded buildings that includes the president's residence, Parliament, and Cabinet offices . . .

[But] Hindus believe that the monkeys are manifestations of the monkey god, Hanuman, and worshippers come to Raisina Hill every Tuesday handing out bananas.

So what has been the progress in resolving these conflicts since 2003? Menon and Kataria (2018), reporting for Reuters News Service, described monkeys continuing to run amok in India's government buildings in New Delhi. The problems were concerning enough that a group of specialists advised parliamentary members on how to prevent monkey attacks by avoiding direct eye contact and not walking between adults and their young. Accustomed to being fed by humans, the monkeys may attack if they are not fed on demand. The macaques chewed optic fiber cables strung along the Ganges River, "derailing" a plan by the prime minister to bring internet to his constituency in Varanasi in 2015. Another monkey gained international attention by taking a ride, along with other commuters, on a New Delhi metro coach. So the interactions between monkeys and people of New Delhi appear to continue, mostly unabated, today.

Thornton and Quinn (2009) evaluated the attitudes of residents in the urban fringe of Calgary in the province of Alberta, Canada, toward cougars. Interactions between humans and cougars have become more frequent as humans have moved into cougar habitat over the last several decades. The cougar population in the study area was high, about 4 per 100 km², and probably expanding, but predation on livestock was uncommon. In the study area, human growth was driven mostly by a need for more rural residential subdivisions. The researchers observed that residents living in cougar habitat sometimes felt anxiety and fear from perceived risks to children, pets, or themselves (Riley and Decker 2000; Teel et al. 2002).

Using a Likert-like seven-point scale, the researchers mailed a survey with 37 closed-end questions to 1,508 randomly selected residents in a stratified design. The survey addressed four areas: attitudes and beliefs about wildlife, including hunting, wildlife rights, education, and government participation; attitudes about cougars, including knowledge and beliefs, measures of risk, and government participation; attitudes about cougar management, ranging from preventing problems to killing problem cougars; and demographic information. Twenty-nine percent of those receiving the surveys responded. The researchers assumed cougars were not of particular interest to the nonrespondents.

About 154 (36%) of the respondents owned cattle or horses, and about one-quarter of these respondents reported livestock losses due to cougars. Most (78%) indicated willingness to adjust husbandry to reduce predation. Overall, about 40% of respondents said they or family members had observed a cougar in the wild. About 43% said that the presence of cougars increased their quality of life, and 65% believed cougars were an acceptable threat to livestock (65%) and humans (54%). Most participants (72%) accepted hunting of cougars and other wildlife if it did not impact the sustainability of the targeted population, and most (87%) believed presence of cougar did not affect opportunities for hunters. Most respondents (71%) wanted to be more involved in government decision-making, most believing wildlife management lacked sufficient public participation. Most respondents (65%) did not feel at personal risk from cougars, and most felt they could accept risks and coexist with the cougars.

Overall, the male respondents tended to accept hunting, and female respondents were more protective of cougars. However, the women felt more at risk and expressed greater reluctance to go into wild areas known to have cougars. Relocation was the most common management action preferred by respondents when a cougar stalked a country skier, injured a hiker on a trail, charged and knocked down a person on a trail (but then left), or killed a person on the trail (but had cubs), wandered repeatedly around a neighborhood, or attacked and killed one or more pets. However, killing the offending cougar was preferred by the respondents if the cougar had attacked and killed a person in the neighborhood, had killed a person on a trail and had a history of aggression, or had killed a person on the trail (with no other information available).

The potential for conflict index (which uses values between 0 and 1, wherein 1 represents greatest conflict and 0 least) analyses showed the highest within-population agreement (0.022) for the statement that it is important to maintain cougar populations for future generations and greatest disagreement (0.52) for the statement "I believe that cougars would attack a human without being provoked." Relatively strong disagreement (0.42) was also found for the statement "Cougars are an acceptable threat to humans."

The information provided in these surveys can be helpful to human-wildlife conflict managers. In this study, the respondents indicated interest in public participation in the management process. The survey showed that respondents who had the least knowledge about cougars and their management—such as the female respondents and respondents newly relocated from urban areas—also had the greatest fear. The researchers suggested that these groups be targeted for educational programs. Thornton and Quinn (2009) also suggested that the tendency toward relocation as the management method of choice might also indicate a "not in my backyard" mentality, with the actual acceptance of cougars being less than respondents implied.

Conspiracy theories provide examples of how beliefs can go wrong with sometimes worldwide impacts. Bale (2007) describes conspiracy theories as "elaborate fantasies that purport to show that various sinister, powerful groups with evil intentions, operating behind the scenes, are secretly controlling the course of world events." Such theories may have the advantage of bringing public attention to an issue, but they can also foster prejudice, genocide, affect health-related choices such as using contraception and being vaccinated, and they can be used as unsupported arguments against science- or other expert-based information on topics such as climate change and genetically modified foods. In many countries, including the United States, conspiracy theories are linked to political activities and behaviors, often fostering a distrust of government.

COVID-19 is seen by the World Health Organization (WHO) not only as a pandemic but also an infodemic ("an over-abundance of information—some accurate and some not—that makes it hard for people to find trustworthy sources and reliable guidance when they need it"; World Health Organization 2020). Fueled by a social media unregulated in many countries, the spread and management of the COVID-19 pandemic worldwide has been strongly influenced not only by a deep and somewhat fortuitous scientific understanding of coronaviruses and technological advances in vaccination but also by conspiracy theories, sometimes amplified rather than disputed by governmental or political leaders. Although technological advances have facilitated a rapid development and deployment of effective vaccines, the conspiracy theories have worked to thwart public health efforts to manage the virus.

In countries where human responses have been evaluated (e.g., Pummerer et al. 2021 in Denmark; Sallam et al. 2020 in Jordan; Marinthe et al. 2020 in France), conspiracy theories have been found to negatively affect the willingness of people to follow scientifically based practices that minimize transfer of the virus (e.g., Bierwiaczonek et al. 2020), including practicing social distancing, undergoing medical diagnoses such as sampling with nasal swabs, and getting vaccinated (WHO 2020). Although the actual cost of these theories to the successful management of the virus is unknown, the landscape level of the beliefs in many regions of the world (e.g., about one-third of US citizens have been shown to believe some COVID-19-related conspiracy theories) as well as the gravity of the behavioral responses suggest an enormous toll has already been taken in lives lost to the coronavirus.

Transparency by governmental agencies, promotion of scientifically based and accurate information by media and social media platforms (WHO 2020), and education in critical thinking have been suggested as tools that might be used to discourage future beliefs that are founded in misinformation and disinformation.

Summary

- Social values, or conceptions of what is good or bad, have a strong genetic base, form early in life, and change little later in life. Values influence group norms and underlie how people view human-wildlife conflicts and their management.

- People with doministic value orientations see humans as being a higher priority than wildlife and accept lethal and other intrusive methods of control. People with mutualistic value orientations see wildlife in a trusting relationship with humans and oppose intrusive methods.
- Beliefs are judgments of what is true or false. Beliefs, whether based on facts or not, strongly influence how people view human-wildlife conflicts and their management.
- Values and beliefs may be rooted in religion. All religions view animals as sentient and foster humane treatment. Intertwined with religious values and beliefs are societal norms, ethics, and cultural or shared values.
- Conspiracy theories, such as those spreading during the COVID-19 pandemic, provide examples of how beliefs can go wrong with sometimes worldwide impacts.
- Attitudes are tendencies to react favorably or unfavorably to a situation, and they can predict support for or opposition to actions surrounding human-wildlife conflicts. Attitudes can be modified by education and experience.
- Levels of risk and emotions influence attitudes toward human-wildlife conflicts. Perceptions of risk, not necessarily actual risk, often drive people's actions regarding human-wildlife conflicts.
- Emotions such as happiness, anger, or disgust relate to specific events and are short-lived. They are sometimes used to persuade people about changes in attitudes or to raise funds.
- Personal involvement and frustration with a human-wildlife conflict tend to increase its importance to a person. Those not suffering from or exposed to the conflict may have negative or nebulous attitudes toward its management.
- Moral decisions, such as taking a personal management action rather than having someone else do it, stimulate the emotional parts of the brain.

Review and Discussion Questions

1. How might a person's genetic composition influence her or his attitudes toward human-wildlife conflicts and their management? Summarize scientific literature supporting your view, including references.
2. Relate Kellert's typologies to value orientations provided by Manfredo et al. (2009). Which typologies and value orientations seem particularly important to human-wildlife conflicts and their management? Explain, with examples.
3. Choose a recent publication on a human-wildlife conflict that would have benefited by the addition of a study on the VBAs of stakeholders. Reference the study. Describe the information needed on VBAs and how the information may have helped the study.
4. What value orientations might lead a community such as Town and Country, Missouri, to recommend translocation of overabundant deer rather than bow hunting by sharp shooters (Beringer et al. 2002)? How might one test for such value orientations? How might one attempt to influence the orientation?
5. Are people in North America shifting from a doministic value orientation to a mutualistic value orientation? State yes or no, and justify your response with examples and references. Explain how such a shift would influence the future of human-wildlife conflicts and their management.

15

Politics and Public Policy

In the world of politics, the needs of practitioners can be facilitated by empathetic individuals and groups or thwarted by those opposed; attention and resources can be quickly brought to bear on human-wildlife conflicts from the local to international levels.

Statement
Human-wildlife conflict practitioners are tied to the political worlds of the societies within which they live and work, and the final decisions about how human-wildlife conflicts are addressed are often political ones.

Explanation
Dye (2008) describes **public policy** as "whatever governments choose to do or not to do" (1). In general, things they choose to do are to regulate conflicts within society; organize conflicts with other societies; distribute rewards and material services to members of society; and extract money from society, mostly as taxes. Public policies are founded in constitutions, legislative acts, and judicial decisions and are embodied in actions, regulations, laws, and funding priorities.

Because wildlife is a public natural resource in many countries and is often managed under the stewardship of public organizations, the resolution (including its private-sector aspects) of human-wildlife conflicts is strongly influenced by public views and opinions.

Public policies are made or used when concerns are successfully raised about a conflict or how the conflict is being managed. For example, concerns for damage to ecosystems or to the commercial fisheries industry by zebra mussels may lead to policies or regulations on movement of ships into ports or through canals, regulations on how ballasts are decontaminated or released, or conditions under which recreational boats can be moved from one lake to another within a local area.

Concerns surrounding human-wildlife conflicts are often based on one or more of these four factors: method effectiveness (i.e., is the method doing what is intended, with benefits that outweigh the costs?); biological or ecological soundness (i.e., are the methods based on sound scientific principles and concepts?); animal welfare or **humane-**

ness (i.e., have efforts been made to minimize the pain and suffering of impacted wildlife?); and animal rights (i.e., have legal and other rights of impacted wildlife been considered?). For instance, individuals or groups may believe the use of steel jaw traps to capture coyotes is unnecessarily painful and oppose use of these traps. Or individuals or groups may believe the use of repellents to reduce coyote depredations on livestock is ineffective and want policies or regulations that allow use of more effective methods.

Successful efforts to deal with the concerns result in new or amended public policies. The processes used to develop new public policies or to amend existing policies vary by type of government (e.g., authoritarian, monarchic, democratic, theocratic, totalitarian) and legal system (i.e., civil, religious, or a combination of these). For example, the National Crop Protection Center in Laguna, Philippines, was established by Presidential Decree No. 936, in May 1976, (after martial-law era) as part of the country's efforts to become self-sufficient in rice production. The center was established to offer scientific information on integrated pest management, including human-wildlife conflicts.

Whereas individuals—e.g., elite such as high-ranking politicians or administrators—can sometimes influence overall public policies, changes come mostly from efforts of the masses (although these may also be initiated by the elite, as we will see) organized as groups. Organized people can gain power and regulate the behaviors of others.

Groups, sometimes based on common values, beliefs, and attitudes (VBAs; see chapter 14) can be formed at any level of society. For instance, some groups whose interests include human-wildlife conflicts are organized to align with governmental levels and units (table 15.1). An illustration of this is the Columbia Audubon Society, aligned with the city of Columbia and one of 11 chapters of the National Audubon Society located in Missouri. Other groups interested in human-wildlife conflicts are organized according to trade or profession. Examples in the United States include the National Sunflower Association and The Wildlife Society. Yet other groups, both national and international, are organized around advocacy for the sustainability of natural resources or wildlife, such as the National Fish and Wildlife Foundation and the National Wildlife Federation. Still others focus on species or other taxa, such as Bat Conservation International and Partners in Flight.

Other groups are organized around specific issues such as animal welfare or animal rights. Advocates of **animal welfare** believe that human-wildlife conflicts

should be managed in a manner that minimizes the pain and suffering of individual animals. They support responsible treatment of animals, whether they be wild, farm, laboratory, or companion animals. An example is the Animal Welfare Institute. Advocates of **animal rights** believe that animals should have legal and other rights equal to those of humans. Some animal rightists believe that animals should never be used for meat, leather, furs, or kept as property, in zoos or circuses, or as pets; they are generally opposed to managing human-wildlife conflicts. An organization representing this category in the United States is PETA, People for the Ethical Treatment of Animals (Wywialowski 1991).

Further, committees and subcommittees are often found within groups, and some may have particular interest in human-wildlife conflicts. One example is the Wildlife Damage Management Working Group of The Wildlife Society. At any given time, any number of groups at any level of various organizations may have an interest in a specific human-wildlife conflict such as coyote predation on pets in a local suburb or European lynx predation on sheep. Indeed, it is the wise human-wildlife conflict manager who assumes such interest exists and adopts transparent policies (i.e., open communication, including all interested individuals and groups from planning to action and evaluation of results) for the formulation and execution of any action surrounding a human-wildlife conflict. In many countries, such transparency is implicit in laws overseeing any governmental actions involving such conflicts.

To exert social pressure, groups often form **coalitions**—agreement between two or more parties to advance common goals and secure common interests—that focus on specific conflicts. Such coalitions often fit into the general concepts such as pluralism (policymaking and decision-making lie within the government but can be influenced by many nongovernmental groups) and models such as group theory (government maintaining a balance between interests of different groups) formulated by political scientists (Dye 2008). One model that applies is the **iron triangle**. In the United States and at its national level, points of power in the triangle are relevant positions and units (depending on the specific issue) of the executive branch (e.g., the Secretary of Agriculture and the Deputy Administrator of Wildlife Services who administers the program; or the director of the National Institute of Food and Agriculture, who administers the federal extension program); relevant positions and units in Congress (e.g., Agricultural

TABLE 15.1   *Examples of groups that sometimes serve as beneficiaries in human-wildlife conflicts*

GOVERNMENT UNIT ORGANIZATIONS, US

Oktibbeha Audubon Society, Starkville, MI
Columbia Audubon Society, Columbia, MO
Michigan Audubon Society
Missouri Audubon Society
National Audubon Society

REGIONAL ORGANIZATIONS, US

SE Association of Fish & Wildlife Agencies

TRADE OR PROFESSION, US

American Fisheries Society
American Sheep Industry Association
American Society of Mammalogists
American Veterinary Association
National Cattlemen's Beef Association
National Cattlemen's Foundation
National Sunflower Association
The Wildlife Society

TRADE OR PROFESSION, OTHER COUNTRIES

Australian Society for Fish Biology
Chinese Veterinary Medical Association
European Deer Farmers Association
Federation of Livestock Farmers Association of Malaysia
National Sunflower Association of Canada
Wildlife Conservation Society, Nepal

NATURAL RESOURCE OR WILDLIFE ADVOCACY, US

National Audubon Society
National Fish & Wildlife Foundation
Natural Resources Defense Council
National Wildlife Federation
Sierra Club
Teaming with Wildlife

NATURAL RESOURCE OR WILDLIFE ADVOCACY, OTHER COUNTRIES

African Wildlife Foundation
Conservation International
Convention on International Trade in Endangered Species in Wild
    Fauna and Flora
Food and Agricultural Organization of the United Nations
International Union for Conservation of Nature
World Organization for Animal Health (OIE)
World Conservation Network
World Wildlife Fund

TAXA, US AND OTHER COUNTRIES

Bat Conservation International
Ducks (Quail, Trout) Unlimited
International Crane (Osprey) Foundation
International Association for Bear Research and Management
National Wild Turkey Federation
North American Wolf Association
Partners in Flight
Pheasants Forever
Purple Martin Conservation Association
Raptor Research Foundation
Rocky Mountain Elk Foundation
Royal Society for Protection of Birds
Timber Wolf Information Network
Whale and Dolphin Conservation Society
Whale Conservation Institute
The Wolverine Foundation
Wolf Recovery Foundation
World Center for Birds of Prey
World Owl Trust

ISSUE—ANIMAL WELFARE, US

American Society for Prevention of Cruelty to Animals
Animal Protection League
Animal Welfare Institute
Defenders of Wildlife
US Humane Society

ISSUE—ANIMAL RIGHTS, US

Animal Liberation Front
Animal Political Action Committee
Animals in Politics
Culture and Animals Foundation
Friends of Animals
Greenpeace
People for the Ethical Treatment of Animals
Trans-Species Unlimited

Appropriations Committee members and many individual senators and representatives); and lobbyists (see table 15.1 for examples). According to the iron triangle concept, policies affecting human-wildlife conflicts at the national level are debated, legislated, implemented, administered, and monitored within the triangle. Analogous triangles occur at all levels of government and in other countries, and legislation is often transferred to other levels or other countries.

Ogden (1971) expanded the concept of iron triangle to a "web of power" that included more groups as play-ers. He described the web as "power clusters" and said it exists in every area of public policy. With power clusters, groups act independently but come together on specific issues that affect their interests at local, state (province, etc.), and national levels. A power cluster has six elements: administrative agencies, legislative committees, special-interest groups, professionals, an attentive public, and a latent public that might be aroused to act on a given issue. Power clusters include behavioral patterns that help shape policymaking: close personal and institutional ties with key people, active communication among cluster elements, internal conflicts among competing interests within clusters, internal cluster decision-making, and a well-developed power structure. For example, Sims (1995) describes the efforts to make available the M-44 and the Livestock Protection Collar in Texas following the federal ban on use of predacides. This statesman mentions the involvement of another US congressman, a Texas state representative, the Texas Department of Agriculture, an extension predator specialist, an inventor, US Wildlife Services (then Animal Damage Control), and the Texas Department of Agriculture. Because Sims was not attempting an exhaustive listing, the power cluster may have included additional individuals and groups.

Other models in political science also apply to managing human-wildlife conflicts. Dye (2008), for example, describes a theory of elites in which policies focused on human-wildlife conflicts come from the interactions of individuals and groups organized as a pyramid. At the bottom of the pyramid is a generally apathetic (but latently powerful) public. Moving upward, there exists a public interested in the issues, a public both interested and active, officials and administrators who have power and responsibility to shape policy, and a few elites (heads of major corporations, associations, etc.) at the top of the pyramid who determine public policy. With this model, groups and the general public have little actual influence on the formation of public policy.

My sense is that none of the models applies exactly but that each—particularly the power cluster model (I mostly use this model for the rest of the chapter) and the elite models—offers insights into how public policy and laws on human-wildlife conflicts are formed in the United States and probably in other countries as well.

How do these groups, individually or as power clusters, affect changes in national public policies? Public policymaking in the United States can be seen as a process with the following steps (fig. 15.1):

- Problem identification, wherein societal problems are publicized, and action is demanded by mass media, interest groups, citizens' initiatives, and the public
- Agenda setting, wherein issues that will be addressed by government are decided by elites, Congress, elective candidates, and mass media
- Policy formulation, wherein proposals to resolve issues are developed by think tanks, the president and executive offices, congressional committees, and interest groups

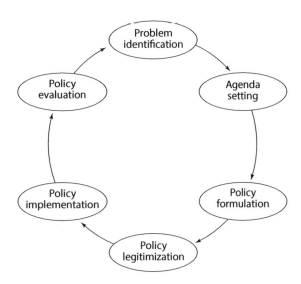

Figure 15.1 Steps in formulating federal public policy in the United States. Illustration by Lamar Henderson, Wildhaven Creative LLC.

- Policy legitimization, wherein a proposal is selected, political support is developed, the proposal is enacted into regulations or law, and its constitutionality is decided by interest groups, the president, Congress, and the courts
- Policy implementation, where departments and agencies are organized, payments or services are provided, or taxes are levied by the president, White House staff, and executive departments and agencies
- Policy evaluation, wherein government programs provide reports, policies are judged, and changes and reforms are proposed by executive departments and agencies, congressional oversight committees, the mass media, and think tanks

Although this description indicates an orderly, stepwise process, the actual process is usually disorganized, with steps occurring in various orders or simultaneously.

Individuals or alliances of groups such as power clusters initiate the process and may be involved in every step. An example is the move of Wildlife Services from the US Fish and Wildlife Service (USFWS) into the US Department of Agriculture (USDA) in 1985 (see the Examples section at the end of the chapter). Groups within the power cluster were involved with the transfer of the unit from its inception, and the power cluster retains active involvement with the unit today.

In their guide to community-based deer management, Decker et al. (2004) provide case studies of power clusters formed at community levels to raise and resolve issues surrounding overabundance of deer. Individuals and groups play varying roles in the policymaking process, which often leads to consensus on a management action rather than a new regulation or law. In Bedford, Massachusetts, for example, a deer manager specialist for the Division of Fisheries and Wildlife both made the community aware it had a problem and advocated for hunting as a solution. In Gettysburg National Military Park, the process, including the development of an environmental impact statement, led to a decision to reduce the number of deer in the park by shooting. The park had previously not allowed hunting within its borders.

Individuals, power clusters, and processes organize to impact managing human-wildlife conflicts in other countries in ways particular to that country. For example, regulations and laws in countries such as Austria, Germany, Switzerland, and the Czech Republic allow individual owners to be compensated for damage to specific forest tree species caused by specific pests (Schaller 2007). These laws were developed through a power cluster that included landowners who hold hunting rights and lease hunting privileges, governmental units who regulate hunting (at the federal level) and enforce complaints law (at the local level), and hunter associations such as the Deutscher Jagdschutzverband.

In the United Kingdom, the Department for Environment, Food and Rural Affairs (DEFRA) described the framework within which public policy is made for resolving human-wildlife conflicts (Government of the United Kingdom 2010). The steps reflect those shown in figure 15.1. DEFRA described the framework as a "process tree." The first step is problem identification (i.e., deciding whether government intervention is justified), followed by informing (gathering and presenting scientifically based information), deciding (choosing which action among options), communicating (informing stakeholders—i.e., persons who are affected by or can affect a problem or its management—of the decision and the reasons for it), implementing, and monitoring. DEFRA cited management of inland great cormorants (*Phalacrocorax carbo*) as an example. Great cormorants compete with anglers and fish farmers who then seek ways to reduce economic losses. DEFRA turned to data from Wetlands Bird Surveys for estimates of cormorant populations and used a model to relate different levels of licensing (i.e., numbers of kill permits) to effects on overall population sizes. By running thousands of iterations, uncertainty in projected future growth of cormorant populations was reduced. Then, by issuing licenses, monitoring population responses, and following an adaptive resource management approach (see chapter 16), the number of licenses could be adjusted annually to changing circumstances. How effectively has this specific process worked? At least one stakeholder is not satisfied. In November 2020, the Angling Trust (2020) argued that the process continues to be too bureaucratic and that DEFRA is unwilling to issue sufficient numbers of permits to adequately manage a still burgeoning cormorant population.

Generally, how effectively does the public policy process work? This is where the logic of the system can pose a challenge. Although the processes just described serve as the norm for developing many public policies, deception (wherein one is given only part of the whole story or the story itself is deliberately misconstrued, hence the term "spin") and expediency (wherein a politician may take the easiest route to re-election rather than the one in the public's best interest) are also part of political life, including both politically based groups and individual politicians.

Whereas most public agencies and politically oriented advocacy groups encourage transparent actions, political methodology sometimes favors sharing only as much information as is needed for a participant to do his or her part, without the participant knowing the full agenda. Further, information may not only be intentionally incomplete but be written with a spin that is intended to mislead. An example is an animal advocacy group promoting a short-term goal for its membership of reducing meat consumption for health and environmental reasons, when the actual long-term goal of the leadership is to ban all meat consumption everywhere, even hamburgers at McDonalds. Ballots or referenda are sometimes intentionally written so that the public is not sure exactly what a "yes" or "no" vote means.

Political campaigns supporting referenda may also include photographs or videos appealing to emotions rather than reason, intending to arouse the latent public and to persuade. For example, a campaign to ban traps as "cruel" may show photographs of animals caught in traps but fail to provide actual information on the humaneness of the targeted trap designs. The campaign might also fail to contrast the relative "cruelty" of the traps with methods left for practitioners if traps are banned. The intent of the campaign is to get public support and votes for the ban, not necessarily to provide objective information.

In these (as in all) situations, it is left to the voter to gather objective information and to make informed assessments and decisions regarding the proposal. The campaigners know that many voters are likely to respond emotionally or to see through the filters of existing ideologies, and they will not bother with more rigorous and time-consuming analytical assessments.

In the United States, bills (i.e., proposed laws) unpopular with the public but important to lobbyists (and sometimes for the re-election of the involved officials) may be attached as riders to publicly popular but unrelated bills. The unpopular rider (which would not be passed on its own) gets passed on the wings of a popular bill, another form of deception. Thus, a rider attached to a US congressional federal budget bill in 2011 returned management of wolf populations to the states of Idaho and Montana, sidestepping the checks and balances that are an intended part of the delisting process under the Endangered Species Act (Kramer 2011).

Political expediency can sometimes short-circuit not only the political process but also the scientific and analytical evaluation of proposed solutions. For example, a recent decision to ban all congressional directives in the United States won votes and gained the support of some constituencies. The "one-size-fits-all" ban, however, eliminated scientific and professional programs in wildlife damage management that included those addressing problems with feral pigs, rodent damage to sugarcane, and academic educational programs. An analytical process that evaluated the need and worth of individual congressional directives would have allowed the separation of "pork" from other contents of the barrel, but this approach was not politically expedient.

While discussion thus far has focused on the United States, the same strengths and problems are seen in other countries, although the details become idiosyncratic. For example, New Zealand farmers distributed rabbit hemorrhagic disease, short circuiting a process of federal evaluation, even as the government of New Zealand determined that the risk was too great for introduction. And fisher people in the Galápagos thought it in their best interests to hold hostage an invasive species eradication program to gain more fishing permits (fig. 15.2).

Once made, public policies set the boundaries within which human-wildlife conflicts are managed. That is, the conflicts are assessed and scientifically investigated, and their management is conducted under the authority of, and within, public policies

Figure 15.2 Threats from fishermen in the Galápagos to restock goats on goat-free islands if demands for increasing fishing licenses are not met. Photo by C. J. Donlan, with permission.

from local to federal levels in all countries where wildlife conflicts occur and are managed. For many countries, the laws, regulations, and policies include those governing the management of natural resources such as wildlife, air, and water; the activities of resource management agencies and the private sector; methods used in managing human-wildlife conflicts; and redress for citizens who believe they have been injured by damage or its management. For example, public regulations and laws often determine the types of oral vaccinations that can be used to manage rabies in Europe and North America, the nature of delivery of the vaccinations (e.g., by aircraft or by hand), and the areas treated.

In the United States, management of human-wildlife conflicts is governed at the national level by many specific laws and their amendments (table 15.2). These laws may be complex in how they are applied. The Federal Insecticide, Fungicide, and Rodenticide Act (FIFRA), for instance, provides for registration, classification, and regulation of use of all pesticides in the United States (see table 15.2). Administration of FIFRA is the responsibility of the US Environmental Protection Agency (EPA). The intent of the legislation is to ensure that benefits derived from use of pesticides outweigh costs to the environment. Pesticides, as interpreted by the EPA, include any substance or mixture of substances intended for preventing, destroying, repelling, or mitigating any pest.

Registration of pesticides by the EPA is based on risk assessment by EPA experts using data provided by

TABLE 15.2   *Selected examples of public policies affecting human-wildlife conflicts in the United States*

Act	Year	Responsible agency	Impact
Airborne Hunting Act	1971	USFWS	Disallows shooting or harassing birds, fish, or other animals from aircraft except for certain reasons, including protection of wildlife, livestock, and human life.
Animal Damage Control Act	1931, 1973, 1987, 1991	USDA	Allows operational and research activity; allows cooperative agreements and exchange of funds.
Animal Welfare Act	1966	USDA	Regulates transport, sale, and handling of dogs, cats, nonhuman primates, guinea pigs, hamsters, and rabbits intended for research or other purposes; requires Animal Care and Use Committees, following USDA guidelines and procedures.
Endangered Species Act	1973	USFWS, NOAA	Prevents extinction of imperiled plants and animals; under Section 7, federal agencies must consult with USFWS to determine impact on imperiled species if any are likely.
Federal Insecticide, Fungicide, and Rodenticide Act (FIFRA)	1910, 1972, 1988	USEPA	Ensures benefits derived from use of pesticides outweigh environmental costs; decisions based on data provided by registrant, following Good Laboratory Practice (GLP) guidelines.
Fish & Wildlife Coordination Act	1934	USFWS	Requires all federal agencies coordinate efforts on environmental issues; authorizes Secretary of the Interior to minimize damage from overabundant species and to control losses from diseases or other causes.
Lacey Act	1900, 2008	USFWS, USDA	Regulates import and export of invasive species, listed as "injurious to human beings, to the interests of agriculture, horticulture, forestry or wildlife resource."
Migratory Bird Treaty Act	1918	USFWS	Coordinates management of game and nongame migratory birds through permitting; exempts some offending blackbirds, grackles, crows, and magpies.
National Defense Authorization Act	2008	DoD	Requires brown tree snake report to congressional defense committees (Section 314).
National Environmental Policy Act	1969	Council on Environmental Quality	Requires an evaluation of environmental risk before any major federal action is taken: usually done progressively as an Initial Environmental Examination, then Environmental Assessment, then Environmental Impact Statement.
Nutria Eradication and Control Act	2003	USDI	Authorizes funds for assistance to Maryland and Louisiana.
Safe, Accountable, Flexible, Efficient Transportation Equity Act		National Highway System	Allows funds for control of terrestrial noxious weeds and aquatic weeds.
U.S. Fish & Wildlife Conservation Act	1980	USFWS	Encourages all federal agencies to conserve and protect nongame wildlife and their habitats, authorizes some wildlife damage action to protect threatened or endangered wildlife.
Water Resources Development Act	2007	Department of Army	Authorizes barrier demonstrations to prevent dispersal of Asian carp, Upper Mississippi.

Abbreviations: DoD, Department of Defense; EPA, US Environmental Protection Agency; NOAA, National Oceanic and Atmospheric Administration; USDA, US Department of Agriculture; USDI, US Department of the Interior; USFWS, US Fish and Wildlife Service.

the proposed registrant. Data are gathered under study plans (protocols) approved by the agency using guidelines called **Good Laboratory Practices** (**GLP**; see below).

Let us assume the registrant is requesting registration of a new chemical toxicant that is formulated as a bait. Assuming the pesticide is a chemical bait, data are likely required (US EPA 2013) on some combination of the following:

- Basic chemistry, including descriptions of color, melting point, freezing point, reactivity, stability in air/soil/water, and product shelf (storage) life.

- Product chemistry, including active and inactive ingredients (see chapter 11).
- Product performance, including toxicity to "target" and other animals, including LD_{50} (the lethal dose that kills 50% of the population), LC_{50} (the lethal concentration when fed as a bait that kills 50% of the population), and dermal toxicity to skin and eyes; and field efficacy, possibly including both confined and open field studies.
- Data from studies that determine hazard to humans, domestic animals, and nontarget organisms including toxicity to other (nontarget) animals, including primary toxicity from feeding on the bait directly and secondary toxicity to prey or animals scavenging an exposed carcass.
- Environmental fate and residue chemistry.

Sprayed pesticides or antimicrobial products would have additional data requirements.

Laboratory and field studies must meet GLP standards to be accepted as data for consideration of registration. These requirements include a chain of custody documenting handling and storage of any materials used in the study; an archive of samples and all original data; documentation of qualification of laboratory and field participants; documentation of the conditions of storage of samples (e.g., that materials stored in a freezer were kept frozen without interruption); documentation of calibration of instruments used in the study; and establishment of approved protocols and standard operating procedures.

Work conducted under GLP guidelines can be very costly. For example, Fagerstone et al. (2008) estimated that total costs for the 48–60 data requirements typically required for registering a new active ingredient can be up to US$2.7 million. This may be a conservative estimate in today's dollars. In their evaluation of potential toxicants for managing mongoose (*Herpestes javanicus auropunctatus*) in Hawaii and US territories, Ruell et al. (2019) considered estimated costs for the federal registration of end products that could be applied below ground and in bait stations as well as aboveground for spot baiting and hand broadcasting. For all uses including the aboveground applications, which require more data for evaluation of risk, the scientists estimated a federal registration cost of at least US$172,000 for one product whose active ingredient is already registered in other products (assuming the current registrant will share the data) to a maximum of US$6.7 million for a product whose active ingredient is not currently registered in the United States.

One consequence of high registration costs has been a reduction in the number of pesticides available for use, including those for managing human-wildlife conflicts. Another has been the formation of data-gathering coalitions, including both public and private-sector entities, to share the costs of registration (Fagerstone and Schafer 1998).

Some laws affecting management of human-wildlife conflicts may either be targeted at specific organisms or situations or embedded within laws that regulate broadly. Noxious weeds, nutria, and brown tree snakes are examples of organisms that have their own acts. The Convention on Illegal Trade in Endangered Species (CITES) influences the movement of invasive species between the United States and other countries (see table 15.2). Other regulations, policies, and laws at state and local levels, including wildlife codes, may also apply. For example, purchase and use of pyrotechnics and firearms are regulated from federal to local levels. The human-wildlife conflict practitioner must, of course, be aware of and compliant with all relevant policies and laws.

What are the legal responsibilities of governments and natural resource agencies when wildlife under their stewardship damage ecosystems or people's properties or cause human, pet, or livestock disease, injury, or death? It depends partly on the country. In the United States, natural resource agencies often have little or no responsibility for the damage itself or for payment of compensation. This is because wildlife species are seen as *ferae naturae* (an ancient common-law doctrine stating that a wild animal cannot be owned by anyone). Legal interest in wildlife mostly lies with states or federal agencies managing lands, which act as a **sovereign** (representing the common interests of its citizens) rather than a **proprietor** (asserting ownership). Because the state has not asserted dominion, the state cannot be held accountable for damage caused by wildlife (Bader and Finstad 2001).

Tan (1990) points to steadfast federal and state courts that refuse awards for restitution of wildlife-inflicted damage when suits are directed at governmental agencies such as public natural resource management agencies. In *Barrett v. State of New York* [220 N.Y. 423, 116 N.E. 99, LRA 1918C 400 (1917)] often used as a reference, a suit was brought against the State of New York in 1917 by a landowner who sustained "considerable commercial destruction" from beavers on his land. The beavers were protected under a New York law stating that no one "shall molest or disturb any wild beaver on the dams, houses, homes or abiding places of same." The claimant argued that since the beavers were "owned" by the state, the state should pay for the damage. The court disagreed, citing ownership of wildlife in a sovereign capacity for the benefit of all people. In another suit, farmers near Horseshoe Lake State Game Preserve, Illinois, sued the state to recover damage to corn and soybeans destroyed in 1946 and 1947 by migratory birds, mostly the Canada goose. Among other arguments, the claimants alleged the federal government was responsible for damage because it had the geese in its possession and control, failed to protect the defendants, took actions in 1946 that stirred up the waterfowl, causing more damage, and is the owner of geese when they are in the United States. The court again struck down the arguments, stating that a claim would have to be based on negligence or wrongful act or omission of an employee of the government while acting within the scope of employment [*Sickman et al. v. United States* (two cases); *Ryal et al. v. United States*, 184 F.2d 616 (7th Cir. 1950)].

While suits against federal or state agencies for wildlife damage itself may be denied, property and human injuries and deaths do occur as a consequence of wildlife. Costs can be high, and liability has been claimed under two circumstances of common law (Bader and Finstad 2001): first, the state has a duty to warn of known dangerous conditions, including those involving wildlife on state property; and second, the state may be required to compensate when it produced artificial conditions that led wildlife to cause harm. In those situations, negligence or mismanagement is presumed to allow wildlife damage that would not otherwise have occurred. Both tort (civil) and criminal lawsuits might be used, constituting another set of laws related to human-wildlife conflicts. Claims may be against public agencies (less likely to be successful) or private entities (more likely to be successful). Successful examples include suits associated with predator attacks or animal-vehicle collisions at airports, where agencies and their staff were found to be negligent or guilty of mismanagement.

In one case, a family was awarded $1.9 million in a wrongful death suit involving a bear attack on an 11-year-old boy that occurred at a US Forest Service campsite in 2007. The bear had attacked other campers 12 hours earlier, and the agencies were found negligent for not warning other campers or closing the campsite (Dobner 2011).

Dale (2009) summarized personal and corporate liability surrounding bird strikes. Because 74% of airstrikes happen at 500 feet or less above ground, airstrike prevention is often the responsibility of the airport and its management. A common liability is failure of the airport manager to take actions that are legally required to maintain a safe operating environment. Wildlife hazard assessments are legally required after these occurrences: an aircraft experiences a multiple wildlife strike; an aircraft experiences substantial damage from striking wildlife; an aircraft experiences an engine ingestion; and wildlife capable of causing damage are observed having access to airport flight patterns or aircraft movement areas (Cleary and Dolbeer 2005). Based on results of the assessment, the US Federal Aviation Agency determines whether a wildlife hazard management plan is needed. If threatened or endangered species are involved, a biological assessment of impacts is also required. Management actions must follow all relevant public regulations and laws.

Failure to take appropriate action can be costly, both in damage following airstrikes and in subsequent litigation. As examples of tort claims, Dale (2009) included

- A November 12, 1975, aborted takeoff of a DC-10 that ingested gulls and caught fire, injuring 30 of 139 people on board, and costing $15 million in suits by the airlines against the US Federal Aviation Administration, the Port Authority of New York and New Jersey, New York City, and several aircraft companies.
- An Air France Concorde that on June 3, 1995, ingested geese at 9 feet above ground upon landing, causing uncontained failure with over $7 million in damage to the Concorde, and settlement out of court for $5.3 million between the French Aviation Authority and the Port Authority of New York and New Jersey.
- An Air France A-320 hitting a flock of gulls on takeoff at Marseille Provence Airport, France, with the impact destroying the engine; the airline awarded $4 million in settlements because of negligence (failure to remove a hedgehog, hit earlier, that attracted the gulls) in operating the airfield.

As examples of criminal charges, Dale (2009) points to

- A January 20, 1995, crash of a Falcon 20 that struck a flock of birds, killing all 10 people on board, with French authorities bringing charges of involuntary manslaughter against the Paris Airport Authority and three of its former officers.

- The September 22, 1995, crash of a US Air Force AWACS B-707 after ingesting geese at Elmendorf Airbase in Alaska, killing all 24 people on board; the senior airport controller and one other controller invoked their Fifth Amendment rights to avoid criminal prosecution.

Lawsuits against wildlife management agencies in the United States are often based on faulty or inadequate address of laws applicable to the proposed action, such as failure to fully comply with the National Environmental Policy Act. Suits often involve wildlife species that the public sees as charismatic, which thus engender public interest. For example, a coalition of conservation and animal protection organizations unsuccessfully sued US Wildlife Services over a program to kill black bears as part of integrated management action to reduce damage to forests in Oregon; the suit targeted the environmental assessment, claiming that it was inadequate (e.g., see Zuckerman 2003). Also in Oregon, four conservation groups sued the Wildlife Services program for its proposed role in killing two wolves at the behest of Oregon Department of Fish and Wildlife (ODFW). The wolves had taken livestock, and nonlethal efforts by the ODFW had failed. Trying to prevent a pattern of killing livestock, the ODFW requested assistance from Wildlife Services and set the terms of the permit so that the alpha female and her four pups were protected. The basis for the suit was that the required environmental analysis was not conducted (Laughlin et al. 2010). This suit was also unsuccessful.

I have focused on policies in the United States, but public policies offer both similarities and differences among countries. As an example, the Government of Japan's Ministry of the Environment describes a basic system for protecting wildlife (Government of Japan 2020) that provides strict protection for about 600 species, classifying them as animals excluded from protection (some marine mammals, rats, and insectivorous mammals), game species (47 species including brown bear, black bear, shika deer, wild boar, Japanese green pheasant, Japanese quail, and tree sparrow), and protected species (the Japanese Redbook lists 47 mammals, 90 birds, 18 reptiles, and 14 amphibia as threatened). Included also are Wildlife Protection Areas, a total of about 3.4 million hectares in about 3,600 locations, where hunting is prohibited and wildlife is protected. Each type of wildlife conservation is enforced by strict federal polices and laws.

Differing public policies can further complicate managing human-wildlife conflicts when the geographic ranges of the wildlife involved extend beyond the borders of the countries that govern how the conflicts are to be managed. Boon et al. (2020) explored policies and regulations for management of carnivores in Sweden, Norway, Finland, Spain, the Netherlands, and Germany. The carnivores included were wolves, bears (*Ursus arctos*), lynx (*L. pardinus*), wolverines, and golden eagles. The policies were explored using scientific literature, policy documents, law texts, and expert knowledge. Policy mixes were evaluated for their overall coherence (two-way communication between all authorities and levels), consistency (whether set policy objectives can be achieved simultaneously and whether the instruments used reinforce one another), and comprehensiveness (whether the policy mix addresses all relevant institutional and systemic failures) within the framework of policies established by the European Union and the Convention on the Conservation of European Wildlife and Natural Habitats (Bern Convention).

The instruments used by one or more of the countries included compensation payments, conservation performance payments, subsidies (for preventive measures), release of livestock owner for reparation when livestock escape was due to wolf presence, culling of problem-causing animals, license or quota hunting for population regulation, maximum acceptable depredation losses, promotion of ecotourism, adaptive hunting quotas, and habitat improvements.

The researchers found "a patchwork of processes, principal plans, and policy instruments that lack a clear uniform underlying guiding strategy that accounts for the multiple temporal and spatial scales and components of the socio-ecological system that the policy mixes aim to address" (Boon et al. 2020, 408). The scientists contended that what is needed is a "holistic, system-based approach with strong collaborative structures across policy boundaries and regions, the inclusion of diverse stakeholders, and constant care and attention to address all objectives simultaneously rather than in isolation" (409).

Examples

TRANSFER OF WILDLIFE SERVICES. The transfer of Wildlife Services (then Animal Damage Control, or ADC) from USFWS back to the USDA in 1985 serves as an example of involvement of a power cluster in the public policy process. The transfer was the consequence of these factors: individual farmers and ranchers complaining that wildlife damage was inadequately addressed in USFWS (problem identification); groups such as the National Cattlemen's Beef Association and the American Sheep Industry expressing concerns to congresspeople (problem identification); congresspeople (particularly Jesse Helms of North Carolina) and 19 western senators expressing concerns and making recommendations to President Reagan (problem identification, agenda setting, and policy formulation); the Department of Justice and the secretaries of agriculture and the interior finding mechanisms for transferring the program (policy formulation and legitimization); and the budgeting power of Congress putting the program's funds into the USDA rather than the USFWS in 1985–1986 (policy implementation). Representatives of 21 members of the power cluster now meet annually (contingent on available funds) as the National Wildlife Services Advisory Committee, which deliberates policies and recommends changes (policy evaluation) and thereby completes steps in the process.

INVASION OF ASIAN BLACK CARP. Simberloff (2005) summarized the politics surrounding risk assessments for biological invasions in the United States. One example was assessment of risks associated with introduction of the Asian black carp in hatcheries within the Mississippi Basin. The completed assessment included concerns regarding introductions of new diseases of fishes and mussels and concerns about foraging on native mollusks, causing competition with native species. The assessment recommended that only sterile triploids be permitted, solely in contained facilities away from open waters.

It soon became apparent, however, that black carp were being held near open water. A consortium of 28 state fish and wildlife agencies subsequently petitioned the USFWS to list black carp as "injurious" under the Lacey Act, and the USFWS began that process. In 2003, however, the USFWS made some procedural changes that left the fish unlisted, removed the USFWS fishery biologist from the listing consortium, and removed its funding support from the consortium. Apparently, an influential state senator was responding to a different power cluster, one that included catfish farmers who saw the carp as valuable for controlling intermediate snail hosts of trematodes. In the words of Simberloff (2005), addressing a more general context, "The problem that seems inadequately treated currently is that a substantial benefit that might accrue to a few has more political weight than a substantial cost that might be borne forever by all" (220).

MANAGEMENT OF LEOPARDS. Marker and Sivamani (2009) summarize the need for policy evaluations and changes for management of leopard (*Panthera pardus*)

attacks on humans in India. In 2006, at least 133 people were killed by leopards. Leopards also prey on dogs, cows, and goats. Leopards have adapted to environments occupied by humans, who continue to move into leopard habitat. People retaliate by injuring or killing the leopards. Indian federal law currently allows two options in managing a leopard that has killed a human: hold it in captivity for its natural life or translocate the animal. Holding the leopard captive is expensive, and translocation can exacerbate problems because leopards are strongly territorial and attempt to return to their original home range; in doing so, they move through human habitations, resulting in further human encounters and attacks.

Based on recommendations from a workshop held in 2007, during which experts on predator management from India and other countries were invited to share information and ideas, the following policy changes were suggested:

- The formation of trained emergency response teams composed of a highly ranked forest conservator, a veterinarian, and at least five support members empowered to declare curfews, provide information to the public, alert police and order ambulances, and decide the fate of the leopard; that Section 11 of the Wildlife Protection Act be amended "to make them more amendable to conflict resolution."
- That laws, such as definition of a "problem animal," be clarified so that officials with responsibility for management actions know that their actions fall within legal bounds, thus avoiding personal liability.

These and other policy recommendations were implemented in a set of guidelines set forth by the Ministry of Environment and Forests in April 2011 (Government of India 2011).

How effectively have the guidelines been implemented? Gubbi et al. (2020) evaluated the impact of these guidelines and responses of field managers to the guidelines during leopard captures in Karnataka state between 2009 and 2016 (to allow evaluation before and after the guidelines). A total of 357 leopards were captured, an increase of about 9.7 per year. Monthly translocations increased threefold. Translocations were mostly to protected areas (85.5%), taken to captivity (10.8%), or died due to capture mortality (3.8%). There were 29 retaliatory killings. In responses to questionnaires, 64% of the managers said they were unaware of the guidelines, and only 1.9% followed them. The researchers concluded that "large-scale improvement is required by bringing in field-level managers, communities, media personnel, and other stakeholders while developing such policies."

TRAP, NEUTER, AND RETURN OF FERAL CATS. Williams (2009) summarized a visit to the University of Hawaii, Oahu, where a 10-year effort to manage feral cats with the trap, neuter, and return (TNR) method was underway. There are at least 1,200 registered caregivers for feral cat colonies and between 100,000 and 300,000 feral cats on the island. The Hawaiian Humane Society has already sterilized over 5,000 feral cats in 2021.

Wildlife agencies and their surrounding power clusters have mostly opposed TNR, instead supporting euthanasia of feral cats. This is partly because feral cats impact other wild species, such as birds and sea otters. The American Bird Conservancy estimates that nationwide about 500 million birds are killed yearly by feral cats. Feral cats probably outnumber other wild predators in North America, and they transmit diseases such as toxoplasmosis, roundworm, and rabies. The evidence accrues that TNR is ineffective (e.g., Longcore et al. 2009). Yet the use of TNR continues to expand rapidly in communities throughout the country. In a briefing published by Alley Cat Allies, Holtz (2013) stated that 331 local governments incorporated TNR into animal management policies and practices, not counting thousands of feral cat groups who privately conducted TNR, and that TNR was "poised to become the predominant method of feral cat management in the United States." Why?

Williams argues the growth of TNR is at least partly because the feral cat lobby is stronger than the wildlife power cluster. The researcher points to cases in which the two clusters have clashed. For example, the feral cat lobby prevented legislation from being enacted that was supported by the wildlife agencies and that would have removed invasive exotic species from National Wildlife Refuges. The lobby was concerned that feral cats might be taken. Alley Cat Allies alone had a staff of at least 10 and a total revenue in fiscal year 2019 of just under US$11 million.

Summary

- Public policy is what governments choose to do or not do, including regulating internal conflicts, organizing conflicts with other societies, distributing rewards and services, and extracting money from society, usually as taxes.
- Public policies are made or used in managing human-wildlife conflicts when citizens are concerned about conflicts or how they are managed. Concerns often focus on effectiveness of methods, their biological or ecological soundness, or animal welfare or rights.
- Individuals sharing common interests gain political strength by organizing into groups, and groups form power clusters (coalitions) around human-wildlife conflicts. Elitist individuals can directly influence public policy.
- Most public policies involving human-wildlife conflicts and their management follow transparent processes, but political deception and expediency also exist. Individuals, groups, and power clusters participate in any or all steps of the public policy process, from increasing awareness of a problem to evaluating effectiveness of new laws. Transfer of the federal Wildlife Services program from USFWS to USDA in 1985 is an example of public policy driven by public interest.
- Once made, public policies in the form of regulations and laws set the legal boundaries within which human-wildlife conflicts are assessed, studied, and managed. Because policies and laws occur at every governmental level, the human-wildlife conflict manager must be aware and compliant with them all.
- Public policy establishes the legal bounds for redress if a citizen believes she or he has been injured by wildlife or its management. Personnel of agencies and private individuals and entities can be held accountable for negligence or creating situations that are unnatural and lead to wildlife damage.

Review and Discussion Questions

1. Find a specific example of how a power cluster was formed on an issue surrounding human-wildlife conflicts, such as the ban on trapping in Colorado. Who were the groups and individuals in the power cluster? Was it based on reason, emotion, or both? Was the general public well informed, or was deception or political expediency involved? How might the process have led to a different outcome?

2. Feral cats are an increasing concern for conservationists and wildlife damage managers. The article by Williams (2009) argues that the feral cat lobby is stronger than that of professional wildlife management. Do you agree? Support your position with specific examples and references.

3. Feral pigs are also an increasing concern for conservationists and wildlife damage managers. Would

you anticipate the same problems for management of feral pigs as Williams (2009) describes for feral cats? Why or why not? What issues would you anticipate in managing the feral pig? Defend your responses with specific examples and references.

4. How much publicity surrounded the use of Compound 1080 to remove Arctic foxes from Kiska Island for the benefit of endangered geese? Do you agree that it is the wise human-wildlife conflict practitioner who assumes interest exists in any management action and adopts a policy of transparency? Does that put the practitioner at a disadvantage with political groups and politicians who practice deceit? Defend your position.

PART VI • **STRATEGIES AND THE FUTURE**

In this part, I consider overall processes and strategies to manage human-wildlife conflicts now and in the future.

16

Operational Procedures and Strategies

Here I explore the use of operational procedures and strategies in managing human-wildlife conflicts.

Statement

Strategies can ensure all aspects of a human-wildlife conflict are considered. An effective strategy resolves the conflict while meeting benchmarks based on criteria such as safety, economy, effectiveness, humaneness, and stakeholder (beneficiary) engagement.

Explanation

A **strategy** is a broad plan of action to achieve a goal. A military concept, the term has been applied more recently in the sense of strategic management. **Holistic** (i.e., comprehensive) strategies are the consequence of careful consideration of every aspect of a human-wildlife conflict (fig. 16.1). Such considerations include history and existing literature on the problem, applicable biological and ecological concepts, and human dimensions; the strategies take advantage of routine plans such as standard operating procedures and applicable decision support systems. Holistic strategies begin with fundamental considerations—for example, whether to use an **autecological** or **synecological** approach.

Autecological approaches are chosen when an individual species—or ecologically closely related species—is the proximate cause of the problem. Suppose, for instance, ring-billed gulls are nesting on air-conditioning systems on a rooftop, fouling the air that is being circulated into rooms and posing a fire hazard with flammable nesting material. An autecological approach would be most appropriate because only one species is causing damage.

Synecological approaches are used in situations where several species might conflict with humans simultaneously, such as at airports or landfills. Here, a broad ecosystem or landscape approach may be needed wherein the habitat is made unsuitable for many or all species, regardless of their ecological niches. Pools of water might be removed to reduce the attraction of the area to wildlife. Grass might be managed at a given height or removed entirely to deter geese, gulls, and other wildlife. Small

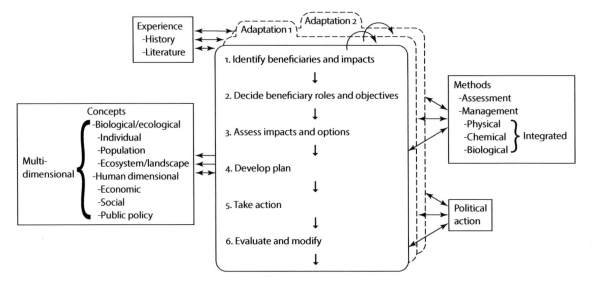

Figure 16.1 Steps in formulating strategies for resolving human-wildlife conflicts. Illustration by Lamar Henderson, Wildhaven Creative LLC.

mammals might be trapped and removed to reduce their presence and attractiveness of the area to raptors and other predators. Fencing might be used to exclude wildlife from critical areas. Human garbage might be covered to reduce its attractiveness to birds, rodents, bears, or other wildlife.

Holistic strategies build on day-to-day operating procedures and tasks. **Standard operating procedures** (SOPs) are written, which are formalized directions to be followed during performance of tasks. As such, SOPs are themselves more like tactics or operational plans than strategies. An SOP ensures that a task, regardless of who does it, can be completed the same way each time it is performed, serving as a form of quality assurance. For this reason, SOPs are routinely included in environmental assessments of many programs required under the National Environmental Policy Act. They are widely used for both research and operational activities in the public and private sectors.

For example, steps for weighing a sample with an analytical balance may be described in a SOP at a research facility. Anyone weighing samples with the balance is expected to follow the detailed directions in the SOP. A written statement or signature may be required, ensuring that the SOP was followed. Steps used to calibrate the balance may be written as a separate SOP. Steps used to freeze and store the weighed samples may have an SOP as well. The procedures for ensuring that the storage device remains below freezing may also be documented as a SOP. A laboratory may there-

fore have hundreds of SOPs, especially if they are following **Good Laboratory Practices** (**GLP**) guidelines for registration of pesticides. In larger laboratories the SOPs are often maintained by a quality assurance officer who also looks after other quality factors such as chain of custody of samples and archival records of laboratory studies. Collectively, these SOPs contribute to a broader strategy of quality control.

Sengl et al. (2008) described the SOP used by Wildlife Services, USDA, in Las Vegas, Nevada, when receiving phone calls from the public seeking help. The SOP describes each step, including

- gathering information on the caller (e.g., contact, location),
- identifying the conflict (e.g., species, type of damage),
- determining the responsible managing agency (e.g., USFWS, Nevada Department of Wildlife, or the USDA's APHIS/Veterinary Services), and
- assessing which public policies may apply (e.g., conducting technical assistance or direct control).

Here the SOP also serves to ensure a high level of quality in response to public calls.

Thousands of SOPs could probably be found at universities and research facilities around the globe that describe procedures to be followed in all aspects of assessing and managing human-wildlife conflicts.

Decision support systems (DSSs) do not make decisions, nor do they constitute stand-alone strategies.

Rather, DSSs help to gather and store information and to analyze and present the information in ways that allow new insights and options for the practitioner. They are incorporated as parts of holistic strategies. Many DSSs are computer based, but they need not be. Hoare (2001), for example, provides a written DSS for managing elephant damage in Africa. Other DSSs are designed as "apps" (applications) for cell phones that can be downloaded and used in developing countries or remote areas where computers may be unavailable (e.g., Eisen et al. 2011).

Computer-based DSSs have evolved from their inception in the 1960s to where many are interactive and provide real-time (virtually instantaneous) information. Most have a database (or knowledge base), a model, and a user interface. DSSs can be particularly useful when options, each involving complex iterative calculations, need to be considered as part of a broader strategy. An early example is the DSS offered by Sterner (2002) for controlling vole damage to alfalfa. The DSS does not decide on a specific strategy for the practitioner, such as when to treat with zinc phosphide; instead, it shows benefits in relation to costs for differing treatments and levels of damage.

DSSs are increasingly available for a wide range of uses. Schneider et al. (1998) describe a DSS that helps to evaluate the level of risk for the spread of zebra mussel in lakes and streams associated with boat use at infested sites in Illinois. Among other revelations, the DSS model predicted that quarantines of lakes with high boater populations could actually increase the risk of infection and that educational efforts could reduce the risk of infestation at larger lakes and their surrounding areas by 80%. Such information is helpful in designing strategies to reduce the movement of zebra mussels between bodies of water.

WeedSOFT was developed by the University of Nebraska as a DSS to help growers, consultants, and extension agents develop strategies for managing weeds. The software provides information according to specific field conditions and considers economic and environmental factors. Photographs help with weed identification, and mapping features allow consideration of groundwater contamination. Programs such as WeedSOFT have many useful applications in managing damage associated with exotic invasive plants.

Eisen et al (2011) describe uses of geographic information systems combined with DSSs to predict outbreaks of vector-borne diseases such as Lyme disease, plague, dengue, malaria, trypanosomiasis, and West Nile virus based on weather conditions and likely distribution of vectors. This research team points particularly to DSS packages that are freely available, can be downloaded on cell phones, provide easy data entry and visual display of analyses, and can be used in resource-limited areas where vector-borne diseases occur.

In their systematic review of mobile apps, Tom-Aba et al. (2018) identified three apps downloadable to cell phones that contained four key variables (surveillance capability, contact tracing, case management, and laboratory data management) needed to facilitate individual and institutional responses to zoonotic diseases such as the West African Ebola outbreak that occurred during 2014–2015. Nikolopoulos et al. (2021) used data from the United Kingdom, United States, India, Germany, and Singapore to evaluate DSSs that could be used to forecast excess demands for products and services at the country level during COVID-19.

Some DSS systems are designed to provide real-time warnings of imminent bird strikes. An example is the radar-based bird strike detection system called MERLIN (https://detect-inc.com/aircraft-birdstrike-avoidance-radar/). The manufacturer claims it is the first production model bird strike radar system. It functions as a DSS by alerting of potentially hazardous bird activity in real time. Such systems can be incorporated into strategies to reduce or eliminate impacts of bird strikes. Huijser et al. (2009) created a DSS that predicts benefits and costs for various methods to mitigate deer-vehicle collisions over a 75-year period.

Whereas SOPs and DSSs can be viewed as tactics that contribute to broader strategies, other approaches can themselves serve as holistic strategies. An ideal holistic strategy can be characterized as integrated (involving a range of management methods), **multidimensional** (including human dimensions as well as biological and ecological concepts), and **adaptive** (wherein one proceeds without a full understanding of the problem and its solution, but the strategy includes sufficient monitoring and experimental data gathering so that post hoc analyses lead to improvements). Most holistic strategies have some combination of these attributes (see fig. 16.1).

Many factors in an individual situation play into holistic strategies. For example, if the problem is deer damage to alfalfa, relevant aspects might include

- Flavor preferences and foraging behavior of deer
- Density of deer in the area
- Ecological season
- Surrounding crops and their closeness to water and shelter, such as forested areas

- Extent of damage
- Availability of suitable fencing, chemical repellents, or kill permits
- Policies of the wildlife agency regarding issuing kill permits
- Attitude of the farmers toward hunters and the wildlife management agency
- Attitude of hunters toward culling operations
- Benefit-to-cost ratio for using fencing, repellents, sharpshooters, or hunters to cull

A **holistic strategy** would take into consideration each of these factors and guide the actions needed to resolve the damage to alfalfa effectively and acceptably. Holistic approaches may include best management practices (BMPs), integrated pest management (IPM; box 16.1), integrated wildlife damage management (IWDM), adaptive management practices (AMPs), and adaptive impact management (AIM).

Best practices, or **best management practices**, are informal standards for techniques or methods that have proven successful over time. BMPs can serve as stand-alone strategies, such as using a trap to remove a raccoon from an attic. BMPs can also be utilized as approved SOPs for entities such as agencies or as part of a broader strategy—e.g., where trapping is part of a national strategy for management of raccoon-based rabies. By considering factors such as efficiency, selec-

tivity, practicality, user safety, and animal welfare, BMPs ensure that every practitioner can follow the best-known technology for a management action, thereby ensuring the best possible outcomes (see, e.g., Furbearer Conservation Technical Work Group 2006). BMPs are widely used in managing human-wildlife conflicts. For instance, Wittenberg and Cock (2001) provide BMPs as part of a tool kit for preventing and managing invasive exotic species.

Wildlife species other than insects are increasingly included as part of IPM programs. For example, Horgan (2017) includes birds and rodents as part of his evaluation of IPM usage in rice. And integrated wildlife damage management (IWDM), a form of IPM, is receiving increased attention by wildlife damage practitioners. As with other IPM strategies, the idea is to reduce reliance on pesticides and to emphasize a mix of methods that work holistically to resolve the problem. The focus is on reducing damage rather than reducing populations of wildlife. The practitioner, following IWDM practices, first determines whether management is needed. If it is, the manager figures out the combination of methods that would work most effectively. SOPs and BMPs are often part of the mix. The approach takes into consideration the factors outlined in figure 16.1. Rodent and bird species, with higher densities and predictable patterns of damage,

---

BOX 16.1 Integrated Pest Management

Integrated pest management (IPM) can stand alone as a strategy or be part of a broader one. IPM was first conceived in the late 1960s for managing insect pests in agriculture by Dr. Raymond Smith and Dr. Perry Adkisson. The notion was to reduce use of pesticides by incorporating a range of nonchemical control options into a single management plan. The plan might call for using pheromonal attractants along with mechanical traps, sterile male releases, natural enemy releases, and bred plant resistance to pests, thereby reducing dependence on chemical pesticides. IPM quickly succeeded in reducing farmer dependence on pesticides—by 50% in the United States. (Drs. Smith and Adkisson received the World Food Prize for this contribution in 1997.)

IPM practices were also quickly expanded beyond agricultural insects. For example, Dr. Smith in the mid-1970s helped to determine the physical, training, and equipment needs to establish the National Crop

Protection Center in Los Baños, Philippines. The center was designed to establish a research and training capability for IPM that included vertebrate species and weeds in addition to insects.

This makes sense because different taxa interact, as was recently demonstrated by Rodenburg et al. (2014) between weeds and Quelea birds in rice in Africa. In their study, more weeds resulted in more bird damage, possibly because of the attraction of weed seeds to the birds. Reducing weeds would therefore reduce bird presence, perhaps reducing the need for avicides or repellents. Interactions were found more recently between weeds and the rodent *Bandicota bengalensis* in the rice paddies of Bago, Myanmar (Htwe et al. 2019).

IPM, incorporating interactions among taxa, has since become a fundamental strategy underlying research and development of pest management practices, including work by major international organizations such as the International Rice Research Institute.

seem particularly amenable to IWDM as an economically beneficial strategy (Sterner 2008).

IWDM has received national attention, as the following examples show. Nugent et al. (2008) used IWDM practices to manage a mix of gull species roosting nightly on Lake Auburn, an unfiltered municipal water source for Auburn, Maine. The gulls were contaminating the drinking water with fecal coliform bacteria. Cotton (2008) described IWDM of a threatened and recovering species, the Louisiana black bear (*Ursus americanus luteolus*) in Louisiana. Campbell and Long (2009) and West et al. (2009) provide excellent reviews of IWDM strategies for managing damage caused by feral pigs worldwide, emphasizing damage and management practices in the United States. Pullins et al. (2018) evaluated the effectiveness of translocation of red-tailed hawks as a first step toward the method being used as part of an IWDM strategy designed to reduce raptor collisions with aircraft. They translocated hawks to various distances from Chicago's O'Hare International Airport, with increasing distances at costs ranging from US$213 to US$426. The researchers found birds older than 1 year were much more likely to return than younger birds. Lischka et al. (2018) developed a conceptual model that integrated social and ecological information for better understanding human-wildlife conflicts and illustrated by using the model on human-black bear conflicts in Durango, Colorado. The researchers pointed to three lessons from the application: the importance of integrating social science throughout the research process, of aligning the scales of social and ecological data, and of understanding human as well as animal behavior.

Ecologically based rodent management (chapter 12; Singleton et al. 1999) is a form of IWDM that reduces use of rodenticides by providing ecologically based, nonchemical methods such as the following: good hygiene that reduces food and cover for rodents; synchronized crop planting; and the trap barrier design system. The latter involves a crop, such as rice, being planted and fenced earlier than the rest of the crop in a small portion of a field; rodents, attracted to the older rice, move through holes placed in the fencing into the patch of older rice and are periodically removed; the method effectively gets rid of rodents without rodenticides. This strategy includes strong socioeconomic components, such as surveys of knowledge, attitudes, and practices of farmers—information that underlies the management plans. This approach has been used successfully in both developing and developed countries, for example, to develop management strategies for intensive organic piggeries and poultry farms in Europe (Singleton et al. 2004), and it continues to gain in use and popularity as an ecologically sound form of rodent management (e.g., Singleton, 2014; Krijger et al. 2017; Liu 2019; Swanepoel et al. 2017).

Use of AMP underscores the belief that in a real and always changing world, one is unlikely to get all the information needed to completely and assuredly solve a human-wildlife conflict. At some point, the practitioner needs to act with the situation, information, and options on hand. The strategy embraces change and uncertainty. It is contrary to the **precautionary principle**, which states that no action should be taken until a problem and all consequences of proposed actions are fully understood, lest one be surprised with an unanticipated form of ecological backlash that causes damage to the environment or human health.

With AMP, the practitioner proceeds with a management action without a full understanding of the situation, monitors the outcomes, and adjusts future actions based on what was learned (see fig. 16.1). By gathering detailed information on an appropriate design while taking the management action, the practitioner learns more about the damage situation and how to improve future actions. In a sense, the manager becomes an amalgam of practitioner and scientist (or involves scientists). A major challenge is achieving a balance between the resources required to acquire new information in a scientifically rigorous manner (Raffaelli and Moller 1999) and the resources needed to resolve the problem (e.g., Roy et at. 2009).

AMPs are increasingly used in wildlife damage management, often as part of IWDM. For instance, Engeman et al. (2007) described use of track plots to adjust the management of feral pig damage to wetlands in Florida. Bryce et al. (2011) described an adaptive approach for eradicating American minks from areas where they compete with native species in and around Cairngorms National Park, Scotland. Treves et al. (2009) suggested that mitigating wolf damage would be facilitated by including clauses in compensation agreements that permit AMPs from the outset. These researchers argued that the clauses should "articulate explicit goals for compensation programs lest the costs skyrocket without measurable success" (16). Roy et al. (2009) stated that AMP could provide a means of filling data gaps for managing invasive species on islands. These scientists argued that some information, such as assessing risk, can be gathered only by carrying out eradications. Madsen et al. (2017) reported the first AMP being used to manage a European migratory waterbird population, the Svalbard pink-footed goose (*Anser brachyrhynchus*).

Although AMP offers ways to reduce uncertainty in managing human-wildlife conflicts, and while AMP is being used increasingly, the profession would benefit from broader adoption of the approach. Richardson et al. (2020), for example, explored the use of AMP in managing invasive non-native species. The scientists evaluated 3,992 articles after searching for "adaptive management" or "adaptive harvest management" on the Web of Science and focusing on managing invasive non-native species. The researchers found that only 56 (about 1%) used AMP, and only 10 of these included all its recommended components.

Some efforts to broaden the use of AMP in Japan within the profession are already being made. Fortin et al. (2020) have proposed a quantitative approach, using predictive models that are revised as new information is gathered on responses of animals to interventions, to serve as a robust framework for finding cost-effective and sustainable strategies for managing human-wildlife conflicts. As a more specific example, Tsunoda and Enari (2020) have suggested that an AMP approach to planning the use of space by reducing urban and farm interfaces with expanding natural ecosystems would be an effective way to optimize human needs, including benefits provided by ecosystem services, while reducing opportunities for human-wildlife conflicts. The space is becoming available because Japan is anticipating a 24% reduction in human population by 2050.

Managing human-wildlife conflicts is a truly multidimensional field, embracing not only its underlying biological and ecological concepts but also concepts related to its human dimensions, including economics, sociality, and public policy. Application of concepts such as **wildlife acceptance capacity** (the mixture of tolerances that can be found among the stakeholders—beneficiaries—in a wildlife damage problem) is as central to solving issues as the selection of appropriate management methods (Decker et al. 2002; see chapter 14). Decker et al. (2002) offer AIM as a variation of AMP. They characterize the approach as both the practitioner and the beneficiaries agreeing that "we don't have all the answers needed for developing a management program that will fix this problem with certainty, but we'll apply what we know, use our best judgment in those things we are less certain about, and commit to learning from the experience of the specific strategy and tactics we employ" (7).

Most recently, Decker et al. (2019) have also proposed a paradigm shift from that of stakeholder to that of public beneficiaries in wildlife conservation,

including managing human-wildlife conflicts. The scientists argue that this shift, though "slight," is important to align the stakeholder concepts with that of wildlife as a public trust and to ensure inclusion of the broadest base of human engagement. They see it as part of a progression from clients (people who pay for services and products) to stakeholders (any people affected by or affecting wildlife or its management) to beneficiaries (the broad public for whom wildlife are managed in a public trust). In the words of Decker et al. (2019), "in our opinion, the future of the profession will be created by those managers and agencies that adopt, refine, and practice the evolving beneficiary approach in conservation and management of public trust wildlife" (517). Further, the researchers argue that such an approach will encourage good governance practices and "make wildlife conservation more relevant to and valued by more citizens." In fundamental agreement with this approach, I have changed the orientation of figure 16.1 and its description from stakeholders to that of beneficiaries.

I summarize by offering a generalized process for the human-wildlife conflict manager (see fig. 16.1). The process is multidisciplinary, integrated, and adaptive. It applies to individual practitioners or those working for agencies and can probably be adapted to specific societies and public policies peculiar to a country. The first step is identifying impacts and beneficiaries, following methods suggested by Decker et al. (2002, 2019). The extent of beneficiary involvement (in each of the steps) needs to be assessed in the second step, as do the specific objectives of any management actions. For example, an objective might be to measurably reduce concerns of homeowners of contracting Lyme disease. Damage surveys might be useful at this time, as might information on the breadth of interested beneficiaries and their attitudes and acceptance capacities (Decker et al. 2002, 2019).

The third step is the process to assess the impacts and consider management options. Here, an overall strategy needs to be chosen (e.g., incorporating or choosing BMPs, IPM, IWDM, AMPs, and/or AIM), as do available DSSs and SOPs. Legal and public policy requirements and media involvement need to be determined. The assessment may take the form of an environmental assessment or an environmental impact statement (both evaluations of environmental risks and benefits, part of the National Environmental Policy Act). For example, effects of concern about Lyme disease might be fear of families to enjoy the outdoors, such as backyards or local parks. Management options might include education of families regarding use of repellents to prevent vector-borne zoonoses; periodic treatment of yards and parks for control of ticks; control of intermediary hosts, such as habitat to reduce the presence of field mice; use of fencing to reduce the presence of white-tailed deer; use of bow hunting or sharpshooters to reduce overabundant deer populations; or some combination of these options.

After all management options are compared and considered and one or more have been selected, a logistical plan of action needs to be developed (step 4). Often, this takes the form of an operational or management plan wherein the details of resources needed, actions, and time frames are provided. Time management plans are sometimes used for complex management actions.

In step 5, the management plan is conducted. Consultations are completed and permits obtained. Equipment and supplies are ordered and personnel brought on board. The media are notified. Management actions are taken. Records of accomplishment are kept in relation to plans in step 4. Sometimes an assessment is conducted before and after a specific action to measure its effectiveness. Additional data may also be collected as part of an adaptive management program. If, for example, education on the use of tick repellents, tick control at local parks, and reduction of overabundant deer by sharpshooting were selected, the permits, equipment, supplies, personnel, and notification of media would be done at this time. Perhaps additional information was needed on abundance of field mice for possible future adaptive actions, and indices of abundance for the mice would be put in place at this time.

Step 6 includes an evaluation of the overall effectiveness of the management action, based on the analysis of information collected during step 5 and on adjustments of the management plan. Since the intent was to reduce the impact of concern for Lyme disease in the community, the effectiveness of the management actions on this factor should be measured. Adjustments in future programs would then be made.

During the process, information can be drawn from history and experience; from basic biological, ecological, and human dimensional concepts; and from methods for assessment and managing of damage (see fig. 16.1). Information gained from the process can be added to the existing base of knowledge.

I describe the process as one that follows a logical, stepwise sequence from identification of an impact to modification of future management actions. Individual

practitioners may see it differently and adjust the content and order of some of the steps according to their own mindsets. I note also that the orderly sequence of events as described on paper is not always the sequence followed in the real world. For example, political support of, or opposition to, the proposed management actions (steps 3 and 4) may necessitate adjustments in plan development, management actions, or deviations from the time sequences for planned management actions (step 5; see fig. 16.1; see chapter 15).

Examples

The Vermont Fish and Wildlife Department (2017) provides an example of BMPs for managing beaver damage. The introductory portion of the guide is organized to provide an overview of beaver management in Vermont, the biology and behavior of beavers, and descriptions of beaver problems. The BMPs are divided into damage prevention techniques, culvert and dam obstructions less than two years old, and damage from older, more established beaver dams. The guide provides information needed to work within applicable federal and state laws.

The BMP begins with a phone call to a local authority, who determines whether the issue can be resolved by preventive techniques. For example, beaver damage to ornamentals might be prevented by fencing until removal or lethal reduction can be completed. Concern for rabies or *Giardia* might be mitigated with educational materials or with removal or lethal reduction of the population. Obstructed culverts and dams less than two years old might be mitigated by monitoring for potential beaver problems; arranging for removal or lethal reduction of the population; installing or maintaining water-control structures and, between June 1 and October 1 (to prevent effects on trout spawning, provided in appendices), notifying anyone who might be affected by a lowering of the water level downstream; installing fencing or a control device; and notifying a fish and wildlife warden or other identified agents. For well-established beaver dams, a site visit may be needed. Here, biologists would be consulted to provide a plan that reduces damage while preserving the value of the wetland.

The manual includes diagrams and drawings of common wire and electric fencing methods, the addresses of people who can help, information on diseases such as giardiasis, discussion of siphons and other control devices and dam removal methods, and information on relevant ordinances, statutes, and regulations along with references and additional reading materials.

Witmer (2007) provides an overview of IWDM as applied to rodent pests in cropland. This researcher suggests that methods need to be selected and put into IWDM strategies based on the species causing problems, the physical environment, and the nature of damage. Correct identification of species causing damage is essential, and the amount and distribution of the damage needs to be known. He emphasizes the importance of knowing the distribution of the rodents within their habitat, how the rodents use the habitat, how the rodents relate to other species, and what human activities exacerbate damage problems. Witmer (2007) suggests that DSSs, such as the Mouser, can sometimes be helpful in determining when rodent control will be cost-effective. He suggests that any potential strategy be evaluated not only in its ecological and biological context but also in the human dimensional world of economics, societal concerns, and public policy and acceptability (see fig. 16.1). This scientist recommends that sufficient monitoring be put in place to allow treatment of the strategy as a large-scale experimental field trial so that its effectiveness can be evaluated in sufficient detail to allow adaptive adjustments and improvements in the future.

Shaffer et al. (2019) reviewed the history of human-elephant conflicts, including the biological and ecological bases for the conflicts, human dimensional factors underlying the conflicts, and current management strategies. The researchers summarized current management strategies as physical separation, mitigation by domestication, translocation, or culling and/or compensation. They argued that while these are important tools, they also have a short-term and site-specific focus, and they fail to deal with the underlying biological, ecological, and human dimensional roots of the conflicts. The researchers proposed a conceptual model that considers water, land, and plant resources and their competitive uses by elephants and people, and that points to conflict hotspots and alternative resource options. The scientists highlighted ecological, anthropological, and geographic information and tools for developing sustainable solutions. The researchers emphasized that while biophysical data may be available for modeling at a landscape level, much ethnographic information—key to addressing the underlying causes of the human-elephant conflicts—is still needed. "Without this knowledge," the scientists conclude, "the task of resolving human-elephant conflict and finding a means for these species to coexist in the Anthropocene is sisyphean" (8).

Summary

- Holistic strategies are the consequence of careful consideration of every aspect of a human-wildlife conflict, including its history and existing literature, applicable biological and ecological concepts, and human dimensions.
- Holistic strategies begin with fundamental consideration, such as whether to use an autoecological approach or a synecological approach.
- Standard operating procedures and decision support systems are incorporated into holistic strategies.
- Ideal holistic strategies are integrated, in that they involve a wide range of management methods; multidimensional, in that they include human-dimensional as well as biological and ecological concepts; and adaptive, in that they include sufficient monitoring and data-gathering to allow post hoc improvements.
- Holistic strategies that can stand alone or be parts of broader strategies include best management practices, integrated pest management, integrated wildlife damage management, adaptive management practices, and adaptive impact management.
- A process that helps the practitioner formulate holistic strategies has the following steps: identify impacts and beneficiaries; determine the extent of beneficiary involvement and specific management objectives; assess impacts and consider management options; develop a logistical plan of action; conduct the plan; evaluate the effectiveness of the management plan; and adjust the plan as needed for improvements.

Review and Discussion Questions

1. How important are SOPs for quality assurance of the responses by the staff who routinely answer calls for assistance with wildlife damage, such as extension, wildlife services, or state wildlife office personnel? How important might SOPs be for registration of a new pesticide? How important might they be for defense of management actions in a courtroom?

2. I state that the precautionary principle is at odds with AMPs in resolving human-wildlife conflicts. Find a literature example where AMPs were used successfully. How about an example where AMPs led to ecological backlash? Which is better, to use AMPs or to follow the precautionary principle?

3. Find a published example of the use of AIMs in resolving a human-wildlife conflict, and provide its

reference. What is meant by impact, and what impacts were involved? Show how the results from the initial management strategy were used to make improvements in subsequent strategies.

4. Choose a specific damage problem, such as muskrat damage to a privately owned pond, or a European lynx attacking lambs at a hotspot, and apply the generalized process (see fig. 16.1) to managing the conflict. Did the process help ensure a compre- hensive consideration of the issue? If not, how might the process be modified to make it more useful?

5. Show how the generalized strategy for resolving human-wildlife conflicts, as seen in figure 16.1, might be applied to management by a public agency or a private sector company, such as reducing damage to cars and buildings and health hazards by starlings in a city roost.

17

Future Directions

This chapter probes the impacts of human population growth, its demographics, and emerging technologies on human-wildlife conflicts. Latest and future directions in information technology as it relates to managing human-wildlife conflicts were covered in chapter 3.

HUMAN POPULATION SIZE AND DEMOGRAPHICS

Statement
Sheer numbers of people and their collective activity will challenge our abilities to manage human-wildlife conflicts over the next 40 or so years.

Explanation
In his own way, Malthus was right. Malthus ([1798] 1993) was one of the first to predict exponential human growth (fig. 17.1). Total human population size reached 7.0 billion on or about October 31, 2011, just 12 years after it reached 6.0 billion. It will exceed 8.0 billion in 2021. The United Nations (Department of Economic and Social Affairs, and Population Division [UNPD] 2019) projects about 11.0 billion people in 2100, above the level projected for 2050 of 9.7 billion. The projections are extrapolations from existing demographic data on individual countries, not from a single equation of the whole population, as created by Malthus. The projections mean an increase of 39% or more during the 39-year period of 2011–2050, adding at least 2.7 billion more people to the planet. This addition alone, 2.7 billion people, is a size that many would argue is the sustainable carrying capacity for the whole human population on the planet, providing for a lifestyle equivalent to middle-class Americans living in the 1990s (Daily et al. 1994) or Europeans in the 21st century (Pimentel et al. 2010) while also providing for sustainable biodiverse ecosystems.

The University of Washington (Institute for Health Metrics and Evaluation [IHME], https://vizhub.healthdata.org/population-forecast/; Vollset et al. 2020) provides a more optimistic analysis, this one also based on population dynamics of individual countries. The differences were due to changes that allowed a more refined prediction of mortality due to all causes to 2100; provided age-specific fertility as a func-

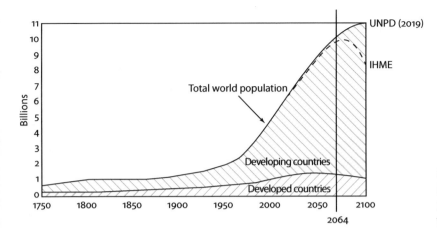

Figure 17.1 A general graph of human population growth. Illustration by Lamar Henderson, Wildhaven Creative LLC.

tion of reproductive age and whether needs for contraception are being met; and incorporated uncertainty for net migration up to 2100. The result was a model that allowed the scientists at IHME to conclude that "because of progress in female educational attainment and access to contraception contributing to declining fertility rates, continued global population growth through the century is no longer the most likely trajectory for the world's population." Instead, these scientists have predicted that the global population would peak at 9.73 billion in 2064, then decline to 8.79 billion in 2100. They explained the difference between theirs and the UNPD forecast as mostly due to 702 million fewer people in sub-Saharan Africa, 584 million fewer in south Asia, and 447 million fewer people in Asia, east Asia, and Oceana.

The researchers suggested that an earlier decline in population was potentially good news for the global environment, but they also warned that "environmental and climate change might still have major and serious consequences in the intervening years unless preventive action and mitigation is vigorously pursued." As evidence, the scientists noted that the current predictions for impacts of climate change by the Intergovernmental Panel on Climate Change are based on a model (the Wittherstein forecast) of human population growth much closer to their model than the one presented by the UNPD.

Although both predictions note that the human growth rate, at 1.1%, has declined from 2.2% in 1960s, both analyses predict the human population will continue to grow for at least the next 46 years. Further, the models are probably conservative in that they do not account for the responses of governments to declining populations and economic concerns for sufficient labor forces to compete in global markets and for

sufficient capital to underwrite programs that support increasing numbers of elderly. For example, in May 2021 the government of China announced that it was increasing the encouraged number of children in families from two to three.

As described by Crist et al. (2017), the sheer increase in the number of people and their demands for goods and services will be accompanied by accelerated pressures on natural ecosystems and their biodiversity. The pressures will be led by food production, including continued efforts to make arable the few remaining natural grasslands and prairies, continued efforts to convert forests to agriculture, and continued efforts to provide food by intensive aquaculture and by overharvest of our freshwater and oceanic coastal and deep-water natural resources. The consequences will be the further decline in size, contiguity, and complexity of natural ecosystems and the earth's natural cycles, whose qualities have already been markedly impacted at global scales by human activity, including profound alterations in nitrogen and phosphorus cycles through use of fertilizers, acidification of the ocean by CO_2 emissions, and greenhouse gases in the atmosphere that have accelerated global warming and climate changes.

Conservation efforts will continue to encourage intensified agriculture rather than expansion of land used for agriculture, efficiencies in food transport, reductions in food waste, and changes in diet as means of reducing impacts on natural ecosystems. But, also as described by Crist et al. (2017), while laudable these efforts on their own are unlikely to succeed. Increased demands for durable housing and other goods, and for transportation, health care, education, and entertainment, and the energy required to produce and deliver them to such a large human population and sometimes in excessive amounts, will further diminish our natural resources and contaminate our natural ecosystems as they negatively impact our entire planet. These efforts will be accompanied by a concomitant increase in the use of our natural ecosystems as a human sink for our sewage, our trash, and our pollutants. Conservation scientists will also continue their efforts to manage these negative impacts of humans on wildlife—from disentangling whales from human trash to mitigating the effects of global warming—but the problems will continue to impact all our wildlife, reaching all corners of the earth at the scales of biomes and the biosphere.

This argument assumes that present levels of lifestyle will continue in developed countries, that lifestyles in some developing countries such as China and India will become more similar to those of other developed countries as part of an emerging global middle class (Ravallion 2010; Kharas 2017), and that energy production is able to meet the increasing demands of all countries for the next 40 or so years. It assumes further that even if the scientific community turns attention back to human overpopulation as the fundamental cause underlying our environmental crisis, and even if it incorporates the needs of all life on the planet into its consideration so that it derives a carrying capacity for humans—say, somewhere between 2 and 5 billion—that can also sustain biodiverse ecosystems, and even if humans, *Homo sapiens*, are sufficiently wise to adopt this limit, underwritten by adopting societal norms such as equality for all and equal rights for women (Crist et al. 2017) and embracing the best approaches that conservation scientists can offer, that it will take generations of this species to save a modest portion of natural ecosystems and wildlife.

This argument also assumes that, in his own way, Malthus was wrong. Malthus assumed that food production would continue to increase in a strictly arithmetic manner; in so stating, he seriously underestimated the human capacity to respond. Foley (2017), who points to this error in Malthus' judgment, puts it this way: "Whether we perceive the finite physical and biological capacity of the planet as hard limits and boundaries, or as the opportunity space for humans to operate and innovate within, reflects deeply held human beliefs and cultural narratives about scarcity and abundance. We can each look at the same planet, and with the same data, with the same laws of physics and biology in mind, and reach different conclusions . . . In other words, whether one considers the planetary glass half full or half empty depends on whom you ask" (252). Foley goes on to add, "But either way, it

would be foolish not to acknowledge the intrinsic limits of the glass" (252). The ingenuity and technological capabilities of humans can greatly facilitate our efforts to slow and perhaps even reverse the tide of declining biodiversity and bide our time while providing for a temporary but unsustainable high quality of life, but these incredible human qualities cannot change the finite limits of the earth itself or the physical or biological laws within which it must function.

So, given a finite planet with an expanding human population exploiting the planet's natural resources, where might those resolving human-wildlife conflicts most strategically place their future investments? One can assume that increased human-wildlife conflicts will occur on all fronts. One challenge will be keeping pace with the quantity and intensity of issues over the next 46 years. Because management resources will be limited, another challenge will be anticipating needs and placing human expertise where and when it is needed most.

Based on the preceding discussion, at least four fronts emerge from the plethora: invasive species, zoonoses, urban areas, and biodiversity.

INVASIVE SPECIES. Most human activities—agriculture, urbanization, production and delivery of goods, education, health care, war, and entertainment—contribute in some manner to invasions by wildlife species. Stressors of ecosystems, for example, not just from agriculture, wars, and global warming, but also from excessive use by all human activities and as sinks for excessive sewage, trash, and other wastes, simplify ecosystems and make them susceptible to invasive species and diseases. Transport of goods and services for consumption, education, health, or entertainment offer unique opportunities for wildlife to stowaway, making novel ecosystems available in hours or days that would have taken millennia or longer for them to access without human assistance. The brown tree snake is an example. It probably arrived on Guam as a stowaway on a military plane and subsequently became a successful invader responsible for significant loss of the island's biodiversity.

Given the broad base of opportunities for dispersal by human transport and the vulnerability of ecosystems simplified by human activity, the numbers of invasive species can be expected to rise impressively in frequency, intensity, and duration during the future crisis. Anticipating and managing these outbreaks would be welcome areas for human-wildlife conflict practitioners to invest their expertise in between now and 2064.

ZOONOSES. In collaboration with their medical colleagues, human-wildlife conflict specialists have played important roles in managing zoonoses, such as detecting and learning the distribution of the diseases in their sylvatic cycles, understanding the nature of the pathogens, and monitoring the courses of the diseases during their spillovers to humans. Facilitated by human activities such as transportation and urbanization, wildlife diseases and zoonoses will also increase in frequency as human populations continue to grow.

Epidemics such as avian influenzas and pandemics such as COVID-19 will likely occur more frequently, so human-wildlife conflict managers will be called on to help resolve these issues with increased frequency over the next four decades. One Health, Eco-Health, and Planetary Health are approaches to managing health that recognize connections between people, plants, animals, and the environment (Lerner and Berg 2017). Such holistic approaches point to similarities in health care among all living things, and particularly EcoHealth has a strong focus on biodiversity. Human-wildlife conflict managers might also consider investing their expertise within the framework of these holistic systems as the approaches gain traction in a future laden with health issues for humans, pets, livestock, and wildlife.

URBAN AREAS. The burgeoning human population will be moving into cities over the next 40 years. The trend toward urbanization will further intensify the need for wildlife damage management and shift its focus. In developed regions, such as Europe, Asia, North America, and Australia income will offer lifestyles for many that are further removed from daily efforts to meet basic needs than lifestyles for many in developing regions. In these developed regions, the trend will continue toward nonconsumptive uses of wildlife for recreation and entertainment such as bird-watching and hiking.

Management goals will of course be led by the needs of agriculture to provide relatively inexpensive and reliable food for city dwellers, and by the need to provide lifestyles in cities free from damage by commensal and other wildlife. Lifestyles here can support a continued shift of value orientations from doministic to mutualistic ones (e.g., Manfredo et al. 2009), so urbanites in these regions will want to consider broader wildlife-conflict issues such as those underlying **recovery** of endangered or threatened species, particularly charismatic ones.

A shift to mutualistic value orientations may also gain momentum in countries like China and India as their middle classes emerge, further encouraging a strong interest in wildlife conservation in these countries. For the wildlife damage practitioner, the shifts

in value orientation suggest that there is wisdom in investing heavily in methods that are acceptable to a public whose sentiments will support only wildlife-friendly methods, particularly nonlethal ones, over tools such as lethal trapping, toxicants, and shooting. Development of additional effective methods that meet criteria for public acceptability would also seem a valuable investment.

In these developed regions, and as humans retreat into cities and as land is abandoned and becomes available for wildlife, one can envision increasing habitat for charismatic species such as mountain lions, pumas, wolves, bobcats, and bears in parts of Europe, North America, Asia (e.g., China), Japan, and South Korea, and opportunities for increased biodiversity of localized ecosystems. Management of human-wildlife conflicts as humans retreat and leave abandoned areas—including old farmlands and repurposed urban entities such as graveyards and amusement parks—will offer demanding new issues for managers of human-wildlife conflicts. Tsunoda and Enari (2020), for example, reported that Japan anticipates increased human-wildlife conflicts as its human population declines by 24% by about 2050, with about 20% of agricultural settlements abandoned.

Xu et al. (2019) reported a dramatic increase in human-wildlife conflicts near the Wolong Nature Reserve, China, consequent to an effort to increase biodiversity in the area. Although this is not an example of abandonment per se, it points to the importance of ensuring an effective plan is in place for managing conflicts with wildlife when biodiversity programs are implemented. Farmers near the Wolong Nature Reserve were required to convert pasture to forest as part of a forest biodiversity program, but the tools used to manage pests, such as guns, traps, and snares, were confiscated and banned. The park is best known for the giant panda (*Ailuropoda melanoleuca*), but farmers reported increased damage from other wildlife including feral pig, Asian black bear (*Ursus thibetanus*), masked palm civet (*Paguma larvata*), hog badger (*Arctonyx collaris*), and Malayan porcupine (*Hystrix brachyura*).

Urbanization in regions of Asia and Africa where populations continue to increase will probably force a different focus on managing wildlife damage. Here, urbanization will occur in the poorest of the cities (Dahl 2005). Because the human effort will be focused on meeting daily needs for survival, management of commensal and feral pests and zoonotic diseases will be heavily emphasized. The outbreak of severe acute respiratory syndrome (SARS) in Hong

Kong in 2002 and 2003 and its rapid spread throughout the globe point to both the increased prospects for emerging zoonoses in densely populated urban areas and to the global connectedness of problems (e.g., Patel and Burke 2009). Investment in the most efficacious and affordable technologies for management of damage caused by commensal and other wildlife species would seem most appropriate for wildlife damage practitioners working in these regions.

In these developing regions, concern and support for all wildlife will also continue to grow, but increasing human densities in many urban areas will force novel approaches if wildlife species and humans are to co-occur or coexist. The leopards of Mumbai, India, are an example. In Mumbai, leopards (*Panthera pardus*) are feared because of their predatory capabilities but coexist with the human residents partly because they manage a feral dog population and reduce the incidences and costs of managing rabies (Braczkowski et al. 2018).

BIODIVERSITY. At the center of human population growth and expansion is the notion that, unless such growth is kept in balance with all life on this planet, human expansion comes at the expense of other species. Given a sustainable human carrying capacity somewhere between 2 billion and 5 billion people, and a present human population approaching 8 billion, this impact is already well underway. Evidence for this, provided by the Living Planet Index, is that the population sizes of mammals, birds, amphibians, reptiles, and fish have declined an average of 68% between 1970 and 2016 (Almond et al. 2021). Given mounting human pressures toward simplification of natural ecosystems, we can anticipate that biodiversity will continue this downward slide. A crucial investment by human-wildlife conflict experts would be into research and operational activities that protect present and future species.

Furthermore, by both UNPD and IHME accounts, sub-Saharan Africa will be a focus of human population pressure over the next 40 years. According to UNPD, most growth will occur in nine countries: Democratic Republic of the Congo, Egypt, Ethiopia, India, Indonesia, Nigeria, Pakistan, the United Republic of Tanzania, and the United States (United Nations 2019; see fig. 17.1). It is the demographic momentum of the youthful population in most of these countries (the increase in the United States is driven mostly by its immigration policies) that drives the upward growth. People in these regions already sustain increasing conflicts with wildlife. Wildlife-safe areas (such as refuges and park reserves) and changing agricultural practices

(such as a shift to sugarcane in parts of India) have increased some wildlife populations. Increasing numbers of people and contact with wildlife will continue to exacerbate conflicts in these regions. Africa and Asia provide habitats for many of the world's most favored wildlife, highlighting the need for conservation-oriented management of wildlife damage in these areas (e.g., Distefano 2005).

Successfully protecting biodiversity will likely be the most challenging of all efforts of future human-wildlife conflict managers, calling for the best investments in strategies and ingenuity. Zoning, wherein areas are designated for use by wildlife, people, or both, and meta-parks, wherein two or more parks are connected by corridors that can collectively provide biological and genetic needs of wildlife, are examples of approaches that might help reduce human-wildlife conflicts in densely populated areas and see increased applications in the future.

Example

Meta-parks seem an increasingly attractive option for survival of wildlife and ecosystem biodiversity as humans encroach. For instance, van Aarde et al. (2006) and van Aarde and Jackson (2007) have proposed the formation of a mega-park that would include a complex of existing ones in several countries, including Kruger National Park in South Africa, as a means of protecting elephant populations. These populations in some of the parks, such as Kruger, are swelling and negatively impacting biodiversity while populations in other parks are low. By connecting them with corridors, as occurs with meta-populations, the scientists suggest that elephants would disperse and redistribute, perhaps reducing or eliminating the need to cull in parks where the megaherbivores are overpopulated.

The mega-park would offer other benefits to elephants, including seasonal access to food and water, and genetic exchange, and help to preserve or improve the overall biodiversity of the system. There are concerns, e.g., that the elephants will not use the corridors sufficiently to disperse or that diseases might be transferred between subpopulations, so the scientists argued for thorough evaluation before such an approach is implemented.

In one recent study, Green et al. (2018) evaluated the use of a corridor by elephants in Kenya. Based on evaluation of over 43,000 photographs with 694 events triggered by elephants, the researchers concluded that the corridor was effective in connecting a southern population to those farther north, also providing safe travel and food and other resources. Such an approach, if successful, would help preserve both elephants and the biodiversity of the parks in the face of mounting human population pressure.

TECHNOLOGICAL ADVANCES

Statement
Human technological advances will fuel unprecedented technical growth in methodology for resolving human-wildlife conflicts. These advances will underwrite how human-wildlife conflict managers can contribute to maintaining the planet's biodiversity while providing continued improvements in methods as judged by such criteria as efficiency, cost-effectiveness, and humaneness.

Explanation
One does not have to look far to find daily examples of major technological advances in virtually every aspect of human enterprise, whether flying a small helicopter on another planet or developing an effective vaccine to COVID-19 in months. Here are some examples of how such advances can be applied to help solve human-wildlife conflicts.

Examples
UNMANNED AERIAL VEHICLES (DRONES). Drones are finding increasing use in managing wildlife in general and human-wildlife conflicts in particular. Companies such as AirShepherd are flying drones in national parks in Zimbabwe, Malawi, and South Africa to help rangers find and catch elephant and rhinoceros poachers. Drones flying along the Klang River in Malaysia gathered data that will be used to assess plastic debris (Geraeds et al. 2019). Wang et al. (2020) found drones about as effective as netting in protecting vineyards from bird damage. Sharkeye, a drone platform, warned beachgoers about the presence of real sharks and rays with over 350 detections in Kiama, Australia (Gorkin et al. 2020). And the SnotBot "is a modified consumer drone that flies through the blow of a whale and collects exhaled 'snot' on petri dishes. This blow contains a treasure trove of valuable biological information: DNA, stress and pregnancy hormones, microbiomes and potentially many other biological compounds/indicators of the animal's health and ecology" (https://whale.org /snotbot/).

BIRD FLIGHT DIVERTERS. Above-ground power lines are increasing globally at about 5% per year (Jenkins et al. 2010). Electricity in distribution lines can be sufficiently high to electrocute birds that alight on distribution lines. But simply flying into the wires kills birds: about 1 million per year in the Netherlands, 175 million per year in the United States, and about 1 billion per year globally. Diverters are colored, spiral or floppy devices placed on powerlines to alert birds to take evasive action or collide. The devices are placed on overhead powerlines using roadcrews in fleets of trucks with buckets (cherry pickers), using helicopters with crew members sitting on the edges of the runners, and, more recently, using diverters designed to be delivered and attached by drones, sometimes with the help of robots that move along the lines.

Ferrer et al. (2020) tested the effectiveness, along roads in southern Spain, of three types of diverters: yellow spiral, orange spiral, and a flapper. The scientists walked along roads and collected all birds fallen beneath wires. They compared the numbers of fallen birds under lines with different diverter types and with roads having no diverters. The researchers found that all diverters reduced the numbers of birds collected under the power lines and that the flapper design provided the most effective overall protection.

REAL-TIME WARNINGS FOR BIRD STRIKES. Collisions between military aircraft and birds show a bimodal distribution, with one peak occurring mostly during takeoffs and another while flying at high speeds and at low altitudes. Damage is related to the size and weight of the birds, but also to the speed of the aircraft and the kinetic energy being expended by the engines (which is high during take-offs and low-altitude flying). Furthermore, most military jets are single engines, so damage to the engine can be disastrous. For these reasons, military authorities have focused on detection of birds and warnings for pilots before planned flights so pilots can be warned or the flights postponed and during flights by warning pilots that a collision with birds might be imminent when flying below a given altitude. Such warning systems, radar-based, have been developed and used since the 1970s.

Gasteren et al. (2019) summarized recent improvements in such systems used in five European countries. The researchers compared the systems used in the Netherlands, Belgium, Germany, Poland, and Israel, countries that conduct military operations within the paths of birds that migrate in huge numbers twice yearly. They compared the frequency of collisions and costs for damage with those of two countries, England and France, that do not have such systems in place.

The systems began in the 1970s in Denmark and the Netherlands with electronic bird intensity counters that were made available on the radar screens of military air surveillance radars, describing mostly flocks of birds as "other targets." Technological improvements came with the addition of weather radar to the systems, called OPERA in Europe. These systems allowed international connections, thereby covering wider areas and displaying larger spatial patterns of migrations. Further improvements came with improved algorithms that provided information on bird density, speed, and direction as a function of altitude.

The algorithm used now, called the Flysafe Bird Avoidance Model, provides updated information every half hour in the airspace of the Netherlands, Germany, and Belgium. Outputs are issued to pilots as BIRDTAM (BIRD-Notice-to-Airmen) warning values from 1–8, based on bird density and level of risk. Values of 5 or higher result in a warning, a prohibition for flying for jet aircraft, or both. For fixed-wing aircraft, which are less vulnerable to damage than jet aircraft, higher values mean slower flight speeds. Measured in numbers of bird strikes per 10,000 hours of flying time and based on comparisons with countries not having warning systems, the systems have reduced damaging airstrikes by 45%, from 7.47 to 4.10. The scientists suggest the systems would have useful applications in civil aviation and in reducing collisions of migrating birds with energy industry and offshore oil and gas production

GENETIC BIOCONTROL. Genetic biocontrol, or **genetic pest management**, involves altering genes of pest populations to either decrease population densities or to render the population harmless. The field was recently reviewed by Clark et al. (2020), who summarized approaches using manipulation of genes, and by Teem et al. (2020), who summarized these as well as more traditional approaches. Three programs have particular potential for managing human-wildlife conflicts: sterile-release, YY males, and gene suppression or drive.

Screw-worms (*Cochliomyia hominivorax*) and the pink bollworm (*Cydia pomonella*) have already been effectively managed using sterile male techniques. By irradiating screw-worm pupae and releasing them in the wild, screw-worm was eradicated from the United States in 1966, and from Mexico and other countries in Central America since 1991 (US Department of Agriculture, undated). The eradication of the pink bollworm from cotton crops in the United States was announced in 2018, following years of policy restrictions and integrated pest management, including sterile male releases.

Sterile male techniques were also used for about 20 years, until 2011, as part of an integrated strategy for managing sea lamprey in the Great Lakes. The high cost of sterilization was one reason for discontinuing the program. Sterile male methods are now being considered for management of American bullfrogs invasive in Europe. Placing the frog eggs under pressure reliably generates triploid chromosomes and makes the gametes sterile. But, as with sterilization of sea lamprey, sterilization of bullfrogs would require a dedicated and large facility, making such a program expensive. At this time the program continues; if it succeeds, it could serve as a model for other programs to manage invasive species or vectors of zoonoses using sterile methods.

YY males, as opposed to normal XY males, produce all male offspring. When YY males are introduced repeatedly into wild areas, the proportion of males in the population increases and the reproductive capability decreases until the population declines and eventually crashes. This approach to population management is being evaluated as a means of eradicating invasive brook trout from the waters of Idaho. Brook trout are a common invasive fish in many states and compete with more desirable species, but attempts to manage the trout have been notably unsuccessful. The YY male program was begun by Idaho Department of Fish and Game in 2008. Traditional hormonal methods using estrogen were used to develop a broodstock over three generations that could provide a supply of YY males. Following a period of research involving the effects of sterile males introduced into natural populations, and field studies on the viability and survivability of such introduced males, field tests have now been initiated in four alpine ponds and seven streams in Idaho. Similar programs have begun in other states, including Washington, Oregon, and New Mexico. If successful, the approach could have broad applications for managing the brook trout and other invasive fishes.

Gene drive or suppression involves editing a gene to include a trait that affects the targeted population. Gene-editing machinery (e.g., CRISPR/Cas9) is used to insert a gene drive allele into a chromosome at a site cut by an enzyme (Redman et al. 2016). In repairing itself, the wild-type allele gene uses the gene drive allele as its template. The result is a chromosome homozygous for the gene inserted by the machinery. The insertion can be made into a gamete or a somatic cell. Because of its homozygosity, the inserted gene will be passed from generation to generation. Although CRISPR/Cas9 is probably the best-known gene editor, others exist that also might be useful in managing invasive species or vectors of wildlife diseases and zoonoses. For example, Teem et al. (2020)

point to Y-CHOP (an acronym for the type of program), a strategy that uses editing machinery to insert an endonuclease that "shreds" the Y-chromosome, resulting in a sterile (XO rather than XY) male.

Unlike techniques using YY and other sterile male methods, where managers can end a program by halting the release of sterile animals (as with the sea lamprey program), gene drives continue on their own after genetically altered organisms are placed within a population. A concern is that the genetically driven gene (the transgene) might escape and move into an unintended population. Such movement was observed at some sites when the bacterial parasite *Wolbachia* was tested for dengue fever control in Australia. The issue has been studied theoretically for other species including the house mouse.

Concern for transgene escape is one reason that the first tests on use of gene drive technology for managing mouse populations will probably occur on islands. On islands, the ocean itself serves as a barrier for mouse emigration. Also, over eons of isolation mice on islands have changed genetically. By inserting the driver gene into a segment of mouse chromosome that is unique to a mouse population on a particular island, translocation of the driver gene to a mouse from another location would be unsuccessful even if a treated mouse were to escape and to mate with an exogenous mouse. The Genetic Biocontrol of Invasive Rodents Programme has begun plans for such a test (Campbell et al. 2019).

Clark et al. (2020) and Teem et al. (2020) summarize a plethora of additional social, technical, and policy considerations that need to be addressed if gene drive or suppression is to see use in managing human-wildlife conflicts.

DE-EXTINCTION. In May 2019, researchers at the Medical Research Council Laboratory of Molecular Biology in Great Britain announced that they had synthesized a form of *Escherichia coli* made entirely from DNA rewritten by humans. The bacteria, named Syn 61 because they were built from 4 million base pairs of DNA based on 61 rather than the usual 64 codons for this type of organism, were reported as alive and well but growing more slowly and developing longer rod-shaped cells than regular *E. coli* (Fredens et al. 2019). These oddities were corrected by later modifications of the code. Although the intent of this human construct was far from a woolly mammoth, the technology demonstrated yet another small step in a growing human technological capability that is on a footpath toward **de-extinction**, i.e., the capability of bringing back walking, talking versions of extinct wildlife. The

contributing technologies, already available, include the now almost routine ability to determine the genetic code of any living organism and to manipulate genetic sequences at will through processes such as CRISPR.

Although the woolly mammoth is not yet available to see at your local zoo or for restoration in a rewilding area, the Harvard-based "Woolly Mammal Revival Project" is hoping to provide a smaller but woolly "proxy" version of such a mammal before long (https://reviverestore.org/projects/woolly-mammoth/).

Darling 58, the genetically engineered version of the American chestnut tree that thumbs its leaves at chestnut blight, is available now. And thanks to more traditional approaches to breeding used to help resolve issues surrounding a different conflict, global warming, so are varieties of coral that, rather than bleaching, continue to thrive and to display their vivid colors in unusually warm waters (Cornwall 2019).

Summary

- The increase in numbers of people, between 2 and 3 billion, from 2021 to 2064 is what many would argue is a realistic carrying capacity for the whole human population if it is to have a modest middle-class lifestyle and provide for the planet's biodiversity.
- At the present size of almost 8 billion people, and with a continued increase to at least 9.7 billion and likely more, demands for goods and services will be accompanied by accelerated pressures on all natural ecosystems and their biodiversity. Consequences will be the further decline in size, contiguity, and complexity of natural ecosystems, the earth's natural cycles, and its biodiversity.
- Human-wildlife conflict specialists can anticipate increased human-wildlife conflicts on all fronts. Practitioners need to focus their future efforts in at least four areas: invasive species, zoonoses, urban human-wildlife interests, and biodiversity.
- Urban interests will depend on whether the population is declining or expanding. People living in areas with declining human populations will continue moving toward a more humanistic value orientation, with some opportunities to contribute positively to biodiversity as abandoned farmlands and urban areas are repurposed.
- People living in areas with increasing populations will focus on more immediate human-wildlife conflicts surrounding commensal rodents and diseases. Because much human growth is anticipated in areas, such as sub-Saharan Africa

and parts of Asia, with many of the world's larger and most charismatic species, maintaining or increasing biodiversity in these areas will be particularly challenging. Human-wildlife conflict specialists should anticipate adjusting their methods to those the people in these regions find acceptable.

- Technological advances are occurring daily and in virtually every aspect of human endeavor. Some of the advances can be applied to managing human-wildlife conflicts, allowing unprecedented opportunities for specialists in this area.

- Unmanned aerial vehicles (drones) are already being used to help rangers find and catch elephant and rhinoceros poachers, to assess plastic trash in rivers, to warn beachgoers of the presence of sharks, and to sample the blow from whales. Applications for drones will continue to expand.

- Bird flight diverters, which are being installed (sometimes by robots and drones) on powerlines, have the potential to greatly reduce the number of birds killed globally by flying into the wires.

- Near real-time warnings of imminent collisions with birds are being provided to military pilots in Europe and other regions of the world. The radar systems that provide this information can be expected to continue to improve and their applications to civil and other uses to greatly expand.

- Genetic biocontrol, or genetic pest management, includes sterile-release, YY males, and gene suppression or drive. One or more of these methods are being evaluated for management of the American bullfrog in Europe, for eradication of brook trout in Idaho and other states, and for eradication of invasive mice on islands. Applications of these technologies could have dramatic consequences in the effectiveness of methods for managing wildlife damage.

- A woolly mammoth is not on the immediate horizon, but the Harvard-based Woolly-Mammal Revival Project is hoping to provide a smaller but woolly "proxy" version before too long. This and other technologies are steps in a growing human technological capability that is on a footpath toward de-extinction—i.e., bringing back walking, talking versions of extinct wildlife.

Review and Discussion Questions

1. Would human-wildlife conflicts exist in a world where the human population was under 2 billion? If your answer is no, defend your view with references. If your answer is yes, describe its nature with references.

2. How important is the planet's biodiversity to the health and well-being of humans? Explain, with references.

3. What specific areas do you think should be the focus of human-wildlife conflict practitioners during the next four decades? Explain your answer using references.

4. List the top five technologies being advanced by humankind. Give a specific example of how each might be applied to manage human-wildlife conflicts.

5. Thinking as a human-wildlife conflict practitioner, what advances would you like to see in technology over the next decade? In information management? In applications of human dimensions?

Glossary

Abatement programs—local, state, or federal public programs that provide consultation services, direct support from agency experts, or subsidies for technologies for abating human-wildlife conflicts

Abiotic—nonliving

Active ingredients—biologically or pharmacologically active components of a pesticide

Acute rodenticides—rodenticides that are effective in a single dose, such as zinc phosphide

Adaptive Management Practices—see **Adaptive strategy**

Adaptive strategy—management or strategy that proceeds without a full understanding of the problem and its solution but includes sufficient monitoring and experimental data gathering so that post hoc analyses lead to improvements in the strategy

Additive mortality—the notion that removal of population members by trapping, hunting, or poisoning might be timed so that it adds to other forms of mortality, thereby facilitating a decreased overall population size

Addling—"loss of development," the destruction of eggs by any physical or chemical means, by puncturing, freezing, or coating with oil such as vegetable oil

Agents—used here as the causes of diseases, often biotic but sometimes abiotic, such as radiation

Altruistic value—the amount an individual is willing to pay to maintain an asset or resource that the individual does not use but wants available for others to use

Animal rights—the belief that wildlife should have legal and other rights similar to those of humans

Animal welfare—the belief that wildlife should be managed in ways that foster health and wellness, and that minimize the pain and suffering of individual animals

Anthropomorphic—described or thought of as having human attributes

Anticoagulant—compounds that reduce the ability of blood to coagulate, used variously as medicine and poison

Apex predator—a predator at the top of its food chain, often also a keystone species

Attenuation—decrease in intensity of a behavioral response to a repeated stimulus

Attitudes—"learned tendencies to react favorably or unfavorably to a situation, individual, object or concept" (Allen et al. 2009, 5)

Attractant—in baits, the inactive ingredient that selectively attracts intended animals

Autecological—ecology of individual organisms or species

Avicides—pesticides used on birds

Baits—mixes of liquids, solids, or powders, parts of organisms, or whole organisms that attract targeted animals; the animals may be lured into traps, or the baits may deliver a management compound such as a pesticide that kills, repels, or contracepts

Bait shyness—avoidance of a bait after consumption of a sublethal amount, based on flavor aversion learning

Basic reproduction number (R_0)—used to predict likelihood of a disease outbreak from a species jump, it is the ratio of secondary cases (i.e., number of individuals in the new population that are infected from primary cases) to primary cases (i.e., number of individuals infected by the jump)

Beliefs—judgments about what is true or false

Benefit-cost analysis (BCA)—an economic analysis that considers all benefits and costs, direct and indirect, present and future (discounted to present value), and wherein total costs are then subtracted from total benefits to derive a net benefit

Bequest value—the value of knowing that something, such as wildlife, will be present for future generations

Best Management Practices (BMPs)—informal standards for techniques or methods that have proven successful over time

Binder—also called a "sticker," in baits, the inactive ingredient used to ensure that the active ingredient adheres to the carrier

Biogeography—study of geographic distribution of organisms in space and time, including landscape ecology

Biological clocks—the intrinsic abilities of organisms to time certain activities according to the rotation of the earth, the movement of the moon around the earth, the earth around the sun, or the juxtaposition of heavenly bodies

Biological control (biocontrol)—using interactions among organisms to limit damage or presence of damaging species, e.g., using lure crops to attract pests away from crops of higher cash value

Biological life cycle—movement of a disease between hosts

Biomagnification—food chain concentration, wherein a substance such as a pesticide moves up the food chain, increasing in concentration by roughly an order of magnitude at each level

Biomes—as defined by Raven and Johnson (1992, 518), "climatically delineated assemblages of organisms that have a characteristic appearance and that are distributed over a wide land area"

Biopesticides—pesticides made from living organisms or their genetic parts

Biosphere—the largest biological community, comprising all life on the shell of the earth

Biophilia—love for living things, an innate emotional connection between humans and other living things driven by a long co-evolutionary history

Biotic—living

Bud caps—paper or plastic coverings placed over the tops of transplants or over buds, usually held in place with one or more staples, to prevent damage

Cable device—also called a "snare," a steel cable with a loop on one end for capturing animals

Camera trap—wherein animals are recognized from photographs taken at stations rather than actually captured, marked, released, and recaptured

Carnivore—organism that eats animals

Carrier—in baits, the matrix in which all of the ingredients are mixed

Carrying capacity (K)—the upper limit of population growth; i.e., the maximum number of population members that an area of habitat can support in a sustained manner

Character displacement—morphological or behavioral changes that occur over time within a population in response to competition with other species

Charismatic megafauna—large animals with widespread popular appeal, such as elephants, tigers, lions, leopards, bears, hawks, and eagles

Chronic rodenticide—a rodenticide, usually an anticoagulant, that requires repeated feedings for physiological or pharmacological effects

Chronobiology—also called "ecophysiology," study of biological cues and rhythms

Circadian rhythms—biological clocks associated with daily rotation of the earth

Circannual rhythms—biological clocks associated with the movement of the earth around the sun, including both seasonal and annual activities

Citizen science—scientific research conducted in whole or in part by amateur scientists

Classical methods—in biocontrol, where the control agent is released one or more times so that the agent becomes established and gradually manages the pest

Climatic climax community—mature seres for a major biome or region, often described by the dominant plant species for that biome or sere

Climax community—stable final stage of ecological succession where energy used equals energy input. Climatic climax communities are final stages for a particular climate, often described as biomes. Edaphic climax communities are most stable communities achieved in areas that are prone to periodic climatic disasters such as flooding or fire

Clumped distribution—the most common distribution of organisms, wherein the distances between neighbors are minimized. Clumped distribution probably occurs because of gathering around limited resources, family or clan grouping, defensive gatherings such as herds, or limitations in movement

Coalition—a group of two or more individuals, factions, organizations, political parties, or other groups that form a partnership, usually temporary, to achieve a common goal

Commensals—uninvited guests that have "come to the table" to eat food intended for humans, including wildlife such as rats and mice

Common chemical sense—stimulation of the pain receptors of the trigeminal system

Community—all the groups of organisms living and interacting in an area at a given time

Community physiognomy—the physical structure of a community or ecosystem, including both its vertical and horizontal dimensions

Compensatory mortality—the notion that some members of a population can be removed by hunting or trapping without overall long-term impacts on population size because an equivalent number of these members would have died anyway from other causes, such as disease or starvation, had they not been hunted or trapped

Competition—when organisms of different species interact, and both species are affected negatively by reduced environmental range or roles, either directly or indirectly by reducing the supply of a needed resource

Competitor release effect—explosive increases in population size when a competitor is removed

Conservation—advocates the wise use of wildlife resources

Conspecific—an individual of the same species

Convergent evolution—species that occupy similar ecological niches may evolve similar characteristics in response to similar selective pressures, thereby becoming more and more similar in appearance, behavior, and physiological attributes

Coprophagy—consuming feces

Corral—a pen for livestock

Corridors—thin runs that connect patches, all having similar habitat

COVID-19—a highly infectious respiratory disease that has species-jumped to humans and is caused by SARS-CoV-2 virus

Crepuscular—active during dawn or dusk

CRISPR—literally meaning "clustered regularly interspaced short palindromic repeats," refers to a technology now used routinely to add genes into chromosome at a desired location

Critical minimum—the minimum population size, below which the population is no longer viable and dies

Crowding—a condition wherein the population density becomes too high, affecting the health of population members when they become stressed

Cull—in wildlife management, to reduce population size by removing, usually by killing, members of the population

Culture—a set of shared values, beliefs, and behaviors that characterize an institution, organization, or group

Damage—an affront, real or perceived, against humans, their property, or something they value (e.g., other wildlife)

Deception—in politics, a method wherein one is given only part of the whole story or the story is deliberately misconstrued to achieve a political objective

Decision support systems (DSSs)—systems, often computer based and involving iterative calculations, that can analyze, organize, and present data in a manner that helps people make decisions

Decoy crops—also called "lure" crops, crops provided as alternatives to attract pests away from crops having higher values

De-extinction—resurrection biology, i.e., generating an organism that either resembles or is an extinct species

Density dependent—in populations, factors that affect the size of the population in a manner that is related to the number of members per unit habitat; e.g., stress due to crowding that causes poorer care and survival of young as the density increases with some species

Depletion trapping—also called "removal trapping," when members of a population are intensely trapped and permanently removed; data gathered can sometimes be used to estimate population size

Despot—a ruler with absolute power, often exercising power in a cruel or oppressive way

Detritus—the dead remains of a formerly living organism

Detritus food circuit—the flow of energy through a food web that begins with consumption of dead organisms and feces

Detrivore—or detritivore, an organism that eats the remains of a dead organism

Diethylstilbestrol—a reproductive inhibitor used to block egg production in some bird species

Digastric—having two-stomachs

Diseases—abnormal conditions that impair the functions of organisms

Displacement behavior—usually self-grooming, touching, or scratching, displayed when an animal has a conflict between two drives; in managing human-wildlife conflicts, displacement behavior can be used to facilitate ingestion of control compounds, such as ingestion of aversive materials during self or heterogrooming

Distress—stress at levels that cause harm

Diurnal—day active

Diversionary feeding—providing food to a pest to protect something of greater value, e.g., feeding bears to keep them from damaging forest trees

Domestication—a special case of symbiosis in which humans artificially select characteristics of plants or animals for qualities that humans desire, and the plants and animals benefit with resources and protection afforded by humans as well as managed harvest with seed collection for the next season

Dominance—social "pecking order" wherein one or more members of a population claim more resources than are made available to others; or the relative importance of one or more species in an ecosystem

Doministic value orientation—a value orientation in which the well-being of humans is seen as a higher priority than that of wildlife

Ebola virus disease—a viral disease with malaria like symptoms, often fatal, first recognized in the Democratic Republic of the Congo in 1976, with bats or other wildlife as reservoirs; formerly known as Ebola hemorrhagic fever

Ecological backlash—where the negative ecological consequences, often unanticipated, of a management action, such as the introduction of an exotic for biological control, outweigh the positive gains of the action

Ecological density—the number of individuals per unit habitat

Ecological equivalents—organisms or species that occupy the same or similar ecological niches in different geographic locations; the species may evolve convergently

Ecological niche—the range of all factors within which the organism must live in a kind of multifunctional array, along with the functions, or roles, of the organism within that environment

Ecology—the study of plants or animals "at home"; i.e., the study of plants or animals interacting with both the living (biotic) and nonliving (abiotic) parts of their environment

Ecophysiology—also called "chronobiology," the study of biological cues and rhythms

Ecosphere—the biosphere as a system with energy flowing through it, and interacting with its components, both living and nonliving

Ecosystem—a biological community interacting with its abiotic components

Ecotone—edges between communities or ecosystems, having some species characterizing each community as well as some unique species of their own

Ectoparasite—vector of a disease that is both smaller than, and lives on the outside of, the host

Edaphic climax community—seres maintained in a stable fashion for prolonged periods of time by recurrent factors such as floods or fires

Emerging zoonosis—a newly recognized or evolved pathogen, or an extant one that has increased in incidence or expanded in geographical, host, or vector range

Emigration rate (e)—the number of members leaving a population per unit time, often expressed as the number of members leaving per 1,000 population members per unit time

Emotions—along with moods, constitute affect, a general class of feelings that humans experience; emotions relate to specific events and are short-lived and conscious, whereas moods are longer-lasting and in the background of consciousness

Endoparasites—a parasite that lives inside the infected organism

Energy subsidies—in ecology, external energy sources that reduce the cost of self-maintenance of an ecosystem, thereby making more energy available for production

Enhancement—increase in intensity of a behavioral response

Enhancer—in baits, an inactive ingredient that encourages consumption of sufficient amounts of the formulation

Enmeshing—tangling fish or other aquatic organisms to catch them, less selective than gill nets

Environmental resistance—collectively, all factors such as starvation, disease, and storms and flooding, that exert pressure against growth of a population

Enzootic cycle—also called the "sylvatic cycle," movement of a disease between wildlife as hosts and vectors; the natural portion of the biological life cycle of a disease

Epidemics—infectious diseases spreading rapidly among large numbers of people

Ethics—sets of moral values defining good and evil, right and wrong, by which people behave

Eury—a prefix used to define a broad range or function; e.g., eurythermal means that an organism can live in a broad range of temperatures

Euryhalic—able to tolerate a broad range of salt concentrations

Euryphagic—able to eat a broad range of foods

Eurythermal—able to live in a broad range of temperatures

Eustress—low levels of stress that might be beneficial

Even distribution—evenly spaced distribution of organisms, such as fields of corn, occasionally found in nature, e.g., creosote bushes. The distribution can be a consequence of competition for resources that are uniformly distributed over an area, such as reproductive territories for penguins

Evenness—in ecology, an index used to measure the distribution of individuals among species in a community

Exclusion—a method of reducing or eliminating damage by separating, physically or otherwise, wildlife from the items that could be damaged

Existence value—value ascribed to something such as wildlife just because it is there, e.g., the value of simply knowing that snow leopards exist in Asia

Exotic—a species appearing in a new habitat or geographic location

Expected value (EV)—a form of economic analysis that can be used when there is uncertainty surrounding the benefits from a management program

Expediency—in politics, taking an easier route to re-election than one in the public's best interest

Exponential growth curve—geometric growth of an unrestricted population, often resembling a "J" shape when graphed with numbers of population members as the ordinate (y-axis) and time as the abscissa (x-axis)

Feeding behaviors—the behaviors associated with eating

Ferae naturae—an ancient common law doctrine stating that a wild animal cannot be owned by anyone

Feral—domesticated wildlife that are released or escape back into the wild, and that live to reproduce and become established as a population, often causing damage to humans, their property, or to other wildlife

First law of thermodynamics—energy can be neither created nor destroyed but can be converted from one form to another, e.g., binding the photonic energy from the sun into a chemically bound form in adenosine triphosphate

Fladry—flags, usually in bright colors like red, hung at regular intervals along the perimeter of an area to deter predators from crossing the boundary; used, e.g., as an ancient method to direct wolves into traps or to keep them away from sheep

Flavor aversion learning—learned association of the flavor of a food with postingestion illness and subsequent avoidance of the flavor

Fleas—wingless insects of the order *Siphonaptera* with mouthparts specialized for sucking

Food chain concentration—also called "biomagnification," concentration of some compounds as they move up a food chain, often increasing by an order of magnitude at each level of transfer, as occurs with, e.g., lipid soluble pesticides

Food hoarding—storing food

Foraging behaviors—the behaviors associated with searching for food

Foregut fermentation—the form of digestion that involves a two-stomach digastric digestive system (also called rumination), elaborate and time consuming but resulting in efficient use of ingested plant material. Used by artiodactyls, and kangaroos, sloths, and colobus monkeys. The microorganisms in the rumen can detoxify poisons.

Forgone opportunity cost—costs accrued when an activity is not conducted in an area, e.g., crops or livestock are purposely not grown, because of concerns for conflicts with wildlife

Foundation species—a dominant primary producer in an ecosystem, such as kelp in an ecosystem with sea otter as the apex predator and keystone species

Fundamental ecological niche—the theoretically greatest potential habitat and ecological roles of an organism; a limiting factor keeps the organism or population within this niche

Fundamental equation for growth—$r = (b + i) - (d + e)$, where r is the population growth rate, b is births (natality), i is immigration, d is deaths (mortality), and e is emigration

Fungicides—pesticides used on fungi

Fyke nets—specialized fishing nets with wings and internal cones to direct fish into a collecting area

Gause's law—If two species occupy the same niche in the same habitat at the same time, one will outcompete the other until it disappears, or one or both species will evolve in divergent manners until their niches are sufficiently separated to reduce or eliminate the competition.

Genetic pest management—when genes of pests are altered to either decrease population densities or render pests harmless

Genetic pollution—mixing genes, thereby causing hybridization or introgression and affecting survival of the species

Genetic resistance—survival by members of a population that are genetically predisposed to resist treatment by a management compound so that the whole population eventually resists treatment, e.g., genetic resistance of some populations of Norway rats to the anticoagulant rodenticide warfarin

Geophagia—feeding drive to eat dirt, soil, earth or starchy things

Giardia—disease caused by parasitic protozoa; sometimes carried by beaver

Gill nets—nets designed so fish or reptiles of a particular range of sizes are caught when they try to pass through the net

Global climate change—any change in climate over time due to either natural variability or human activity

Global warming—increase in the earth's air and oceanic temperatures

Good Laboratory Practices (GLP)—standardized practices that need to be followed in the conduct of studies used for registration of pesticides and drugs and that will stand up to both scientific and legal challenges

Granivorous—feeding on grains

Grazing energy circuit—the major path of energy flow in systems where sunlight is the direct source of energy, such as a grassland or a savanna

Growth rate (*r*)—the change in numbers of population members over time; the sum of four other rates: natality (birth, *b*), mortality (death, *d*), immigration (*i*, movement into), and emigration (*e*, movement out of); see fundamental equation for growth

Gustation—sampling food for taste

Habitat—the place one goes to find an organism, i.e., the environment of the organism including both living and nonliving components

Habitat fragmentation—when a habitat needed by wildlife is made smaller and smaller or is isolated from previously interconnected habitats

Herbicides—pesticides used on plants

Herbivore—organism that eats plants

Herbivory—eating plants

Hindgut fermentation—also called "monogastric fermentation," this digestive system is used by perissodactyls, elephants, lagomorphs, and rodents. The system uses one stomach, and is fast but inefficient. Toxins ingested in the food are absorbed into the blood, and the liver is relied on to detoxify them.

Holistic strategy—comprehensive strategies that are the consequence of careful consideration of every aspect of a problem

Home range—the area within which an animal conducts its daily activities

Hormone—substance produced by an organ of an animal and released into the bloodstream; the substance then affects "target" organs, stimulating them to either produce more or less of another substance, helping the animal to regulate a physiological process

Hosts—with diseases, organisms that harbor pathogens

Human dimensions—the human aspects surrounding an issue, including economic factors, human perceptions of the issue and responses to it, and politics and public policies related to the issue

Humaneness—often measured as the extent of effort to minimize pain and suffering of animals, as with a management method such as trapping

Human tolerance—in wildlife management, beliefs surrounding acceptable risks associated with wildlife

Human-wildlife conflicts—all negative interactions between people and wildlife, including those caused by people as well as those for which wildlife is the proximate cause

Human-wildlife interactions—all types of interactions between people and wildlife, those perceived by people as both positive and negative ones

Immigration rate (*i*)—the number of new members recruited into a population, often expressed as the number of new members per 1,000 population members per unit time

Inactive ingredients—ingredients in a bait that do not have biological or pharmacological activity but serve other functions, such as binding, attracting, enhancing consumption, masking flavors, or preserving

Instantaneous growth rate—dN/dt, or, the instantaneous change in the rate of growth, dN, for a population with size N, at the instantaneous change in time, dt

Instinct—unlearned response to an environmental stimulus consisting of encoded, stereotyped behaviors, often observed with insects, amphibians, reptiles, and birds

Integrated pest management (IPM)—management that is effective, involves a variety of methods, and minimizes the use of pesticides

Intermediate hosts—also called "secondary hosts," vectors that carry parasitic pathogens for prolonged periods of time

Interspecific interactions—interactions between an organism and other species

Inundative method—in biocontrol, when large numbers of agents are released periodically throughout a single season so that sheer numbers overwhelm the problem population

Invader species—the first organisms moving into an area devoid of life

Invasive species—non-native species that adversely affect the habitat they invade

Iron triangle—a model formulated by political scientists that ascribes three foci of political power held together in an irrevocably bound system; in the US federal government, the foci are relevant positions and units of the executive branch, Congress, and lobbyists

Isocline—a curve through points at which the function's slope will always be the same, e.g., in Lotka-Volterra models, the sloped line where $dN_1/dt = 0$

J-shaped growth curve—see exponential growth curve

K—see carrying capacity

K/2—the inflection point of a sigmoidal population growth curve wherein overall natality is high, and density dependent control factors are still minimal; the point is sometimes viewed as the point of maximum sustainable yield for the population, and targeted in game management plans

Kevlar cord—a para-aramid synthetic fiber string often used in grid systems to repel wildlife such as geese, ducks, or gulls from bodies of water

Keystone species—a species that has a much greater effect on its environment than would be indicated by its abundance. Examples include jaguars in Central and South America, and grizzly bear, prairie dog, and beaver in North America

K-growth species—species that characterize mature seres of ecosystems, and that tend to have relatively fewer offspring, but invest parental care in the survival of the offspring

Landscape ecology—the geographic description of communities and ecosystems, often based today on remote sensing, imagery, and layering provided by geographic information systems

Learning—complex behavioral responses that are modified by experience

Lice—wingless obligate parasites, of which there are over 3,000 species

Liebig's law of the minimum—the factor in the environment that occurs in the least quantity in relation to need will limit the growth of the organism or its population

Life table, cohort or horizontal—a table that provides information on factors such as natality, mortality, or survivorship rates as functions of age, and that is based on all members of a population that were born within the same period of time

Life table, static or vertical—a table that provides information on factors such as natality, mortality, or survivorship rates as functions of age, and that is based on individuals of different ages, but all sampled within one period of time

Lincoln-Petersen Index—a method for estimating population size based on capturing and marking population members, releasing them, and then recapturing a portion of them; today physical capturing may be unnecessary because individuals might be identifiable using photographs or DNA samples, e.g. from hair or feces

Lingering—said of a wildlife disease that has been known since ancient times, such as rabies

Livestock protection dogs (LPDs)—dogs used to protect livestock from predators

Local knowledge—usually means a knowledge system found in the cultural traditions of a local community

Lotka-Voltera models—a series of mathematical equations that describes the theoretical interactions of competing species, including those between predator and prey

Lunar rhythms—biological clocks tied to the rotation of the moon around the earth

Lure (decoy) crops—crops provided as alternatives to attract pests away from crops having higher values

Lyme disease—borreliosis, a disease caused by bacteria of the genus *Borrelia*, transmitted by ticks of the genus *Ixodes*, and causing flulike symptoms in humans, eventually affecting joints, heart, and nervous system if left untreated

Macroparasites—parasites that can be seen with the naked eye, e.g., protozoa and helminths

Malthusian (growth curve)—see exponential growth curve

Management—getting something or someone to do what you want them to do

Marginal value theorem—foraging models that are based on energy intake versus energy used to forage, such as where food available in one patch is diminished to the point where an animal must decide whether to keep foraging at greater and greater energetic costs, or to invest the time and energy into locating a new patch with more food

Market value—what a good sells for in the local market

Masking agent—in baits, an inactive ingredient that overshadows or otherwise hides flavors in baits that might induce avoidance

Mast—seeds and fruit, such as acorns and other nuts, of woody plants that wildlife can eat

Matrix—in landscape ecology, a broader ecosystem or habitat that surrounds a smaller one

Maximum growth rate—growth rate under ideal conditions

Maximum sustainable yield—see $K/2$

Mean—in mathematics, the average

Median—the middle value of all values when ranked from lowest to highest, e.g., median damage of 5.7% (third value) in corn fields with damage ranging from 0.3% (first value) to 15.7% (fifth value)

Mesopredator release—dramatic increase in numbers of middle-level predators as a consequence of reduction in numbers of apex predators

Metapopulations—spatially separated populations of the same species that have some genetic interaction, often considered management units for species existing within fragmented landscapes

Microhabitat—a habitat that varies only slightly, but with describable differences, from the habitat of a similar species

Microparasites—microscopic parasites that can complete life cycles within one organism, e.g., viruses, fungi, obligate intracellular bacteria (e.g., *Rickettsia*)

Midges—tiny "no-see-ums," or sand flies

Minimum distance sampling—when a point is randomly selected within a field and a distance is measured between that point and the nearest damage, then the distance is measured between that damage and the nearest second point of damage, then between the second and third points, and so on, providing data with an accurate measure of damage

Mites—along with ticks, small arthropods of the order Acarina

Molluscicides—pesticides used on snails

Monocultural agroecosystems—agroecosystems simplified to produce a single crop

Monogastric—having a single stomach

Mortality rate (*d***)**—death rate, often expressed as the number of deaths per 1,000 population members per unit time

Multidimensional strategy—a management strategy based on human dimensional as well as biological and ecological concepts

Mutualistic value orientation—a value orientation in which wildlife is seen in trusting relationships with humans, as part of our extended family

Mylar tape—brightly colored plastic tape, often red on one side and silver on the other, used to repel birds

Myopathy—any disease that affects voluntary movement in the body. Capture myopathy in wildlife results from extreme exertion and struggling as a result of capture but can result from natural causes of stress.

Myxomatosis—a disease of the European rabbit that causes skin lumps, puffiness, and sometimes conjunctivitis and blindness and reduces resistance to bacterial infections such as pneumonia; used successfully in the past to manage European rabbits in Australia

Natality rate (*b***)**—birth rate, often expressed as number of births per 1,000 population members per unit time, or number of births per 1,000 reproductive aged females per unit time

Nectivorous—feeding on nectar

Neophobia—fear of something novel, such as a new food

Net benefit—the benefit derived by considering all benefits and costs, direct and indirect, present and future (discounted to present value), and then subtracting total costs from total benefits

Net present value (NPV)—present and future year expenditures versus benefits for the life of a management program, expressed as a present net, where future years are incorporated into present values by using a future discount rate, e.g., inflation rate

Nidus—a "nest" for a disease, i.e., a long-term host for the pathogen of a disease

Nocturnal—active at night

Omnivore—eating a broad range of foods

Operational programs—a term often used when professional experts manage human-wildlife conflicts, e.g., for landowners, farmers, ranchers, homeowners, or businesses

Optimal foraging theorem—a model of foraging which theorizes that organisms maximize their energy intake per unit time; the model is usually based on three categories of issues; type of food or prey; the currency used, such as time or energy; and constraints on time, such as risk of predation

Option value—the value of knowing that one will have the opportunity to use or enjoy something such as wildlife, at a future time of one's choice

Oral rabies vaccination—vaccinations that are delivered to targeted wildlife in massive baiting campaigns and that have successfully reduced the incidences of rabies in wildlife species, first in Europe and later in North America

Outbreaks—when infectious diseases occur at higher frequencies than expected

Overabundant—populations with sizes or densities that threaten human life or livelihood, impact densities of (other) favored wildlife species, have too many members for their own good, or impact ecosystems and cause their dysfunction

Overshadowing—when two or more conditioned stimuli are presented along with a single unconditional stimulus, an organism associates the unconditional stimulus only with the most salient conditioned stimulus

Pandemics—when the spread of diseases are larger than epidemics and potentially global

Parasite—see parasitism

Parasitism—one organism negatively affects another organism of a different species wherein the parasite, usually the smaller of the two species, benefits by consuming the food, nutrients, or a physical part of some of the other species, the host

Patches—areas of relatively homogeneous habitat surrounded by a broader but different habitat, the matrix

Pathogens—biological agents that cause illnesses or diseases

Peer review—critical evaluation of a piece of work by individuals well qualified to conduct the evaluation. In scientific method, information is published in most scientific journals only after passing through peer review.

Pest—the individual, population, or species that is causing damage

Pesticide—any mix of substances (or devices) that affect the behavior of pest plants or animals

Pheromones—substances, usually highly volatile ones, secreted into the environment by an individual, that have behavioral or physiological effects on other individuals of the same species

Photoperiod—length of daylight daily; changes in photoperiod are detected by many organisms, and used as a cue to time certain activities

Photosynthesis—the process of converting carbon dioxide into organic compounds, often sugars, using energy from sunlight

Physiognomy—landscape or ecosystem structure

Piscicides—pesticides used on fishes

Plague—a disease caused by the bacterium *Yersinia pestis*, and manifested as bubonic, septicemic, or pneumonic

Population density—the number of population members per unit area of habitat

Population growth rate (r)—the change in numbers of population members as a function of time—the sum of birth (or natality) rate (b) and immigration rate (i) minus the sum of mortality (or death) rate (d) and emigration rate (e); or $r = (b + i) - (d + e)$

Population index—an estimate of the relative change in population size, without necessarily knowing the actual population size; often relatively inexpensive to determine and can be useful for managing human-wildlife conflicts

Populations—groups of actually or potentially interbreeding organisms living in the same place at the same time

Population size (N)—is the number of population members at a given moment in time

Prebaiting—providing a bait, or its flavor, without the active ingredient so that the animal becomes familiar with the flavor, sometimes improving initial acceptance of the bait when the active ingredient is added back

Precautionary principle—states that no action should be taken until a problem and all consequences of proposed actions are fully understood

Predacides—pesticides used on predators

Predation—attacking another animal; wherein the predator, usually the larger of the two species, often benefits by consuming some of the other species, the prey

Preservative—in baits, an inactive ingredient that prevents degradation of the bait and protects it from the weather

Primary flavor aversion—unlearned avoidance of some flavors, such as "bitter"

Primary scientific literature—original reports of studies, following the scientific method, and related observational or experimental designs, or both, and reviewed by peers prior to publication

Primary succession—succession that occurs when organisms enter an area devoid of life, such as after a volcanic eruption or a prolonged major flood

Prions—abnormally folded protein bodies, possibly the cause of bovine spongiform encephalopathy (i.e., "mad cow disease") in cattle, chronic wasting disease in cervids, and scrapie in sheep

Propagule pressure—the likelihood of a successful wildlife invasion, based partly on the number of individuals emigrating and partly on the frequency of invasions

Proprietor—someone acting as an owner

Protists—any organism whose cell or cells contain a nucleus or nuclei that is not an animal, a plant, or a fungus

Public policy—"whatever governments choose to do or not to do" (Dye 2008, 1)

Pyrotechnics—use of explosives; in managing human-wildlife conflicts, this might involve a broad range of activities, from use of "whistlers" and "bangers," i.e., fireworks to scare wildlife, or plastic or other explosives to blow up beaver dams

Quadrat sampling—where a grid of specific dimensions, e.g., a 10×10 grid of cornstalks, is used to sample damage, starting with a randomly determined coordinate and continuing until a pre-established number of quadrats are sampled

Rabbit hemorrhagic disease virus (RHDV)—causes hemorrhagic disease, a highly contagious disease specific to both domestic and wild rabbits; used to manage rabbits in Australia and New Zealand

Rabies—a viral neuroinvasive disease that causes acute encephalitis in warm-blooded animals

Random distribution—least common type of distribution of organisms that show no clear pattern, and in which the position of one organism is independent of the position of another; e.g., distribution of tropical fig trees that are pollinated by winds

Range of tolerance—any organism must live within a range for each of the physical components of its environment, and cannot survive in environments where any of these ranges are exceeded

Realized ecological niche—the actual niche occupied by an organism or population, less than the fundamental ecological niche due to interactions between the organism or population and other biota

Reasoning—behavioral responses based on rational thought and strategy

Recovery—in conservation, allowing ecosystem changes to occur by ecological succession over time without further human intervention

Recruitment ($b + i$)—natality (b) plus immigration (i), additions to population over time; see fundamental equation for growth

Reflex—the response of an organism, or part of an organism, to an environmental stimulus, more sophisticated than a taxis, modifiable by learning

Removal trapping—also called "depletion trapping," when members of a population are intensely trapped and permanently removed; data gathered can sometimes be used to estimate population size

Reptilicides—pesticides used on reptiles

Reservoir—with diseases, a long-term host for a disease

Restoration—in conservation, when humans attempt to speed up the recovery of an ecosystem by assisting natural ecological processes

R-growth species—species characterizing early seral stages of ecosystems, which tend to produce large numbers of offspring but make little parental investment in their survival

Richness—a diversity index that measures the number of species and distribution of individuals among species

Rickettsia—obligate intracellular bacterial microparasites

Rinderpest—cattle plague, a major infectious viral disease of cattle, domestic buffalo, and some other ungulates, declared eradicated by the World Organization for Animal Health in 2011

Risks—the probability of occurrence of damage and the severity of the consequences

Rodenticides—pesticides used on rodents and other small mammals

Rumination—or "foregut fermentation," elaborate and time consuming but efficient digestion using a two-stomach system; used by artiodactyls, kangaroos, sloths, and colobus monkeys. Microorganisms in the rumen detoxify poisons important for managing human-wildlife conflicts.

Salience—in flavor aversion learning, likelihood of being associated with an illness

Salt drive—urge to eat salt

Sand fleas—small decapod crustaceans

Sand flies—blood sucking dipterans

Sanguinivorous—diet restricted to eating blood

Scientific method—a method that attempts to minimize opportunities for personal or other biases that can lead to misinterpretation of results, and to maximize the odds that results observed are actually due to the factors being controlled and studied. Scientific method is usually a series of steps followed sequentially (in theory, often not in practice): gather background information and current literature; form a hypothesis and its negative form, the "null hypothesis"; develop an experimental or observational design to test the hypotheses; run the experiment, gathering data;

analyze thc data following appropriate statistical methods; interpret the results; and communicate the findings to the rest of the world.

Secondary hosts—also called "intermediate hosts," vectors that carry parasitic pathogens for prolonged periods of time

Secondary literature—accumulations and summaries of primary scientific literature, but including also other experiential and sometimes anecdotal information, that often lead to recommendations for specific management actions. With human-wildlife conflicts, this is often a place where the art as well as the science can be found.

Secondary succession—succession that occurs when an ecosystem is pushed back to an earlier successional stage such as by fire, herbicide, or intensive grazing

Second law of thermodynamics—some energy is lost as heat whenever energy is converted from one form to another

Seines—nets to capture fish and other aquatic organisms

Semiochemical—odor of another species, often a predator

Seral stage—see sere

Sere—each successional community within an ecosystem. In an ecosystem over time, seral stages replace each other in an orderly and predictable manner; i.e., the presence of one community alters the ecosystem in such a manner that it is replaced by the next seral stage until a mature, climax community is achieved.

Shelford's law of tolerance—the factor in the environment that occurs in excessive quantity in relation to need will limit the growth of the organism or its population

Sigmoidal growth curve—growth of a population under conditions where resistance to growth increases with increasing population density; the shape of the curve often resembles an "S" when graphed with numbers of population members as the ordinate (*y*-axis) and time as the abscissa (*x*-axis)

Snare—also called a "cable device," a steel cable with a loop on one end for capturing animals

Social facilitation—when the presence of one animal improves the performance of another

Social values—conceptions of what is good and bad, desirable and undesirable

Sovereign—a state or federal agency representing the common interests of citizens

Species jump—when a pathogen overcomes a species barrier and infects a new organism

Specific hungers—eating preferences for a specific item, such as soil or salt

S-shaped growth curve—see sigmoidal growth curve

Stakeholders—persons who are affected by or can affect a problem or its management

Standard operating procedures (SOPs)—written, formalized directions to be followed during performance of tasks

Steno—able to live in only a narrow range or function, e.g., stenothermal means that an organism can survive only in a narrow ranges of temperatures

Stenohalic—able to survive only in narrow ranges of salt concentrations

Stenophagic—has a diet restricted to few foods or food types

Stenothermal—able to survive only in narrow ranges of temperatures

Stereotyped behaviors—used to describe a carefully orchestrated and rigidly set sequence of behavioral and physiological events

Strategy—a broad plan of action to achieve a goal

Succession—orderly replacement of a community of organisms with another, starting with invader organisms if the area has not had prior living communities, and ending with a stable climax community where energy used by the system balances energy input. Each distinctive community is called a seral stage, or sere.

Sweepstakes route—in evolution, a hazardous or accidental dispersal mechanism by which an organism can move from place to place, such as rats moving from island to island on a log or raft

Sylvatic cycle—also called the "enzootic cycle," movement of a disease between wildlife as hosts and vectors; the natural portion of the biological life cycle

Sympatric—species that occupy the same or similar ecological niches in the same geographic location at the same time

Synecological—a holistic approach to ecology, involving groups of organisms or landscapes

Taxis—tropism, but in a decidedly directed manner

Territorial behavior—when one member, usually a male, of a population establishes a geographical area and defends it against other conspecifics

Theoretical ecological niche—the theoretically greatest potential habitat and ecological roles of an organism, a limiting factor keeps the organisms or populations within this niche

Toxoplasmosis—a parasitic infection caused by the protozoan *Toxoplasma gondii*; felids including domestic, wild, and feral cats are definitive hosts while many warm-blooded animals serve as intermediate hosts

Trammel nets—nets designed for species not easily caught in gill nets, like flatfish and sturgeon

Transect survey—a path along which one counts and records occurrences, such as bird sightings or calls

Trawl nets—a long bag or sock-shaped net that is pulled through the water to capture aquatic organisms

Trophic level—means to nourish, and refers to the organism's position in the food chain

Tropism—the general attraction or avoidance of some environmental stimulus such as temperature or light intensity

Tubing—tubes made of paper or plastic that are used to protect seedlings from herbivores such as gophers, mice, rabbits, mountain beavers, beavers, and deer

Turbo-fladry—electrified fladry

Uniform distribution—see even distribution

Value orientations—values that characterize cultures, a form of ideology

Vectors—with diseases, carriers of diseases

Vertebrate pest control—the aspect of wildlife damage management that deals with vertebrates, usually mostly birds and mammals, but sometimes including reptiles, amphibia, and fishes

Warfarin—a first-generation anticoagulant rodenticide named after the Wisconsin Alumni Research Foundation

West Nile virus—a virus of the family *Flaviviridae* that infects humans, pets, and wildlife, often transmitted with the bite of a mosquito

Whirling disease—a protozoan disease that infects salmon and trout species

Wilderness discount—the portion of damage that is prevented or managed by a natural ecosystem, e.g., wilderness offering alternative prey for predators, thereby reducing livestock losses

Wildlife—any nondomesticated plant or animal

Wildlife acceptance capacity—the mixture of tolerances for a wildlife damage problem that can be found among its stakeholders

Wildlife conservation—see conservation

Wildlife crossings—structures that allow wildlife to cross human-made structures safely, including overpasses, underpasses, tunnels, fish ladders, culverts, canopy bridges, and green roofs

Wildlife damage—a real or perceived threat to humans, their property, or something that they value

Wildlife damage management—the science and art of diminishing the negative aspects of wildlife while maintaining or enhancing their positive aspects

Wildlife disease—an abnormal condition that impairs bodily functions, that is associated with specific symptoms and signs, and that is carried by nondomesticated organisms

Willingness to pay—estimate of the value of something based on how much a person is willing to pay for the activity, such as a hunting, fishing, or hiking trip

Zeitgebers—German for "time giver"; internal cues and rhythms adjusted by external cues

Zoning—in conservation, a management practice that sets aside some geographic areas for human activity, others for wildlife, and some for both.

Zoonoses—diseases that are transmitted between wildlife and humans, or their livestock or pets

References

AAMI. 2019. "AAMI Animal Car Accidents Data 2019." *AAMI Informed* (blog), May 27. https://www.aami.com.au/aami-informed/on-the-road/safe-driving/aami-reveals-peak-periods-for-animal-collisions.html.

AAMI. 2020. "AAMI Data Reveals Australia's Animal Collision Hotspots." *AAMI Informed* (blog), May 20. https://www.aami.com.au/aami-informed/on-the-road/safe-driving/aami-animal-car-accidents-data.html.

Abbott, K. L. 2005. "Supercolonies of the Invasive Yellow Crazy Ant, *Anoplolepis gracilipes*, on an Oceanic Island: Forager Activity Patterns, Density and Biomass." *Insectes Sociaux* 52 (3): 266–273. https://doi.org/10.1007/s00040-005-0800-6.

Abra, F., B. Granziera, M. Huijser, et al. 2019. "Pay or Prevent? Human Safety, Costs to Society and Legal Perspectives on Animal-Vehicle Collisions in São Paulo State, Brazil." *PLoS One* 14 (4): e0215152. https://doi.org/10.1371/journal.pone.0215152.

Accipiter Radar Technologies. 2017. "Accipiter: Environmental Protection." Product brochure. Orchard Park, NY. https://www.accipiterradar.com/wp-content/uploads/2017/05/Accipiter_Environmental_Brochure-V1.10417_Web.pdf.

Agaba, P., T. Asiimwe, T. G. Moorhouse, et al. 2009. "Biological Control of Water Hyacinth in the Kagera River Headwaters of Rwanda: A Review through 2001." Martinez, CA: Clean Lakes, Inc. https://www.cleanlake.com/images/rwanda_bio_paper.pdf.

Aguirre, A. A., T. Longcore, M. Barbieri, et al. 2019. "The One Health Approach to Toxoplasmosis: Epidemiology, Control, and Prevention Strategies." *EcoHealth* 16 (2): 378–390. https://doi.org/10.1007/s10393-019-01405-7.

Aguirre, W., and S. G. Poss. 2000. Non-indigenous Species in the Gulf of Mexico: "*Cyprinus carpio* Linnaeus 1758." Ecosystem-Gulf States Marine Fisheries Commission (GSMFC). https://web.archive.org/web/20021028090157/http://www.gsmfc.org/nis/nis/Cyprinus_carpio.html.

Ahmed, M., and L. Fiedler. 2002. "A Comparison of Four Rodent Control Methods in Philippine Experimental Rice Fields." *International Biodeterioration and Biodegradation* 49 (2–3): 125–132. https://doi.org/10.1016/S0964-8305(01)00112-3.

Allan, J. 2002. "The Costs of Bird Strikes and Bird Strike Prevention." In *Human Conflicts with Wildlife: Economic Considerations: Proceedings of the Third NWRC Special Symposium*, August 1–3, 2000, Fort Collins, CO, edited by L. Clark, 147–153. Fort Collins, CO: National Wildlife Research Center. https://digitalcommons.unl.edu/nwrchumanconflicts/18/.

Allan, J., J. Bell, and V. Jackson. 1999. "An Assessment of the World-Wide Risk to Aircraft from Large Flocking Birds." In *Proceedings of the 1999 Bird Strike Committee USA/Canada, 1st Joint Meeting*, Vancouver, BC. https://digitalcommons.unl.edu/birdstrike1999/4/.

Allen, L., R. Engeman, and H. Krupa. 1996. "Evaluation of Three Relative Abundance Indices for Assessing Dingo Populations." *Wildlife Research* 23 (2): 197–205. https://doi.org/10.1071/WR9960197.

Allen, S., D. Wickwar, F. Clark, et al. 2009. "Values, Beliefs, and Attitudes Technical Guide for Forest Service Land and Resource Management, Planning, and Decision-Making." Portland, OR: US Department of Agriculture, Forest Service, Pacific Northwest Research Station. https://doi.org/10.2737/PNW-GTR-788.

Almond, R. E. A., M. Grooten, and T. Petersen, eds. 2021. "Living Planet Report 2020: Bending the Curve of Biodiversity Loss." Gland, Switzerland: World Wildlife Federation. https://oursharedseas.com/wp-content/uploads/2020/10/WWF_Living-Planet-Report-2020.pdf.

Aloo, P., J. Njiru, J. S. Balirwa, and C. S. Nyamweya. 2017. "Impacts of Nile Perch, *Lates niloticus*, Introduction on the Ecology, Economy and Conservation of Lake Victoria, East Africa." *Lakes and Reservoirs: Science, Policy and Management for Sustainable Use* 22 (4): 320–333. https://doi.org/10.1111/lre.12192.

Alroy, J. 2001. "A Multispecies Overkill Simulation of the End-Pleistocene Megafaunal Mass Extinction." *Science* 292 (5523): 1893–1896. https://doi.org/10.1126/science.1059342.

Alves, J., M. Carneiro, J. Cheng, et al. 2019. "Parallel Adaptation of Rabbit Populations to Myxoma Virus." *Science* 363 (6433): 1319–1326. https://doi.org/10.1126/science.aau7285.

Amano, T., K. Ushiyama, G. Fujita, et al. 2004. "Alleviating Grazing Damage by White-Fronted Geese: An Optimal

Foraging Approach." *Journal of Applied Ecology* 41: 675–688. https://doi.org/10.1111/j.0021-8901.2004.00923.x.

Amstrup, S., G. York, T. McDonald, et al. 2004. "Detecting Denning Polar Bears with Forward-Looking Infrared (FLIR) Imagery." *BioScience* 54 (4): 337–344. https://doi.org/10.1641/0006-3568(2004)054[0337:DDPBWF]2.0.CO;2.

Andelt, W., K. Burnham, and D. Baker. 1994. "Effectiveness of Capsaicin and Bitrex Repellents for Deterring Browsing by Captive Mule Deer." *Journal of Wildlife Management* 58 (2): 330–334. https://doi.org/10.2307/3809398.

Anderson, A., C. A. Lindell, K. M. Moxcey, et al. 2013. "Bird Damage to Select Fruit Crops: The Cost of Damage and the Benefits of Control in Five States." *Crop Protection* 52: 103–109. https://doi.org/10.1016/j.cropro.2013.05.019.

Ang, B., T. Lim, and L. Wang. 2018. "Nipah Virus Infection." *Journal of Clinical Microbiology* 56 (6): e01875-17. https://doi.org/10.1128/JCM.01875-17.

Angling Trust. 2020. "Government Rejects Plea to Add Cormorants to the General Licence." Campaigns, November 16. https://anglingtrust.net/2020/11/16/government-rejects-plea-to-add-cormorants-to-the-general-licence/.

Anthony, R. M., J. Evans, and G. Lindsey. 1986. "Strychnine-Salt Blocks for Controlling Porcupines in Pine Forests: Efficacy and Hazards." In *Proceedings of the 12th Vertebrate Pest Conference*, March 4–6, San Diego, CA, edited by T. P. Salmon, 191–195. Davis: University of California. https://digitalcommons.unl.edu/vpc12/4/.

Apprill, A., C. Miller, M. Moore, et al. 2017. "Extensive Core Microbiome in Drone-Captured Whale Blow Supports a Framework for Health Monitoring." *MSystems* 2 (5): e00119-17. https://doi.org/10.1128/mSystems.00119-17.

Aramini, J. J., C. Stephen, and J. P. Dubey. 1998. "*Toxoplasma gondii* in Vancouver Island Cougars (*Felis concolor vancouverensis*): Serology and Oocyst Shedding." *Journal of Parasitology* 84 (2): 438–440. https://doi.org/10.2307/3284508.

Arjo, W. 2003. "Mountain Beaver: The Little Rodent with a Large Appetite." *Western Forester*, July/August. https://digitalcommons.unl.edu/icwdm_usdanwrc/194/.

Arlinghaus, R., and T. Mehner. 2003. "Socio-economic Characterisation of Specialised Common Carp (*Cyprinus carpio* L.) Anglers in Germany, and Implications for Inland Fisheries Management and Eutrophication Control." *Fisheries Research* 61 (1–3): 19–33. https://doi.org/10.1016/S0165-7836(02)00243-6.

Arnett, E., C. Hein, M. Schirmacher, et al. 2013. "Evaluating the Effectiveness of an Ultrasonic Acoustic Deterrent for Reducing Bat Fatalities at Wind Turbines." *PLoS One* 8 (6): e65794. https://doi.org/10.1371/journal.pone.0065794.

Asa, C., and A. Moresco. 2019. "Fertility Control in Wildlife: Review of Current Status, Including Novel and Future Technologies." In *Reproductive Sciences in Animal Conservation*, edited by P. Comizzoli, J. L. Brown, and W. V. Holt, 507–543. Cham, Switzerland: Springer International. https://doi.org/10.1007/978-3-030-23633-5_17.

Atkinson, C. T., K. L. Woods, Robert J. Dusek, et al. 1995. "Wildlife Disease and Conservation in Hawaii: Pathogenicity of Avian Malaria (*Plasmodium relictum*) in Experi-

mentally Infected Iiwi (*Vestiaria coccinea*)." *Parasitology* 111 (S1): S59–S69. https://doi.org/10.1017/S003118200007582X.

Avery, M. 1996. "Food Avoidance by Adult House Finches, *Carpodacus mexicanus*, Affects Seed Preferences of Offspring." *Animal Behaviour* 51 (6): 1279–1283. https://doi.org/10.1006/anbe.1996.0132.

Avery, M., J. Humphrey, and D. Decker. 1997. "Feeding Deterrence of Anthraquinone, Anthracene, and Anthrone to Rice-Eating Birds." *Journal of Wildlife Management* 61 (4): 1359–1365. https://doi.org/10.2307/3802138.

Avery, M., and M. Lowney. 2016. "Vultures." Wildlife Damage Management Technical Series. Ft. Collins, CO: USDA, APHIS, Wildlife Services, National Wildlife Research Center. https://digitalcommons.unl.edu/nwrcwdmts/5.

Avery, M., M. Pavelka, D. Bergman, et al. 1995. "Aversive Conditioning to Reduce Raven Predation on California Least Tern Eggs." *Colonial Waterbirds*, 18 (2): 131–138. https://doi.org/10.2307/1521474.

Avery, M., and S. Werner. 2017. "Frightening Devices." In *Ecology and Management of Blackbirds (Icteridae) in North America*, edited by G. M. Linz, M. L. Avery, and R. A. Dolbeer, 159–174. Boca Raton, FL: CRC Press. https://digitalcommons.unl.edu/icwdm_usdanwrc/2001.

Babbie, E. 2001. *The Practice of Social Research*. Belmont, CA: Wadsworth/Thomson Learning.

Bader, H., and G. Finstad. 2001. "Conflicts between Livestock and Wildlife: An Analysis of Legal Liabilities Arising from Reindeer and Caribou Competition on the Seward Peninsula of Western Alaska." *Environmental Law* 31 (3): 549–580.

Bagavathiannan, M. V., and R. C. Van Acker. 2008. "Crop Ferality: Implications for Novel Trait Confinement." *Agriculture, Ecosystems and Environment* 127 (1): 1–6. https://doi.org/10.1016/j.agee.2008.03.009.

Baker, C., A. Gordon, and M. Bode. 2017. "Ensemble Ecosystem Modeling for Predicting Ecosystem Response to Predator Reintroduction." *Conservation Biology* 31 (2): 376–384. https://doi.org/10.1111/cobi.12798.

Baker, H. 1965. "Characteristics and Modes of Origin of Weeds." In *Genetics of Colonizing Species: Proceedings of the First International Union of Biological Sciences Symposia on General Biology*, edited by H. G. Baker and G. L. Stebbins, 147–172. New York: Academic Press.

Balciunas, J. K. 2000. "A Proposed Code of Best Practices for Classical Biological Control of Weeds." In *Proceedings of the X International Symposium on Biological Control of Weeds*, July 4–14, 1999, edited by N. R. Spencer, 435–436. Bozeman: Montana State University.

Bale, J. 2007. "Political Paranoia v. Political Realism: On Distinguishing between Bogus Conspiracy Theories and Genuine Conspiratorial Politics." *Patterns of Prejudice* 41 (1): 45–60. https://doi.org/10.1080/00313220601118751.

Bank, F.G., S. Hagood, B. Ruediger, et al. 2002. *Wildlife Habitat Connectivity across European Highways*. Publication No. FHWA-PL-02-011 HPIP/08-02(7M)EW. Washington, DC: Office of International Programs FHWA/US Department of Transportation (HPIP). https://international.fhwa.dot.gov/Pdfs/wildlife_web.pdf.

Barnes, E. 2005. *Diseases and Human Evolution*. Albuquerque: University of New Mexico Press.

Barras, S. 2004. "Double-Crested Cormorants in Alabama." USDA National Wildlife Research Center: Staff Publications, 77. https://digitalcommons.unl.edu/icwdm_usdanwrc/77.

Barras, S., R. Dolbeer, R. Chipman, et al. 2000. "Bird and Small Mammal Use of Mowed and Unmowed Vegetation at John F. Kennedy International Airport, 1998 to 1999." In *Proceedings of the 19th Vertebrate Pest Conference*, March 6–9, San Diego, CA, edited by T. P. Salmon and A. C. Crabb, 31–36. Davis: University of California Agriculture and Natural Resources. https://doi.org/10.5070/V419110070.

Baxter, A. 2007. "Laser Dispersal of Gulls from Reservoirs near Airports." In *Proceedings of the 2007 Bird Strike Committee USA/Canada, 9th Annual Meeting*, Kingston, Ontario. https://digitalcommons.unl.edu/birdstrike2007/2/.

Baxter, J. 2016. *The Disney Conservation Fund: Carrying Forward a Conservation Legacy*. Los Angeles: Disney Editions.

Baxter, J., and D. Hancock. 2020. *The Disney Conservation Fund: Carrying Forward a Conservation Legacy*. Disney Editions. https://thewaltdisneycompany.com/app/uploads/static/conservation-fund-book/index.html.

Beasley, J., and O. Rhodes Jr. 2008. "Relationship between Raccoon Abundance and Crop Damage." *Human–Wildlife Conflicts* 2 (2): 248–259. https://doi.org/10.26077/g4bp-bd34.

Beckmann, J., A. Clevenger, M. Huijser, et al., eds. 2010. *Safe Passages: Highways, Wildlife, and Habitat Connectivity*. Illustrated edition. Washington, DC: Island Press.

Behnke, R. J. 1992. *Native Trout of Western North America*. American Fisheries Society Monograph No. 6. Bethesda, MD: American Fisheries Society.

Belant, J. 1997. "Gulls in Urban Environments: Landscape-Level Management to Reduce Conflict." *Landscape and Urban Planning* 38 (3–4): 245–258. https://doi.org/10.1016/S0169-2046(97)00037-6.

Belmain, S. R., M. S. Ali, A. K. Azad, et al. 2008. "Scientific Assessment Report on Bamboo Flowering, Rodent Outbreaks and Food Security: Rodent Ecology, Pest Management, and Socio-economic Impact in the Chittagong Hill Tracts, Bangladesh." Bangladesh: United Nations Development Programme. https://doi.org/10.13140/RG.2.2.20184.55047.

Belshe, R. 2009. "Implications of the Emergence of a Novel H1 Influenza Virus." *New England Journal of Medicine* 360 (25): 2667–2668. https://doi.org/10.1056/nejme0903995.

Bendell, L. I. 2015. "Favored Use of Anti-Predator Netting (APN) Applied for the Farming of Clams Leads to Little Benefits to Industry While Increasing Nearshore Impacts and Plastics Pollution." *Marine Pollution Bulletin* 91 (1): 22–28. https://doi.org/10.1016/j.marpolbul.2014.12.043.

Bennett, D. 1995. *A Little Book of Monitor Lizards: A Guide to the Monitor Lizards of the World and Their Care in Captivity*. Aberdeen: Viper Press.

Bennett, D., and R. Timm. 2016. "The Dogs of Roman Vindolanda, Part II: Time-Stratigraphic Occurrence, Ethnographic Comparisons, and Biotype Reconstruction," *Archaeofauna* 25: 107–126.

Beringer, J., L. P. Hansen, J. A. Demand, et al. 2002. "Efficacy of Translocation to Control Urban Deer in Missouri: Costs, Efficiency, and Outcome." *Wildlife Society Bulletin* 30 (3): 767–774. https://www.jstor.org/stable/3784230.

Berio Fortini, L., L. Kaiser, and D. LaPointe. 2020. "Fostering Real-Time Climate Adaptation: Analyzing Past, Current, and Forecast Temperature to Understand the Dynamic Risk to Hawaiian Honeycreepers from Avian Malaria." *Global Ecology and Conservation* 23: e01069. https://doi.org/10.1016/j.gecco.2020.e01069.

Berube, D. 2020. "Mosquitoes Bite: A Zika Story of Vector Management and Gene Drives." In *Synthetic Biology 2020: Frontiers in Risk Analysis and Governance*, edited by B. D. Trump, C. L. Cummings, J. Kuzma, and I. Linkov, 143–163. Cham, Switzerland: Springer International. https://doi.org/10.1007/978-3-030-27264-.7_7.

Beschta, R., L. Painter, and W. Ripple. 2018. "Trophic Cascades at Multiple Spatial Scales Shape Recovery of Young Aspen in Yellowstone." *Forest Ecology and Management* 413: 62–69. https://doi.org/10.1016/j.foreco.2018.01.055.

Bhat, M., R. Huffaker, and S. Lenhart. 1993. "Controlling Forest Damage by Dispersive Beaver Populations: Centralized Optimal Management Strategy." *Ecological Applications* 3 (3): 518–530. https://doi.org/10.2307/1941920.

Bhattarai, B., D. Morgan, and W. Wright. 2021. "Equitable Sharing of Benefits from Tiger Conservation: Beneficiaries' Willingness to Pay to Offset the Costs of Tiger Conservation." *Journal of Environmental Management* 284: 112018. https://doi.org/10.1016/j.jenvman.2021.112018.

Bierwiaczonek, K., J. Kunst, and O. Pich. 2020. "Belief in COVID-19 Conspiracy Theories Reduces Social Distancing over Time." *Applied Psychology: Health and Well-Being* 12 (4): 1270–1285. https://doi.org/10.1111/aphw.12223.

Blackburn, T., J. Lockwood, and P. Cassey. 2015. "The Influence of Numbers on Invasion Success." *Molecular Ecology* 24 (9): 1942–1953. https://doi.org/10.1111/mec.13075.

Blackwell, B., D. Felstul, and T. Seamans. 2013. "Managing Airport Stormwater to Reduce Attraction to Wildlife." USDA National Wildlife Research Center: Staff Publications, 1451. https://digitalcommons.unl.edu/icwdm_usdanwrc/1451.

Blackwell, B., T. Seamans, B. Washburn, et al. 2006. "Use of Infrared Technology in Wildlife Surveys." In *Proceedings of the 22nd Vertebrate Pest Conference*, March 6–9, Berkeley, CA, edited by R. A. Timm and J. M. O'Brien, 467–472. Davis: University of California–Davis. https://doi.org/10.5070/V422110116.

Blaustein, A., and J. Kiesecker. 2002. "Complexity in Conservation: Lessons from the Global Decline of Amphibian Populations." *Ecology Letters* 5 (4): 597–608. https://doi.org/10.1046/j.1461-0248.2002.00352.x.

Blejwas, K., C. Williams, G. Shin, et al. 2006. "Salivary DNA Evidence Convicts Breeding Male Coyotes of Killing Sheep." *Journal of Wildlife Management* 70 (4): 1087–1093. https://doi.org/10.2193/0022-541X(2006)70[1087:SDECBM]2.0.CO;2.

Bodenchuk, M., J. R. Mason, and W. Pitt. 2002. "Economics of Predation Management in Relation to Agriculture, Wildlife, and Human Health and Safety." In *Human Conflicts with Wildlife: Economic Considerations: Proceedings* of the Third NWRC Special Symposium, August 1–3, 2000, Fort Collins, CO, edited by L. Clark, J. Hone, J. A. Shivik, et al., 80–90. Fort Collins, CO: National Wildlife Research Center. https://digitalcommons.unl.edu /nwrchumanconflicts/9.

Bongomin, F., R. Kwizera, and D. Denning. 2019. "Getting Histoplasmosis on the Map of International Recommendations for Patients with Advanced HIV Disease." Journal of Fungi 5 (3): 80. https://doi.org/10.3390/jof5030080.

Bonnell, M., and S. Breck. 2017. "Using Resident-Based Hazing Programs to Reduce Human-Coyote Conflicts in Urban Environments." Human–Wildlife Interactions 11 (2): 5. https://doi.org/10.26077/ab7k-6j25.

Boon, A., C. Sandström, U. Arbieu, et al. 2020. "Governing Dual Objectives within Single Policy Mixes: An Empirical Analysis of Large Carnivore Policies in Six European Countries." Journal of Environmental Policy and Planning 23 (4): 399–413. https://doi.org/10.1080 /1523908X.2020.1841614.

Booth, T. W. 1994. "Bird Dispersal Techniques." In Prevention and Control of Wildlife Damage, edited by S. E. Hygnstrom, R. M. Timm, and G. E. Larson, E19–E24. Lincoln: University of Nebraska Cooperative Extension. https:// digitalcommons.unl.edu/icwdmhandbook/58/.

Borer, C., S. Sapp, and L. Hutchinson. 2013. "Flowering Dogwood (Cornus florida L.) as Mediator of Calcium Cycling: New Insights Are Revealed by Analysis of Foliar Partitioning." Trees 27 (4): 841–849. https://doi.org/10.1007 /s00468-012-0838-9.

Borer, E., P. Hosseini, E. Seabloom, et al. 2007. "Pathogen-induced Reversal of Native Dominance in a Grassland Community." Proceedings of the National Academy of Sciences of the United States of America 104 (13): 5473–5478. https://doi.org/10.1073/pnas.0608573104.

Boyle, C. M. 1960. "Case of Apparent Resistance of Rattus norvegicus Berkenhout to Anticoagulant Poisons." Nature 188 (4749): 517. https://doi.org/10.1038/188517a0.

Braczkowski, A., C. O'Bryan, M. Stringer, et al. 2018. "Leopards Provide Public Health Benefits in Mumbai, India." Frontiers in Ecology and the Environment 16 (3): 176–182. https://doi.org/10.1002/fee.1776.

Bradford, W. 1856. History of Plymouth Plantation. Vol. 33. Boston: Massachusetts Historical Society.

Bradshaw, L., and D. M. Waller. 2016. "Impacts of White-Tailed Deer on Regional Patterns of Forest Tree Recruitment." Forest Ecology and Management 375: 1–11. https://doi.org/10.1016/j.foreco.2016.05.019.

Bragard, C., K. Dehnen-Schmutz, F. Di Serio, et al. 2020. "Commodity Risk Assessment of Oak Logs with Bark from the US for the Oak Wilt Pathogen Bretziella fagacearum under an Integrated Systems Approach." EFSA Journal 18 (12): e06352. https://doi.org/10.2903/j.efsa.2020.6352.

Breck, S., N. Lance, and P. Callahan. 2006. "A Shocking Device for Protection of Concentrated Food Sources from Black Bears." Wildlife Society Bulletin 34 (1): 23–26. https://www.jstor.org/stable/3784930.

Breck, S., R. Williamson, C. Niemeyer, et al. 2002. "Non-Lethal Radio Activated Guard for Deterring Wolf Depredation in Idaho: Summary and Call for Research."

USDA Wildlife Services: Staff Publications, 467. https://digitalcommons.unl.edu/icwdm_usdanwrc/467/.

Breck, S., K. Wilson, and D. Andersen. 2003. "Beaver Herbivory and Its Effect on Cottonwood Trees: Influence of Flooding along Matched Regulated and Unregulated Rivers." River Research and Applications 19 (1): 43–58. https://digitalcommons.unl.edu/cgi/viewcontent.cgi ?article=1080&context=icwdm_usdanwrc.

Breed, D., L. Meyer, J. Steyl, et al. 2020. "Conserving Wildlife in a Changing World: Understanding Capture Myopathy—a Malignant Outcome of Stress during Capture and Translocation." Conservation Physiology 7 (1): coz027. https://doi.org/10.1093/conphys/coz027.

Brewster, R., S. Henke, B. Turner, et al. 2019. "Cost-Benefit Analysis of Coyote Removal as a Management Option in Texas Cattle Ranching." Human–Wildlife Interactions 13 (3): 10. https://doi.org/10.26077/2hd9-1v35.

Britton-Simmons, Kevin H., and Karen C. Abbott. 2008. "Short- and Long-term Effects of Disturbance and Propagule Pressure on a Biological Invasion." Journal of Ecology 96 (1): 68–77. https://doi.org/10.1111/j.1365-2745 .2007.01319.x.

Brochu, C., and G. Storrs. 2012. "A Giant Crocodile from the Plio-Pleistocene of Kenya, the Phylogenetic Relationships of Neogene African Crocodylines, and the Antiquity of Crocodylus in Africa." Journal of Vertebrate Paleontology 32 (3): 587–602. https://doi.org/10.1080/02724634.2012.652324.

Brock, J. 2005. "Four Surveyors of the Gods: In the XVIII Dynasty of Egypt—New Kingdom c. 1400 B.C." Ancient Egypt, 15.

Bromley, C., and E. Gese. 2001a. "Effects of Sterilization on Territory Fidelity and Maintenance, Pair Bonds, and Survival Rates of Free-Ranging Coyotes." Canadian Journal of Zoology 79 (3): 386–392. https://doi.org/10.1139/z00-212.

Bromley, C., and E. Gese. 2001b. "Surgical Sterilization as a Method of Reducing Coyote Predation on Domestic Sheep." Journal of Wildlife Management 65 (3): 510–519. https://doi.org/10.2307/3803104.

Bruce, H. 1959. "An Exteroceptive Block to Pregnancy in the Mouse." Nature 184 (4680): 105. https://doi.org/10.1038 /184105a0.

Bruggers, R. L., J. E. Brooks, R. A. Dolbeer, et al. 1986. "Responses of Pest Birds to Reflecting Tape in Agriculture." All-India Co-Ordinated Research Project on Economic Ornithology. Wildlife Society Bulletin 14 (2): 161–170. https://www.jstor.org/stable/3782066.

Bruggers, R. L., and C. Elliott. 1989. Quelea quelea: Africa's Bird Pest. New York: Oxford University Press.

Bruinderink, G., and E. Hazebroek. 1996. "Ungulate Traffic Collisions in Europe." Conservation Biology 10 (4): 1059–1067. https://www.jstor.org/stable/2387142.

Bruns, A., M. Waltert, and I. Khorozyan. 2020. "The Effectiveness of Livestock Protection Measures against Wolves (Canis lupus) and Implications for Their Co-existence with Humans." Global Ecology and Conservation 21: e00868. https://doi.org/10.1016/j.gecco.2019.e00868.

Bryce, R., M. Oliver, L. Davies, et al. 2011. "Turning Back the Tide of American Mink Invasion at an Unprecedented Scale through Community Participation and Adaptive

Management." *Biological Conservation* 144 (1): 575–583. https://doi.org/10.1016/j.biocon.2010.10.013.

Budy, P., and J. W. Gaeta. 2017. "Brown Trout as an Invader: A Synthesis of Problems and Perspectives in North America." In *Brown Trout: Biology, Ecology and Management*, edited by J. Lobón-Cerviá and N. Sanz, 525–544. Chichester, UK: John Wiley & Sons. https://doi.org/10.1002/9781119268352.ch20.

Bullard, R., S. Shumake, D. Campbell, et al. 1978. "Preparation and Evaluation of a Synthetic Fermented Egg Coyote Attractant and Deer Repellent." *Journal of Agricultural and Food Chemistry* 26 (1): 160–163. https://doi.org/10.1021/jf60215a037.

Burbidge, A. A., and K. D. Morris. 2002. "Introduced Mammal Eradications for Nature Conservation on Western Australian Islands: A Review." In *Turning the Tide: The Eradication of Invasive Species*, edited by C. R. Veitch and M. N. Clout, 64–70. Gland, Switzerland: IUCN SSC Invasive Species Specialist Group. http://issg.org/database/species/reference_files/TurTid/Burbidge.pdf.

Burridge, M. 2005. "Controlling and Eradicating Tick Infestations on Reptiles." *Compendium on Continuing Education for the Practising Veterinarian (North American Edition)* 27 (5): 371–376. https://www.vetfolio.com/learn/article/controlling-and-eradicating-tick-infestations-on-reptiles.

Butchko, P. 1990. "Predator Control for the Protection of Endangered Species in California." In *Proceedings of the 14th Vertebrate Pest Conference 1990*, March 6–8, Sacramento, CA, edited by L. R. Davis and R. E. Marsh, 237–240. Davis: University of California. https://digitalcommons.unl.edu/vpc14/11/.

Butler, J., J. Shanahan, and D. Decker. 2001. "Wildlife Attitudes and Values: A Trend Analysis." Issue 1, part 4, HDRU Series. Ithaca, NY: New York State College of Agriculture and Life Sciences. Human Dimensions Research Unit, Fish and Wildlife Reference Service, Cornell University.

CABI. 2021. "*Discula destructiva* (Anthracnose of Dogwood)." *Invasive Species Compendium*. Wallingford, UK: CAB International. https://www.cabi.org/isc/datasheet/20079#REF-DDB-183911.

Cadi, A., and P. Joly. 2004. "Impact of the Introduction of the Red-Eared Slider (*Trachemys scripta elegans*) on Survival Rates of the European Pond Turtle (*Emys orbicularis*)." *Biodiversity and Conservation* 13 (13): 2511–2518. https://doi.org/10.1023/B:BIOC.0000048451.07820.9c.

Cai, J., X. Zhou, X. Yan, et al. 2019. "Top 10 Species Groups in Global Aquaculture 2017." Food and Agriculture Organizations of the United Nations. http://www.fao.org/3/ca5224en/CA5224EN.pdf.

Campbell, E. 1999. "Barriers to Movements of the Brown Treesnake (*Boiga irregularis*)." In *Problem Snake Management: The Habu and the Brown Treesnake*, edited by G. H. Rodda, Y. Sawai, and H. Tanaka, 306–312. Ithaca, NY: Cornell University Press. https://doi.org/10.7591/9781501737688-030.

Campbell, K., and C. J. Donlan. 2005. "Feral Goat Eradications on Islands." *Conservation Biology* 19: 1362–1374. https://doi.org/10.1111/j.1523-1739.2005.00228.x.

Campbell, K. J., J. Saah, P. Brown, et al. 2019. "A Potential New Tool for the Toolbox: Assessing Gene Drives for Eradicating Invasive Rodent Populations." In *Island Invasives: Scaling Up to Meet the Challenge*, edited by C. R. Veitch, M. N. Clout, A. R. Martin, et al., 6–14. Gland, Switzerland: IUCN.

Campbell, T., and D. Long. 2009. "Feral Swine Damage and Damage Management in Forested Ecosystems." *Forest Ecology and Management* 257 (12): 2319–2326. https://digitalcommons.unl.edu/icwdm_usdanwrc/890/.

Castillo-Huitrón, N., E. Naranjo, D. Santos-Fita, et al. 2020. "The Importance of Human Emotions for Wildlife Conservation." *Frontiers in Psychology* 11: 1277. https://doi.org/10.3389/fpsyg.2020.01277.

Caudell, J., S. Shwiff, and M. Slater. 2010. "Using a Cost-Effectiveness Model to Determine the Applicability of OvoControl G to Manage Nuisance Canada Geese." *Journal of Wildlife Management* 74 (4): 843–848. https://doi.org/10.2193/2008-470.

Caughley, G. 1981. "Overpopulation." In *Problems in Management of Locally Abundant Wild Mammals*, edited by P. A. Jewell and S. Holt, 7–19. New York: Academic Press.

Caut, S., J. Casanovas, E. Virgos, et al. 2007. "Rats Dying for Mice: Modelling the Competitor Release Effect." *Austral Ecology* 32 (8): 858–868. https://doi.org/10.1111/J.1442-9993.2007.01770.X.

Chalkowski, K., C. Lepczyk, and S. Zohdy. 2018. "Parasite Ecology of Invasive Species: Conceptual Framework and New Hypotheses." *Trends in Parasitology* 34 (8): 655–663. https://doi.org/10.1016/j.pt.2018.05.008.

Chapron, G., P. Kaczensky, J. Linnell, et al. 2014. "Recovery of Large Carnivores in Europe's Modern Human-Dominated Landscapes." *Science* 346 (6216): 1517–1519. https://doi.org/10.1126/science.1257553.

Chapron, G., S. Legendre, R. Ferrière, et al. 2003. "Conservation and Control Strategies for the Wolf (*Canis lupus*) in Western Europe Based on Demographic Models." *Comptes Rendus Biologies* 326 (6): 575–587. https://doi.org/10.1016/s1631-0691(03)00148-3.

Charter, M., I. Izhaki, Y. Mocha, et al. 2016. "Nest-Site Competition between Invasive and Native Cavity Nesting Birds and Its Implication for Conservation." *Journal of Environmental Management* 181: 129–134. https://doi.org/10.1016/j.jenvman.2016.06.021.

Cheke, R., and M. Sidatt. 2019. "A Review of Alternatives to Fenthion for Quelea Bird Control." *Crop Protection* 116: 15–23. https://doi.org/10.1016/j.cropro.2018.10.005.

China Daily Staff. 2003. "Monkeys Terrorize India Workers, Tourists." *China Daily*, November 3. https://www.chinadaily.com.cn/en/doc/2003-11/03/content_277874.htm.

Chipman, R., T. DeVault, D. Slate, et al. 2008. "Non-Lethal Management to Reduce Conflicts with Winter Urban Crow Roosts in New York: 2002–2007," In *Proceedings of the 23rd Vertebrate Pest Conference*, March 17–20, San Diego, CA, edited by R. M. Timm and M. B. Madon, 88–93. Davis: University of California. https://doi.org/10.5070/V423110418.

Choquenot, D., and J. Hone. 2002. "Using Bioeconomic Models to Maximize Benefits from Vertebrate Pest Control:

Lamb Predation by Feral Pigs." In *Human Conflicts with Wildlife: Economic Considerations: Proceedings of the Third NWRC Special Symposium*, August 1–3, 2000, Fort Collins, CO, edited by L. Clark, J. Hone, J. A. Shivik, et al., 65–79. Fort Collins, CO: National Wildlife Research Center. https://digitalcommons.unl.edu /nwrchumanconflicts/8/.

Christian, J. J., and D. E. Davis. 1964. "Endocrines, Behavior, and Population: Social and Endocrine Factors Are Integrated in the Regulation of Growth of Mammalian Populations." *Science* 146 (3651): 1550–1560. https://doi.org /10.1126/science.146.3651.1550.

Chua, K. B. 2010. "Risk Factors, Prevention and Communication Strategy during Nipah Virus Outbreak in Malaysia." *Malaysian Journal of Pathology* 32 (2): 75–80.

Cilliers, C. J. 1991. "Biological Control of Water Hyacinth, *Eichhornia crassipes* (Pontederiaceae), in South Africa." *Agriculture, Ecosystems and Environment* 37 (1–3): 207–217. https://doi.org/10.1016/0167-8809(91)90149-R.

Cilliers, D. 2003. "South African Cheetah Compensation Fund." *Carnivore Damage Prevention News* 6: 15–16.

Clark, L., C. Clark, and S. Siers. 2018. "Brown Tree Snakes Methods and Approaches for Control." In *Ecology and Management of Terrestrial Vertebrate Invasive Species in the United States*, edited by W. C. Pitt, J. C. Beasley, and G. W. Witmer, 107–134. Boca Raton, FL: CRC Press. https:// digitalcommons.unl.edu/icwdm_usdanwrc/2032.

Clark, L., J. Eisemann, J. Godwin, et al. 2020. "Invasive Species Control and Resolution of Wildlife Damage Conflicts: A Framework for Chemical and Genetically Based Management Methods." In *GMOs: Implications for Biodiversity Conservation and Ecological Processes*, edited by A. Chaurasia, D. L. Hawksworth, and M. Pessoa de Miranda, 193–222. Cham, Switzerland: Springer International. https://doi.org/10.1007/978-3-030-53183-6_9.

Clark, L., and J. Hall. 2006. "Avian Influenza in Wild Birds: Status as Reservoirs, and Risks to Humans and Agriculture." *Ornithological Monographs* 60: 3–29. https://doi.org /10.2307/40166825.

Clark, L., and J. Shivik. 2002. "Aerosolized Essential Oils and Individual Natural Product Compounds as Brown Treesnake Repellents." *Pest Management Science* 58 (8): 775–783. https://doi.org/10.1002/ps.525.

Cleary, E., and R. Dolbeer. 2005. *Wildlife Hazard Management at Airports: A Manual for Airport Personnel*. 2nd ed. Washington, DC: Airports Program, Animal and Plant Health Inspection Service, US Department of Agriculture, and US Federal Aviation Administration. https:// digitalcommons.unl.edu/icwdm_usdanwrc/133/.

Clergeau, P., J. Savard, G. Mennechez, et al. 1998. "Bird Abundance and Diversity along an Urban-Rural Gradient: A Comparative Study between Two Cities on Different Continents." *The Condor* 100 (3): 413–425. https://doi.org /10.2307/1369707.

Clover, M. R. 1954. "A Portable Deer Trap and Catch-Net." *California Fish and Game* 40 (4): 367–373.

Cobián-Rojas, D., J. Schmitter-Soto, C. Betancourt, et al. 2018. "The Community Diversity of Two Caribbean MPAs Invaded by Lionfish Does Not Support the Biotic

Resistance Hypothesis." *Journal of Sea Research* 134: 26–33. https://doi.org/10.1016/j.seares.2018.01.004.

Coleman, T. 2020. "Forest Insect and Disease Leaflet 162: Gypsy Moth." FS-1159 (Revised). Washington, DC: US Department of Agriculture, Forest Service, State & Private Forestry. https://www.fs.fed.us/foresthealth/docs /fidls/FIDL-162-Ldispardispar.pdf.

Conner, M., M. Ebinger, and F. Knowlton. 2008. "Evaluating Coyote Management Strategies Using a Spatially Explicit, Individual-Based, Socially Structured Population Model." *Ecological Modelling* 219 (1–2): 234–247. https://doi.org/10 .1016/j.ecolmodel.2008.09.008.

Connor, M. J. 1992. "The Red-eared Slider, *Trachemys scripta elegans*." *Tortuga Gazette* 28 (4): 1–3. https://tortoise.org /archives/elegans.html.

Conover, G., R. Simmonds, and M. Whalen. 2007. "Management and Control Plan for Bighead, Black, Grass, and Silver Carps in the United States." Washington, DC: Aquatic Nuisance Species Task Force, Asian Carp Working Group. https://www.asiancarp.org/Documents/Carps _Management_Plan.pdf.

Conover, M. 2019. "Numbers of Human Fatalities, Injuries, and Illnesses in the United States Due to Wildlife." *Human–Wildlife Interactions* 13 (2): 12. https://doi.org/10 .26077/r59n-bv76.

Conover, M., and R. Vail. 2014. *Human Diseases from Wildlife*. Baton Rouge: Taylor & Francis.

Cornwall, W. 2019. "Researchers Embrace a Radical Idea: Engineering Coral to Cope with Climate Change." *Science News*, March 21. https://doi.org/10.1126/science.aax4091.

Côté, S., T. Rooney, J. Tremblay, et al. 2004. "Ecological Impacts of Deer Overabundance." *Annual Review of Ecology, Evolution, and Systematics* 35: 113–147. https://doi .org/10.1146/annurev.ecolsys.35.021103.105725.

Cotton, W. 2008. "Resolving Conflicts between Humans and the Threatened Louisiana Black Bear." *Human–Wildlife Conflicts* 2 (2): 151–152. https://digitalcommons.unl.edu /hwi/44/.

Courtenay, W. R., Jr., D. A. Hensley, J. N. Taylor, et al. 1984. "Distribution of Exotic Fishes in the Continental United States." In *Distribution, Biology and Management of Exotic Fishes*, edited by W. R. Courtenay Jr. and J. R. Stauffer Jr., 41–77. Baltimore: Johns Hopkins University Press.

Courtenay, W. R., Jr., and J. D. Williams. 2004. "Snakeheads (Pisces, Channidae)—A Biological Synopsis and Risk Assessment." US Geological Survey Circular 1251. Denver, CO: US Geological Survey. https://doi.org/10.3133/cir1251.

Crist, E., C. Mora, and R. Engelman. 2017. "The Interaction of Human Population, Food Production, and Biodiversity Protection." *Science* 356 (6335): 260–264. https://doi.org/10 .1126/science.aal2011.

Cruz, F., C. J. Donlan, K. Campbell, et al. 2005. "Conservation Action in the Galapagos: Feral Pig (*Sus scrofa*) Eradication from Santiago Island." *Biological Conservation* 121 (3): 473–478. https://doi.org/10.1016/j.biocon.2004.05 .018.

Cucchi, T., and J.-D. Vigne. 2006. "Origin and Diffusion of the House Mouse in the Mediterranean." *Human Evolution* 21 (2): 95. https://doi.org/10.1007/s11598-006-9011-z.

Cucchi, T., J.-D. Vigne, and J.-C. Auffray. 2005. "First Occurrence of the House Mouse (*Mus musculus domesticus* Schwarz & Schwarz, 1943) in the Western Mediterranean: A Zooarchaeological Revision of Subfossil Occurrences." *Biological Journal of the Linnean Society* 84 (3): 429–445. https://doi.org/10.1111/j.1095-8312.2005.00445.x.

Cummings, J., J. Guarino, C. E. Knittle, et al. 1987. "Decoy Plantings for Reducing Blackbird Damage to Nearby Commercial Sunflower Fields." *Crop Protection* 6 (1): 56–60. https://doi.org/10.1016/0261-2194(87)90029-9.

Curtis, P. 2020. "After Decades of Suburban Deer Research and Management in the Eastern United States: Where Do We Go from Here?" *Human–Wildlife Interactions* 14 (1): 16. https://doi.org/10.26077/k7ye-k912.

Cuthbert, R., and G. Hilton. 2004. "Introduced House Mice *Mus musculus*: A Significant Predator of Threatened and Endemic Birds on Gough Island, South Atlantic Ocean?" *Biological Conservation* 117 (5): 483–489. https://doi.org/10.1016/j.biocon.2003.08.007.

Cutler, D., and L. Summers. 2020. "The COVID-19 Pandemic and the $16 Trillion Virus." *JAMA* 324 (15): 1495–1496. https://doi.org/10.1001/jama.2020.19759.

Dabritz, H., M. Miller, E. R. Atwill, et al. 2007. "Detection of *Toxoplasma gondii*-like Oocysts in Cat Feces and Estimates of the Environmental Oocyst Burden." *Journal of the American Veterinary Medical Association* 231 (11): 1676–1684. https://doi.org/10.2460/javma.231.11.1676.

Dadam, D., R. Robinson, A. Clements, et al. 2019. "Avian Malaria-Mediated Population Decline of a Widespread Iconic Bird Species." *Royal Society Open Science* 6 (7): 182197. https://doi.org/10.1098/rsos.182197.

Daehler, C. 1998. "The Taxonomic Distribution of Invasive Angiosperm Plants: Ecological Insights and Comparison to Agricultural Weeds." *Biological Conservation* 84 (2): 167–180. https://doi.org/10.1016/S0006-3207(97)00096-7.

Dahl, R. 2005. "Population Equation: Balancing What We Need with What We Have." *Environmental Health Perspectives* 113 (9): A598–A605. https://doi.org/10.1289/ehp.113-a598.

Daily, G., A. Ehrlich, and P. Ehrlich. 1994. "Optimum Human Population Size." *Population and Environment* 15 (6): 469–475. https://doi.org/10.1007/BF02211719.

Dale, L. 2009. "Personal and Corporate Liability in the Aftermath of Bird Strikes: A Costly Consideration." *Human–Wildlife Conflicts* 3 (2): 216–225. https://doi.org/10.26077/24re-wa34.

Darrow, P., and J. Shivik. 2009. "Bold, Shy, and Persistent: Variable Coyote Response to Light and Sound Stimuli." *Applied Animal Behaviour Science* 116 (1): 82–87. https://doi.org/10.1016/j.applanim.2008.06.013.

Daskin, J., and R. Pringle. 2018. "Warfare and Wildlife Declines in Africa's Protected Areas." *Nature* 553 (7688): 328–332. https://doi.org/10.1038/nature25194.

Davidson, W. R., and G. L. Doster. 1997. "Health Characteristics and White-Tailed Deer Population Density in the Southeastern United States." In *The Science of Overabundance: Deer Ecology and Population Management*, edited by W. McShea, H. Underwood, and J. Rappole, 164–184. Washington, DC: Smithsonian Books.

Davis, D. E. 1966. *Integral Animal Behavior*. New York: Macmillan.

Davis, N., D. O'Dowd, P. Green, et al. 2008. "Effects of an Alien Ant Invasion on Abundance, Behavior, and Reproductive Success of Endemic Island Birds." *Conservation Biology* 22 (5): 1165–1176. https://doi.org/10.1111/j.1523-1739.2008.00984.x.

Dayoub, M., R. Birech, M. Haghbayan, et al. 2021. "Co-Design in Bird Scaring Drone Systems: Potentials and Challenges in Agriculture." In *Proceedings of the International Conference on Advanced Intelligent Systems and Informatics 2020*, edited by A. E. Hassanien, A. Slowik, V. Snášel, et al., 598–607. Cham, Switzerland: Springer International. https://doi.org/10.1007/978-3-030-58669-0_54.

De Beer, W., S. Marincowitz, T. Duong, et al. 2017. "*Bretziella*, a New Genus to Accommodate the Oak Wilt Fungus, *Ceratocystis fagacearum* (Microascales, Ascomycota) [2017]." *MycoKeys* 27: 1–19. https://doi.org/10.3897/mycokeys.27.20657.

DeCalesta, D. 1992. "Impact of Deer on Species Diversity of Allegheny Hardwood Stands." *Proceedings of the Northeastern Weed Science Society Abstracts* 46: 135. https://www.newss.org/proceedings/proceedings_1992_vol46.pdf.

DeCalesta, D. 1997. "Deer and Ecosystem Management," In *The Science of Overabundance: Deer Ecology and Population Management*, edited by W. McShea, H. Underwood, and J. Rappole, 267–279. Washington, DC: Smithsonian Books.

DeCapita, M. 2000. "Brown-headed Cowbird Control on Kirtland's Warbler Nesting Areas in Michigan, 1972–1995." In *Ecology and Management of Cowbirds and Their Hosts*, edited by J. N. M. Smith, T. L. Cook, S. L. Rothstein, et al., 333–341. Austin: University of Texas Press.

Decker, D., A. Forstchen, W. Siemer, et al. 2019. "Moving the Paradigm from Stakeholders to Beneficiaries in Wildlife Management." *Journal of Wildlife Management* 83 (3): 513–518. https://doi.org/10.1002/jwmg.21625.

Decker, D. J., T. B. Lauber, and W. F. Siemer. 2002. "Human-Wildlife Conflict Management: A Practitioner's Guide." Ithaca, NY: Northeast Wildlife Damage Management Research and Outreach Cooperative. https://hdl.handle.net/1813/40557.

Decker, D. J., D. B. Raik, and W. F. Siemer. 2004. "Community-Based Deer Management: A Practitioner's Guide." Ithaca, NY: Northeast Wildlife Damage Management Research and Outreach Cooperative. https://hdl.handle.net/1813/40558.

Decker, S., A. Bath, A. Simms, et al. 2010. "The Return of the King or Bringing Snails to the Garden? The Human Dimensions of a Proposed Restoration of European Bison (*Bison bonasus*) in Germany." *Restoration Ecology* 18 (1): 41–51. https://doi.org/10.1111/j.1526-100X.2008.00467.x.

Deevey, E. S., Jr. 1947. "Life Tables for Natural Populations of Animals." *Quarterly Review of Biology* 22 (4): 283–314. https://doi.org/10.1086/395888.

DeLiberto, S., and S. Werner. 2016. "Review of Anthraquinone Applications for Pest Management and Agricultural Crop Protection." *Pest Management Science* 72 (10): 1813–1825. https://doi.org/10.1002/ps.4330.

Delibes-Mateos, M., M. Farfán, C. Rouco, et al. 2018. "A Large-Scale Assessment of European Rabbit Damage to Agriculture in Spain." *Pest Management Science* 74 (1): 111–119. https://doi.org/10.1002/ps.4658.

Dellamano, F. 2006. "Controlling Birds with Netting: Blueberries, Cherries and Grapes." *New York Fruit Quarterly* 14 (2): 3–5. http://dev.nyshs.org/wp-content /uploads/2016/10/Controlling-Birds-with-Netting -Blueberries-Cherries-and-Grapes.pdf.

DeNicola, E., O. Aburizaize, A. Siddique, et al. 2016. "Road Traffic Injury as a Major Public Health Issue in the Kingdom of Saudi Arabia: A Review." *Frontiers in Public Health* 4: 215. https://doi.org/10.3389/fpubh.2016 .00215.

Dethier, V. G., and E. Stellar. 1964. *Animal Behavior: Its Evolutionary and Neurological Basis.* Englewood Cliffs, NJ: Prentice-Hall.

Dev, N. S. Gogul, K. S. Sreenesh, et al. 2019. "IoT Based Automated Crop Protection System." In *2019 2nd International Conference on Intelligent Computing, Instrumentation and Control Technologies (ICICICT)*, vol. 1: 1333–1337. https://doi.org/10.1109/ICICICT46008.2019.8993406.

DeVault, T., J. Beasley, L. Humberg, et al. 2007. "Intrafield Patterns of Wildlife Damage to Corn and Soybeans in Northern Indiana." *Human–Wildlife Conflicts* 1 (2): 205–213. https://doi.org/10.26077/0j2d-d311.

DeVault, T., B. Blackwell, T. Seamans, et al. 2016. "Identification of off Airport Interspecific Avian Hazards to Aircraft." *Journal of Wildlife Management* 80 (4): 746–752. https://doi.org/10.1002/jwmg.1041.

Dhammi, A., J. Van Krestchmar, L. Ponnusamy, et al. 2016. "Biology, Pest Status, Microbiome and Control of Kudzu Bug (Hemiptera: Heteroptera: Plataspidae): A New Invasive Pest in the U.S." *International Journal of Molecular Sciences* 17 (9): 1570. https://doi.org/10.3390/ijms17091570.

Diagne, C., B. Leroy, R. E. Gozlan, et al. 2020. "InvaCost, a Public Database of the Economic Costs of Biological Invasions Worldwide." *Scientific Data* 7 (1): 277. https://doi .org/10.1038/s41597-020-00586-z.

Diagne, C., B. Leroy, A.-C. Vaissière, et al. 2021. "High and Rising Economic Costs of Biological Invasions Worldwide." *Nature* 592 (7855): 571–576. https://doi.org/10.1038 /s41586-021-03405-6.

Diamond, J. 1997. *Guns, Germs, and Steel: The Fates of Human Societies.* New York: W.W. Norton.

Diamond, S., R. Giles Jr, R. L. Kirkpatrick, et al. 2000. "Hard Mast Production before and after the Chestnut Blight." *Southern Journal of Applied Forestry* 24 (4): 196–201. https://doi.org/10.1093/sjaf/24.4.196.

Di Castri, F. 1990. "On Invading Species and Invaded Ecosystems: The Interplay of Historical Chance and Biological Necessity." In *Biological Invasions in Europe and the Mediterranean Basin*, edited by F. Di Castri, A. J. Hansen, and M. Beussche, 3–16. Dordrecht, Netherlands: Kluwer Academic.

Ding, J., R. Mack, P. Lu, et al. 2008. "China's Booming Economy Is Sparking and Accelerating Biological Invasions." *BioScience* 58 (4): 317–324. https://doi.org/10 .1641/B580407.

Distefano, E. 2005. "Human-Wildlife Conflict Worldwide: Collection of Case Studies, Analysis of Management Strategies and Good Practices." Rome: Food and Agricultural Organization of the United Nations (FAO), Sustainable Agriculture and Rural Development Initiative (SARDI). https://www.fao.org/3/au241e/au241e.pdf.

Dobner, J. 2011. "Judge Awards $1.9M to Family of Boy Killed by Bear." Associated Press, May 3. https://www.heraldnet .com/news/judge-awards-1-9m-to-family-of-boy-killed-by -bear/.

Doherty-Bone, T. M., A. A. Cunningham, M. C. Fisher, et al. 2020. "Amphibian Chytrid Fungus in Africa—Realigning Hypotheses and the Research Paradigm." *Animal Conservation* 23 (3): 239–244. https://doi.org/10.1111/acv.12538.

Dolbeer, R. 1998. "Population Dynamics: The Foundation of Wildlife Damage Management for the 21st Century." In *Proceedings of the 18th Vertebrate Pest Conference*, March 2–5, Costa Mesa, CA, edited by R. O. Baker and A. C. Crabb, 2–11. Davis: University of California. https://digitalcomm ons.unl.edu/vpc18/9/.

Dolbeer, R., M. Begier, P. Miller, et al. 2021. "Wildlife Strikes to Civil Aircraft in the United States, 1990–2019." Office of Airport Safety and Standards, Serial Report No. 26, Washington, DC: US Department of Transportation, Federal Aviation Administration. https://rosap.ntl.bts.gov /view/dot/58293.

Dolbeer, R., J. Belant, and J. Sillings. 1993. "Shooting Gulls Reduces Strikes with Aircraft at John F. Kennedy International Airport." *Wildlife Society Bulletin* 21 (4): 442–450. https://www.jstor.org/stable/3783417.

Dolbeer, R., and R. B. Chipman. 1999. "Shooting Gulls to Reduce Strikes with Aircraft at John F. Kennedy International Airport, 1991–1998." Special report for the Port Authority of New York and New Jersey by US Department of Agriculture. Sandusky, OH: National Wildlife Research Center.

Dolbeer, R., A. R. Stickley Jr., and P. P. Woronecki. 1979. "Starling, *Sturnus vulgaris*, Damage to Sprouting Wheat in Tennessee and Kentucky, USA." *Protection Ecology* 1: 159–169.

Dolbeer, R., P. Woronecki, and R. Bruggers. 1986. "Reflecting Tapes Repel Blackbirds from Millet, Sunflowers, and Sweet Corn." *Wildlife Society Bulletin (1973–2006)* 14 (4): 418–425. https://www.jstor.org/stable/3782281.

Dolbeer, R., S. Wright, and E. Cleary. 2000. "Ranking the Hazard Level of Wildlife Species to Aviation." *Wildlife Society Bulletin* 372–378. https://www.jstor.org/stable /3783694.

Dollinger, A. 2000. "Egyptian Vermin." *An Introduction to the History and Culture of Pharaonic Egypt.* https://web.archive .org/web/20090814160653/http://nefertiti.iwebland.com /timelines/topics/pests.htm.

Donlan, J. 2005. "Re-wilding North America." *Nature* 436 (7053): 913–914. https://doi.org/10.1038/436913a.

Dorr, B., and D. Fielder. 2017. "Double-Crested Cormorants: Too Much of a Good Thing?" USDA National Wildlife Research Center: Staff Publications, 1977. https:// digitalcommons.unl.edu/icwdm_usdanwrc/1977.

Dorrestein, A., C. Todd, D. Westcott, et al. 2019. "Impacts of an Invasive Ant Species on Roosting Behavior of an Island

Endemic Flying-Fox." *Biotropica* 51 (1). 75–83. https://doi.org/10.1111/btp.12620.

Dove, C., R. Snow, M. Rochford, et al. 2011. "Birds Consumed by the Invasive Burmese Python (*Python molurus bivittatus*) in Everglades National Park, Florida, USA." *Wilson Journal of Ornithology* 123 (1): 126–131. https://doi.org/10.1676/10-092.1.

Drake, D., S. Dubay, and M. Allen. 2021. "Evaluating Human–Coyote Encounters in an Urban Landscape Using Citizen Science." *Journal of Urban Ecology* 7 (1): juaa032. https://doi.org/10.1093/jue/juaa032.

Durland Donahou, A., W. Conard, K. Dettloff, et al. 2019. "*Faxonius rusticus* (Girard, 1852)." In *Nonindigenous Aquatic Species Database*. Gainesville, FL: US Geological Survey. https://nas.er.usgs.gov/queries/factsheet.aspx?SpeciesID=214.

Duron, Q., A. Shiels, and E. Vidal. 2017. "Control of Invasive Rats on Islands and Priorities for Future Action." *Conservation Biology* 31 (4): 761–771. https://doi.org/10.1111/cobi.12885.

Dye, T. 2008. *Understanding Public Policy*. 12th ed. Englewood Cliffs, NJ: Pearson/Prentice Hall.

Eason, C., L. Shapiro, S. Ogilvie, et al. 2017. "Trends in the Development of Mammalian Pest Control Technology in New Zealand." *New Zealand Journal of Zoology* 44 (4): 267–304. https://doi.org/10.1080/03014223.2017.1337645.

Ebert, C., F. Knauer, B. Spielberger, et al. 2012. "Estimating Wild Boar Sus Scrofa Population Size Using Faecal DNA and Capture-Recapture Modelling." *Wildlife Biology* 18 (2): 142–152. https://doi.org/10.2981/11-002.

Egan, M., C. Day, T. Katzner, et al. 2020. "Relative Abundance of Coyotes (*Canis latrans*) Influences Gray Fox (*Urocyon cinereoargenteus*) Occupancy across the Eastern United States." *Canadian Journal of Zoology* 99 (2): 63–72. https://doi.org/10.1139/cjz-2019-0246.

Eisemann, J., B. Petersen, and K. Fagerstone. 2003. "Efficacy of Zinc Phosphide for Controlling Norway Rats, Roof Rats, House Mice, *Peromyscus* spp., Prairie Dogs and Ground Squirrels: A Literature Review (1942–2000)." In *Proceedings of the 10th Wildlife Damage Management Conference*, April 6–9, Hot Springs, AR, edited by K. A. Fagerstone and G. W. Witmer, 229–236. Bethesda, MD: The Wildlife Society. https://digitalcommons.unl.edu/icwdm_usdanwrc/212.

Eisen, L., M. Coleman, S. Lozano-Fuentes, et al. 2011. "Multi-Disease Data Management System Platform for Vector-Borne Diseases." *PLoS Neglected Tropical Diseases* 5 (3): e1016. https://doi.org/10.1371/journal.pntd.0001016.

Elfekih, S., S. Metcalfe, T. Walsh, et al. 2021. "Genomic Insights into a Population of Introduced European Rabbits *Oryctolagus cuniculus* in Australia and the Development of Genetic Resistance to Rabbit Hemorrhagic Disease Virus." *Transboundary and Emerging Diseases*, February 9. https://doi.org/10.1111/tbed.14030.

Emerton, L. 1999. "Balancing the Opportunity Costs of Wildlife Conservation for Communities around Lake Mburo National Park, Uganda." Evaluating Eden Series Discussion Paper No. 5. London: International Institute for Environment and Development (IIED). https://pubs.iied.org/sites/default/files/pdfs/migrate/7798IIED.pdf.

Engeman, R. 2002. "Economic Considerations of Damage Assessment." In *Human Conflicts with Wildlife: Economic Considerations: Proceedings of the Third NWRC Special Symposium*, August 1–3, 2000, Fort Collins, CO, edited by L. Clark, J. Hone, J. A. Shivik, et al., 36–41. Fort Collins, CO: National Wildlife Research Center. https://digitalcommons.unl.edu/nwrchumanconflicts/4/.

Engeman, R., and L. Allen. 2000. "Overview of a Passive Tracking Index for Monitoring Wild Canids and Associated Species." *Integrated Pest Management Reviews* 5 (3): 197–203. https://doi.org/10.1023/A:1011380314051.

Engeman, R., and D. Campbell. 1999. "Pocket Gopher Reoccupation of Burrow Systems Following Population Reduction." *Crop Protection* 18 (8): 523–525. https://doi.org/10.1016/s0261-2194(99)00055-1.

Engeman, R., B. Constantin, S. Shwiff, et al. 2007. "Adaptive and Economic Management Methods for Feral Hog Control in Florida." *Human–Wildlife Conflicts* 1 (2): 178–185. https://doi.org/10.26077/gvrc-sb22.

Engeman, R., and M. Linnell. 1998. "Trapping Strategies for Deterring the Spread of Brown Tree Snakes from Guam." *Pacific Conservation Biology* 4 (4): 348–353. https://digitalcommons.unl.edu/icwdm_usdanwrc/633/.

Engeman, R., M. Pipas, K. Gruver, et al. 2000. "Monitoring Coyote Population Changes with a Passive Activity Index." *Wildlife Research* 27 (5): 553–557. https://doi.org/10.1071/WR98090.

Engeman, R., A. Shiels, and C. Clark. 2018. "Objectives and Integrated Approaches for the Control of Brown Tree Snakes: An Updated Overview." *Journal of Environmental Management* 219: 115–124. https://doi.org/10.1016/j.jenvman.2018.04.092.

Engeman, R., S. Shwiff, B. Constantin, et al. 2002a. "An Economic Analysis of Predator Removal Approaches for Protecting Marine Turtle Nests at Hobe Sound National Wildlife Refuge." *Ecological Economics* 42 (3): 469–478. https://doi.org/10.1016/S0921-8009(02)00136-2.

Engeman, R., S. Shwiff, H. Smith, et al. 2002b. "Monetary Valuation Methods for Economic Analysis of the Benefit-Costs of Protecting Rare Wildlife Species from Predators." *Integrated Pest Management Reviews* 7 (3): 139–144. https://digitalcommons.unl.edu/icwdm_usdanwrc/88/.

Engeman, R., and G. Witmer. 2000. "IPM Strategies: Indexing Difficult to Monitor Populations of Pest Species." In *Proceedings of the 19th Vertebrate Pest Conference*, March 6–9, San Diego, CA, edited by T. P. Salmon and A. C. Crabb, 184–189. Davis: University of California. https://doi.org/10.5070/V419110013.

Engle, C., T. Christie, B. Dorr, et al. 2021. "Principal Economic Effects of Cormorant Predation on Catfish Farms." *Journal of the World Aquaculture Society* 52 (1): 41–56. https://doi.org/10.1111/jwas.12728.

Escobar, L., J. Escobar-Dodero, and N. D. Phelps. 2018. "Infectious Disease in Fish: Global Risk of Viral Hemorrhagic Septicemia Virus." *Reviews in Fish Biology and Fisheries* 28 (3): 637–655. https://doi.org/10.1007/s11160-018-9524-3.

Fagerstone, K. A. 2007. "Mitigating Impacts of Terrestrial Invasive Species." *Encyclopedia of Pest Management*, vol. 2, edited by D. Pimentel, 347–352. Boca Raton, FL: CRC Press.

Fagerstone, K. A., and G. Keirn. 2012. "Wildlife Services—A Leader in Developing Tools and Techniques for Managing Carnivores." In *Proceedings of the 14th Wildlife Damage Management Conference*, April 18–21, 2011, Nebraska City, NE, edited by S. N. Frey, 44–55. https://digitalcommons.unl.edu/icwdm_usdanwrc/1133/.

Fagerstone, K. A., L. Miller, J. Eisemann, et al. 2008. "Registration of Wildlife Contraceptives in the United States of America, with OvoControl and GonaCon Immunocontraceptive Vaccines as Examples." *Wildlife Research* 35 (6): 586–592. https://doi.org/10.1071/WR07166.

Fagerstone, K. A., L. Miller, G. Killian, et al. 2010. "Review of Issues Concerning the Use of Reproductive Inhibitors, with Particular Emphasis on Resolving Human-Wildlife Conflicts in North America." *Integrative Zoology* 5 (1): 15–30. https://doi.org/10.1111/j.1749-4877.2010.00185.x.

Fagerstone, K. A., and E. W. Schafer Jr. 1998. "Status of APHIS Vertebrate Pesticides and Drugs." In *Proceedings of the 18th Vertebrate Pest Conference*, edited by R. O. Baker and A. C. Crabb, 319–324. Davis: University of California. https://doi.org/10.5070/V418110320.

Farri, T. A., and R. A. Boroffice. 1999. "An Overview on Lhe Status and Control of Water Hyacinth in Nigeria." In *Proceedings of the First IOBC Global Working Group Meeting for the Biological and Integrated Control of Water Hyacinth*, November 16–19, 1998, Harare, Zimbabwe, edited by M. P. Hill, M. H. Julien, and T. D. Center, 182. South Africa: ARC Weeds Research Division.

Fausch, K., and R. White. 1981. "Competition between Brook Trout (*Salvelinus fontinalis*) and Brown Trout (*Salmo trutta*) for Positions in a Michigan Stream." *Canadian Journal of Fisheries and Aquatic Sciences* 38 (10): 1220–1227. https://doi.org/10.1139/f81-164.

Feldhamer, G., L. Drickamer, S. Vessey, et al. 2007. *Mammalogy: Adaptation, Diversity, Ecology*. Baltimore: Johns Hopkins University Press.

Feldman, R., M. Stanton, D. Borys, et al. 2019. "Medical Outcomes of Bromethalin Rodenticide Exposures Reported to US Poison Centers after Federal Restriction of Anticoagulants." *Clinical Toxicology* 57 (11): 1109–1114. https://doi.org/10.1080/15563650.2019.1582776.

Fernandez, A., R. Richardson, D. Tschirley, et al. 2009. "Wildlife Conservation in Zambia: Impacts on Rural Household Welfare." Food Security Collaborative Working Papers 55053, Michigan State University, Department of Agricultural, Food, and Resource Economics. https://doi.org/10.22004/ag.econ.55053.

Fernandez-Duque, F., R. Bailey, and D. Bonter. 2019. "Egg Oiling as an Effective Management Technique for Limiting Reproduction in an Invasive Passerine." *Avian Conservation and Ecology* 14 (2): 20. https://doi.org/10.5751/ACE-01491-140220.

Ferraz, K., M. Lechevalier, H. Couto, et al. 2003. "Damage Caused by Capybaras in a Corn Field." *Scientia Agricola* 60 (1): 191–194. https://doi.org/10.1590/S0103-90162003000100029.

Ferreira, S., C. Greaver, and C. Simms. 2017. "Elephant Population Growth in Kruger National Park, South Africa,

under a Landscape Management Approach." *Koedoe* 59 (1): 1–6. https://doi.org/10.4102/koedoe.v59i1.1427.

Ferrer, M., V. Morandini, R. Baumbusch, et al. 2020. "Efficacy of Different Types of 'Bird Flight Diverter' in Reducing Bird Mortality Due to Collision with Transmission Power Lines." *Global Ecology and Conservation* 23: e01130. https://doi.org/10.1016/j.gecco.2020.e01130.

Ficetola, G., C. Coïc, M. Detaint, et al. 2007. "Pattern of Distribution of the American Bullfrog *Rana catesbeiana* in Europe." *Biological Invasions* 9 (7): 767–772. https://doi.org/10.1007/s10530-006-9080-y.

Ficetola, G., W. Thuiller, and E. Padoa-Schioppa. 2009. "From Introduction to the Establishment of Alien Species: Bioclimatic Differences between Presence and Reproduction Localities in the Slider Turtle." *Diversity and Distributions* 15 (1): 108–116. https://doi.org/10.1111/j.1472-4642.2008.00516.x.

Fine, P. 2002. "The Invasibility of Tropical Forests by Exotic Plants." *Journal of Tropical Ecology* 18 (5): 687–705. https://doi.org/10.1017/S0266467402002456.

Finlayson, B., R. Schnick, R. Cailteux, et al. 2000. *Rotenone Use in Fisheries Management: Administrative and Technical Guidelines Manual*. Bethesda: MD: American Fisheries Society.

Fisher, A., C. Mills, M. Lyons, et al. 2021. "Remote Sensing of Trophic Cascades: Multi-temporal Landsat Imagery Reveals Vegetation Change Driven by the Removal of an Apex Predator." *Landscape Ecology* 36, 1341–1358. https://doi.org/10.1007/s10980-021-01206-w.

Fitzwater, W. 1972. "Barrier Fencing in Wildlife Management." In *Proceedings of the 5th Vertebrate Pest Conference*, March 7–9, Fresno, CA, edited by R. E. Marsh, 49–55. Davis: University of California. https://digitalcommons.unl.edu/vpc5/11/.

Fleishman, E., N. McDonal, R. Mac Nally, et al. 2003. "Effects of Floristics, Physiognomy and Non-Native Vegetation on Riparian Bird Communities in a Mojave Desert Watershed." *Journal of Animal Ecology* 72 (3): 484–490. https://doi.org/10.1046/j.1365-2656.2003.00718.x.

Flowers, R. 1986. "Supplemental Feeding of Black Bear in Tree Damaged Areas of Western Washington." In *Symposium Proceedings: Animal Damage Management in Pacific Northwest Forests*, March 25–27, Spokane, WA, edited by D. M. Baumgartner, R. Mahoney, J. Evans, 147–148. Pullman: Washington State University.

Flueck, W. T., Smith-Flueck, J.A.M., and Naumann, C. M. 2003. "The Current Distribution of Red Deer (*Cervus elaphus*) in Southern Latin America." *Zeitschrift für Jagdwissenschaft* 49: 112–119. https://doi.org/10.1007/BF02190451.

Foley, J. 2017. "Living by the Lessons of the Planet." *Science* 356 (6335): 251–252. https://doi.org/10.1126/science.aal4863.

Forrester, J., T. Weiser, and J. Forrester. 2018. "An Update on Fatalities Due to Venomous and Nonvenomous Animals in the United States (2008–2015)." *Wilderness and Environmental Medicine* 29 (1): 36–44. https://doi.org/10.1016/j.wem.2017.10.004.

Forsyth, D., D. Wilson, T. Easdale, et al. 2015. "Century-Scale Effects of Invasive Deer and Rodents on the Dynamics of

Forests Growing on Soils of Contrasting Fertility." *Ecological Monographs* 85 (2): 157–180. https://doi.org/10.1890/14-0389.1.

Fortin, D., C. Brooke, P. Lamirande, et al. 2020. "Quantitative Spatial Ecology to Promote Human-Wildlife Coexistence: A Tool for Integrated Landscape Management." *Frontiers in Sustainable Food Systems* 4: 600363. https://doi.org/10.3389/fsufs.2020.600363.

Forys, E., J. Campo, E. Silva, et al. 2020. "A Comparison of 2 Methods to Deter Fish Crows from Depredating Seabird Eggs." *Wildlife Society Bulletin* 44 (4): 670–676. https://doi.org/10.1002/wsb.1139.

Franklin, W., and K. Powell. 1994. "Guard Llamas: A Part of Integrated Sheep Protection." PM-1527. Ames: Iowa State University, University Extension. https://web.archive.org/web/20140911114749/https://store.extension.iastate.edu/Product/pm1527-pdf.

Fredens, J., K. Wang, D. de la Torre, et al. 2019. "Total Synthesis of *Escherichia coli* with a Recoded Genome." *Nature* 569 (7757): 514–518. https://doi.org/10.1038/s41586-019-1192-5.

Freed, L., R. Cann, M. Goff, et al. 2005. "Increase in Avian Malaria at Upper Elevation in Hawai'i." *The Condor* 107 (4): 753–764. https://doi.org/10.1093/condor/107.4.753.

Freedman, A., and R. Wayne. 2017. "Deciphering the Origin of Dogs: From Fossils to Genomes." *Annual Review of Animal Biosciences* 5: 281–307. https://doi.org/10.1146/annurev-animal-022114-110937.

Frei, B., J. Nocera, and J. Fyles. 2015. "Interspecific Competition and Nest Survival of the Threatened Red-Headed Woodpecker." *Journal of Ornithology* 156 (3): 743–753. https://doi.org/10.1007/s10336-015-1177-6.

Frenkel, J. K., J. P. Dubey, and Nancy L. Miller. 1970. "*Toxoplasma gondii* in Cats: Fecal Stages Identified as Coccidian Oocysts." *Science* 167 (3919): 893–896. https://doi.org/10.1126/science.167.3919.893.

Fu, Z. 1997. "Rabies and Rabies Research: Past, Present and Future." *Vaccine* 15 (S1): S20–S24. https://doi.org/10.1016/s0264-410x(96)00312-x.

Fuller, P. L., A. J. Benson, and M. E. Neilson. 2019a. "*Channa marulius* (Hamilton, 1822)." In *Nonindigenous Aquatic Species Database*. Gainesville, FL: US Geological Survey. https://nas.er.usgs.gov/queries/FactSheet.aspx?speciesID=2266.

Fuller, P. L., J. Larson, A. Fusaro, et al. 2019b. "*Oncorhynchus mykiss* (Walbaum, 1792)." In *Nonindigenous Aquatic Species Database*. Gainesville, FL: US Geological Survey. https://nas.er.usgs.gov/queries/factsheet.aspx?SpeciesID=910.

Fuller, P. L., E. Maynard, D. Raikow, et al. 2019c. "*Morone americana* (Gmelin, 1789)." In *Nonindigenous Aquatic Species Database*. Gainesville, FL: US Geological Survey. https://nas.er.usgs.gov/queries/FactSheet.aspx?speciesID=777.

Fuller, P. L., and M. Neilson. 2021. "*Ictalurus furcatus* (Valenciennes in Cuvier and Valenciennes, 1840)." In *Nonindigenous Aquatic Species Database*. Gainesville, FL: US Geological Survey. https://nas.er.usgs.gov/queries/FactSheet.aspx?SpeciesID=740.

Furbearer Conservation Technical Work Group. 2006. "Best Management Practices for Trapping in the United States: Introduction." Association of Fish & Wildlife Agencies. https://www.fishwildlife.org/application/files/5015/2104/8473/Introduction_comp.pdf.

Gabriel, M., L. Diller, J. Dumbacher, et al. 2018. "Exposure to Rodenticides in Northern Spotted and Barred Owls on Remote Forest Lands in Northwestern California: Evidence of Food Web Contamination." *Avian Conservation and Ecology* 13 (1): 2. https://doi.org/10.5751/ACE-01134-130102.

Gabriel, M., L. Woods, G. Wengert, et al. 2015. "Patterns of Natural and Human-Caused Mortality Factors of a Rare Forest Carnivore, the Fisher (*Pekania pennanti*) in California." *PLoS One* 10 (11): e0140640. https://doi.org/10.1371/journal.pone.0140640.

Galef, B. 1980. "Diving for Food: Analysis of a Possible Case of Social Learning in Wild Rats (*Rattus norvegicus*)." *Journal of Comparative and Physiological Psychology* 94 (3): 416–425. https://doi.org/10.1037/h0077678.

Galef, B., and P. Henderson. 1972. "Mother's Milk: A Determinant of the Feeding Preferences of Weaning Rat Pups." *Journal of Comparative and Physiological Psychology* 78 (2): 213–219. https://doi.org/10.1037/h0032186.

Gallardo, B., and D. Aldridge. 2013. "The 'Dirty Dozen': Socio-Economic Factors Amplify the Invasion Potential of 12 High-Risk Aquatic Invasive Species in Great Britain and Ireland." *Journal of Applied Ecology* 50 (3): 757–766. https://doi.org/10.1111/1365-2664.12079.

Gao, C., Y. Wang, X. Gu, et al. 2020. "Association between Cardiac Injury and Mortality in Hospitalized Patients Infected with Avian Influenza A (H7N9) Virus." *Critical Care Medicine* 48 (4): 451–458. https://doi.org/10.1097/CCM.0000000000004207.

Gao, G. F., and L. Wang. 2021. "Perspectives: COVID-19 Expands Its Territories from Humans to Animals." *China CDC Weekly* 3, no. 41 (2021): 855–858. https://doi.org/10.46234/ccdcw2021.210.

Garcia, J., W. Hankins, and K. Rusiniak. 1974. "Behavioral Regulation of the Milieu Interne in Man and Rat." *Science* 185 (4154): 824–831. https://doi.org/10.1126/science.185.4154.824.

García-García, M. 2020. "A History of Mouse Genetics: From Fancy Mice to Mutations in Every Gene." In *Animal Models of Human Birth Defects*, edited by A. Liu, 1–38. Singapore: Springer. https://doi.org/10.1007/978-981-15-2389-2_1.

Garden, P, P McClelland, and K Broome. 2019. "The History of the Aerial Application of Rodenticide in New Zealand." In *Island Invasives: Scaling Up to Meet the Challenge*, edited by C. R. Veitch, M. N. Clout, A. R. Martin, et al., 114–119. Gland, Switzerland: IUCN.

Garshelis, D., S. Baruch-Mordo, A. Bryant, et al. 2017. "Is Diversionary Feeding an Effective Tool for Reducing Human–Bear Conflicts? Case Studies from North America and Europe." *Ursus* 28 (1): 31–55. https://doi.org/10.2192/URSU-D-16-00019.1.

Gasteren, H. van, K. Krijgsveld, N. Klauke, et al. 2019. "Aeroecology Meets Aviation Safety: Early Warning Systems in Europe and the Middle East Prevent Collisions between Birds and Aircraft." *Ecography* 42 (5): 899–911. https://doi.org/10.1111/ecog.04125.

Gaynor, K., K. Fiorella, G. Gregory, et al. 2016. "War and Wildlife: Linking Armed Conflict to Conservation." *Frontiers in Ecology and the Environment* 14 (10): 533–542. https://doi.org/10.1002/fee.1433.

Gehring, T., K. VerCauteren, and J.-Marc Landry. 2010. "Livestock Protection Dogs in the 21st Century: Is an Ancient Tool Relevant to Modern Conservation Challenges?" *BioScience* 60 (4): 299–308. https://doi.org/10.1525/bio.2010.60.4.8.

Geraeds, M., T. van Emmerik, R. de Vries, et al. 2019. "Riverine Plastic Litter Monitoring Using Unmanned Aerial Vehicles (UAVs)." *Remote Sensing* 11 (17): 2045. https://doi.org/10.3390/rs11172045.

Germano, J., K. Field, R. Griffiths, et al. 2015. "Mitigation-Driven Translocations: Are We Moving Wildlife in the Right Direction?" *Frontiers in Ecology and the Environment* 13 (2): 100–105. https://doi.org/10.1890/140137.

Gese, E. M. 2004. "Survey and Census Techniques for Canids." USDA National Wildlife Research Center: Staff Publications, 337. https://digitalcommons.unl.edu/icwdm_usdanwrc/337/.

Giefer, M., and L. An. 2020. "Synthesizing Remote Sensing and Biophysical Measures to Evaluate Human–Wildlife Conflicts: The Case of Wild Boar Crop Raiding in Rural China." *Remote Sensing* 12 (4): 618. https://doi.org/10.3390/rs12040618.

Gilsdorf, J., S. Hygnstrom, K. VerCauteren, et al. 2004. "Evaluation of a Deer-Activated Bio-Acoustic Frightening Device for Reducing Deer Damage in Cornfields." *Wildlife Society Bulletin* 32 (2): 515–523. https://doi.org/10.2193/0091-7648(2004)32[515:EOADBF]2.0.CO;2.

Glahn, J., G. Ellis, P. Fioranelli, et al. 2000. "Evaluation of Moderate and Low-Powered Lasers for Dispersing Double-Crested Cormorants from Their Night Roosts." In *Proceedings of the 9th Wildlife Damage Management Conference*, State College, PA, edited by M. C. Brittingham, J. Kays, and R. McPeake, 34–45. Bethesda, MD: The Wildlife Society. https://digitalcommons.usu.edu/wdmconference/2000/session1/4/.

González-Muñoz, N., C. Bellard, C. Leclerc, et al. 2015. "Assessing Current and Future Risks of Invasion by the 'Green Cancer' *Miconia calvescens*." *Biological Invasions* 17: 3337–3350. https://doi.org/10.1007/s10530-015-0960-x.

Gorkin, R., K. Adams, M. Berryman, et al. 2020. "Sharkeye: Real-Time Autonomous Personal Shark Alerting via Aerial Surveillance." *Drones* 4 (2): 18. https://doi.org/10.3390/drones4020018.

Gosselink, T., T. Van Deelen, R. Warner, et al. 2003. "Temporal Habitat Partitioning and Spatial Use of Coyotes and Red Foxes in East-Central Illinois." *Journal of Wildlife Management*, 67 (1): 90–103. https://doi.org/10.2307/3803065.

Goswami, V., M. Madhusudan, and K. Karanth. 2007. "Application of Photographic Capture–Recapture Modelling to Estimate Demographic Parameters for Male Asian Elephants." *Animal Conservation* 10 (3): 391–399. https://doi.org/10.1111/j.1469-1795.2007.00124.x.

Götz, T., and V. Janik. 2015. "Target-Specific Acoustic Predator Deterrence in the Marine Environment." *Animal Conservation* 18 (1): 102–111. https://doi.org/10.1111/acv.12141.

Gough, P. M., and J. W. Beyer. 1981. "Bird-Vectored Diseases." In *Proceedings of the 5th Great Plains Wildlife Damage Control Workshop*, October 13–15, Lincoln, NE, edited by R. M. Timm and R. J. Johnson, 260–272. Lincoln, NE: Institute of Agriculture and Natural Resources, University of Nebraska, 1982. https://digitalcommons.unl.edu/gpwdcwp/125/.

Government of India, Ministry of Environment and Forests. 2011. "Guidelines for Human-Leopard Conflict Management." https://www.conservationindia.org/wp-content/uploads/guidelines-human-leopard-conflict-management1.pdf.

Government of Japan, Ministry of the Environment. 2020. "Section 6: Wildlife Protection." In *Nature & Parks: Nature Conservation in Japan*. Updated February 4. https://www.env.go.jp/en/nature/npr/ncj/section6.html.

Government of the United Kingdom, Department for Environment, Food and Rural Affairs (DEFRA). 2010. *Wildlife Management in England: A Policy Making Framework for Resolving Human-Wildlife Conflicts*. Bristol: Crown Copyright. https://www.yumpu.com/en/document/read/11857903/wildlife-management-in-england-archive-defra.

Gregory, N. G., L. M. Milne, A. T. Rhodes, et al. 1998. "Effect of Potassium Cyanide on Behaviour and Time to Death in Possums." *New Zealand Veterinary Journal* 46 (2): 60–64. https://doi.org/10.1080/00480169.1998.36057.

Green, S., Z. Davidson, T. Kaaria, et al. 2018. "Do Wildlife Corridors Link or Extend Habitat? Insights from Elephant Use of a Kenyan Wildlife Corridor." *African Journal of Ecology* 56 (4): 860–871. https://doi.org/10.1111/aje.12541.

Grimm, B., B. Lahneman, P. Cathcart, et al. 2013. "Autonomous Unmanned Aerial Vehicle System for Controlling Pest Bird Population in Vineyards." In *Proceedings of ASME 2012 International Mechanical Engineering Congress and Exposition*, Vol. 4: *Dynamics, Control and Uncertainty, Parts A and B*, November 9–15, Houston, TX, 499–505. American Society of Mechanical Engineers Digital Collection. https://doi.org/10.1115/IMECE2012-89528.

Grosman, P., J. Jaeger, P. Biron, et al. 2009. "Reducing Moose–Vehicle Collisions through Salt Pool Removal and Displacement: An Agent-Based Modeling Approach." *Ecology and Society* 14 (2): 17. http://www.ecologyandsociety.org/vol14/iss2/art17/.

Gubbi, S., A. Kolekar, and V. Kumara. 2020. "Policy to On-Ground Action: Evaluating a Conflict Policy Guideline for Leopards in India." *Journal of International Wildlife Law and Policy* 23 (2): 127–140. https://doi.org/10.1080/13880292.2020.1818428.

Guerisoli, M.L.M., and J. Pereira. 2020. "Deer Damage: A Review of Repellents to Reduce Impacts Worldwide." *Journal of Environmental Management* 271: 110977. https://doi.org/10.1016/j.jenvman.2020.110977.

Guerreiro, A. 2019. "Local Ecological Knowledge about Human-Wildlife Conflict: A Portuguese Case Study." *Portuguese Journal of Social Science* 18 (2): 189–211. https://doi.org/10.1386/pjss_00005_1.

Gustavson, C., D. Kelly, M. Sweeney, et al. 1976. "Prey-Lithium Aversions. I: Coyotes and Wolves." *Behavioral Biology* 17 (1): 61–72. https://doi.org/10.1016/S0091-6773(76)90272-8.

IIaas, R., M. Thomas, and G. Towns. 2003. *An Assessment of Potential Use of Gambusia for Mosquito Control in Michigan.* Fisheries Division Technical Report 2003-2. State of Michigan Department of Natural Resources. https://quod .lib.umich.edu/cache/5/0/2/5026207.0001.001/00000001 .tif.18.pdf.

Hagle, S., K. Gibson, and S. Tunnock. 2003. *A Field Guide To Diseases and Insect Pests of Northern and Central Rocky Mountain Conifers.* Report No. R1-03-08. Missoula, MT: US Department of Agriculture, Forest Service, State and Private Forestry, Northern and Intermountain Regions. http://dnrc.mt.gov/divisions/forestry/docs/assistance /pests/fieldguide/complete-field-guide/fg-full-temp.pdf.

Haight, R., F. Homans, T. Horie, et al. 2011. "Assessing the Cost of an Invasive Forest Pathogen: A Case Study with Oak Wilt." *Environmental Management* 47 (3): 506–517. https://doi.org/10.1007/s00267-011-9624-5.

Haim, Abraham, Uri Shanas, Ora Brandes, et al. 2007. "Suggesting the Use of Integrated Methods for Vole Population Management in Alfalfa Fields." *Integrative Zoology* 2 (3): 184–190. https://doi.org/10.1111/j.1749-4877 .2007.00054.x.

Handegard, L. 1988. "Using Aircraft for Controlling Black-birds/Sunflower Depredations." In *Proceedings of the 13th Vertebrate Pest Conference,* Monterey, CA, March 1–3, edited by A. C. Crabb and R. E. Marsh, 293–294. Davis: University of California. https://digitalcommons.unl.edu /vpcthirteen/59/.

Haney, J. C. 2007. "Wildlife Compensation Schemes from around the World: An Annotated Bibliography." Washington, DC: Conservation Science and Economics Program, Defenders of Wildlife. https://defenders.org /sites/default/files/publications/wildlife_compensation _schemes_from_around_the_world.pdf.

Haney, J. C., T. Kroeger, F. Casey, et al. 2007. "Wilderness Discount on Livestock Compensation Costs for Imperiled Gray Wolf *Canis lupus.*" In *Science and Stewardship to Protect and Sustain Wilderness Values: Eighth World Wilderness Congress Symposium; September 30–October 6, 2005; Anchorage, AK,* vol. 49, A. Watson, J. Sproull, L. Dean, eds., 141–151. Proceedings RMRS-P-49. Fort Collins, CO: US Depart-ment of Agriculture, Forest Service, Rocky Mountain Research Station. https://www.fs.usda.gov/treesearch /pubs/31021.

Hansen, M. 2010. "The Asian Carp Threat to the Great Lakes." Letter dated February 9, 2010. Great Lakes Fishery Commission to the U.S. House Committee on Transporta-tion and Infrastructure, Subcommittee on Water Resources & Environment. http://www.michigantu.org/images /pdffiles/asian_carp/Hansen%20written%20testimony%20 about%20Asian%20carp%202-9-10%20FINAL.pdf.

Hardy, A., J. Fuller, M. Huijser, et al. 2007. "Evaluation of Wildlife Crossing Structures and Fencing on US Highway 93 Evaro to Polson Phase I: Preconstruction Data Collection and Finalization of Evaluation Plan." FHWA/ MT-06-008/1744-1. Bozeman, MT: Western Transporta-tion Institute, Montana State University. https://www .mdt.mt.gov/other/webdata/external/research/docs /research_proj/wildlife_crossing/final_report.pdf.

Hart, D., and R. W. Sussman. 2005. *Man the Hunted: Primates, Predators, and Human Evolution.* Boulder, CO: Westview Press.

Hartup, B., L. Schneider, J. M. Engels, et al. 2014. "Capture of Sandhill Cranes Using Alpha-Chloralose: A 10-Year Follow-Up." *Journal of Wildlife Diseases* 50 (1): 143–145. https://doi.org/10.7589/2013-06-140.

Håstein, T., B. J. Hill, and J. R. Winton. 1999. "Successful Aquatic Animal Disease Emergency Programmes." *Revue Scientifique et Technique* 18 (1): 214–227. https://doi.org/10 .20506/rst.18.1.1161.

Havelaar, A. 2007. "Methodological Choices for Calculating the Disease Burden and Cost-of-Illness of Foodborne Zoonoses in European Countries." Report No. 07-002. Maisons-Alfort, France: Network for the Prevention and Control of Zoonoses, Med-Vet-Net Administration Bureau. https://web.archive.org/web/20090121024116/http://www .medvetnet.org/pdf/Reports/Report_07-002.pdf.

Heinrich, J., and S. Craven. 1990. "Evaluation of Three Damage Abatement Techniques for Canada Geese." *Wildlife Society Bulletin* 18 (4): 405–410. https://www.jstor.org/stable /3782739.

Helle, E., and K. Kauhala. 1993. "Age Structure, Mortality, and Sex Ratio of the Raccoon Dog in Finland." *Journal of Mammalogy* 74 (4): 936–942. https://doi.org/10.2307 /1382432.

Herráez, P., A. E. Monteros, A. Fernández, et al. 2013. "Capture Myopathy in Live-Stranded Cetaceans." *Veterinary Journal* 196 (2): 181–188. https://doi.org/10.1016/j.tvjl.2012 .09.021.

Hess, S., D. Van Vuren, and G. Witmer. 2017. "Feral Goats and Sheep." In *Ecology and Management of Terrestrial Vertebrate Invasive Species in the United States,* 289–310. CRC Press. https://doi.org/10.1201/9781315157078-14.

Hoare, R. E. 2001. *A Decision Support System for Managing Human-Elephant Conflict Situations in Africa.* African Elephant Specialist Group. Gland, Switzerland: International Union for Conservation of Nature (IUCN). https://www.iucn.org /sites/dev/files/import/downloads/hecdssen.pdf.

Hoare, R. E. 2003. "Technical Brief: Review of Compensation Schemes for Agricultural and Other Damage Caused by Elephants." African Elephant Specialist Group. Gland, Switzerland: International Union for Conservation of Nature (IUCN). https://www.iucn.org/sites/dev/files /import/downloads/heccomreview.pdf.

Hobbs, R., and L. Hinds. 2018. "Could Current Fertility Control Methods Be Effective for Landscape-Scale Management of Populations of Wild Horses (*Equus caballus*) in Australia?" *Wildlife Research* 45 (3): 195–207. https://doi.org/10.1071/WR17136.

Hockings, K., and M. McLennan. 2012. "From Forest to Farm: Systematic Review of Cultivar Feeding by Chimpanzees—Management Implications for Wildlife in Anthropogenic Landscapes." *PLoS One* 7 (4): e33391. https://doi.org/10.1371/journal.pone.0033391.

Holcomb, L. 1976. "Experimental Use of AV-Alarm for Repelling Quelea from Rice in Somalia." In *Proceedings 7th Bird Control Seminar,* 9–11 November, Bowling Green, OH, edited by W. B. Jackson, 275–278. Bowling Green,

OH: Environmental Studies Center, Bowling Green State University. https://digitalcommons.unl.edu/icwdmbirdcontrol/83/.

Holtz, H. 2013. "Trap-Neuter-Return Ordinances and Policies in the United States: The Future of Animal Control." Law and Policy Brief. Bethesda, MD: Alley Cat Allies. http://s3.amazonaws.com/tzi/resources/attachments/f48f203fe6ec6ab750c99668edb045904a342aeb.pdf?1386700458.

Horgan, F. G. 2017. "Integrated Pest Management for Sustainable Rice Cultivation: A Holistic Approach." In *Achieving Sustainable Cultivation of Rice*, Vol. 2: *Cultivation, Pest and Disease Management*, T. Sasaki, 309–342. Cambridge: Burleigh Dodds Science. https://doi.org/10.19103/AS.2016.0003.23.

Howells, R., and G. Garrett. 1992. "Status of Some Exotic Sport Fishes in Texas Waters." *Texas Journal of Science* 44 (3): 317–324.

Hoyer, I., E. Blosser, C. Acevedo, et al. 2017. "Mammal Decline, Linked to Invasive Burmese Python, Shifts Host Use of Vector Mosquito towards Reservoir Hosts of a Zoonotic Disease." *Biology Letters* 13 (10): 20170353. https://doi.org/10.1098/rsbl.2017.0353.

Htwe, N., G. Singleton, and D. Johnson. 2019. "Interactions between Rodents and Weeds in a Lowland Rice Agro-Ecosystem: The Need for an Integrated Approach to Management." *Integrative Zoology* 14 (4): 396–409. https://doi.org/10.1111/1749-4877.12395.

Huddle Insurance. 2019. "The Roo Report: An Inside Look at Australia's Kangaroo Collision Problem." https://huddle.com.au/assets/pdf/Huddle-Roo-Report.pdf.

Huijser, M., J. Duffield, A. Clevenger, et al. 2009. "Cost–Benefit Analyses of Mitigation Measures Aimed at Reducing Collisions with Large Ungulates in the United States and Canada: A Decision Support Tool." *Ecology and Society* 14 (2): 15. http://www.ecologyandsociety.org/vol14/iss2/art15/.

Huijser, M. P., P. McGowen, J. Fuller, et al. 2007. *Wildlife-Vehicle Collision Reduction Study: Report to Congress.* FHWA-HRT-08-034. Washington DC: US Department of Transportation, Federal Highway Administration. https://www.fhwa.dot.gov/publications/research/safety/08034/08034.pdf.

Humair, F., L. Humair, F. Kuhn, et al. 2015. "E-commerce Trade in Invasive Plants." *Conservation Biology* 29 (6): 1658–1665. https://doi.org/10.1111/cobi.12579.

Hutson, C., K. Lee, J. Abel, et al. 2007. "Monkeypox Zoonotic Associations: Insights from Laboratory Evaluation of Animals Associated with the Multi-state US Outbreak." *American Journal of Tropical Medicine and Hygiene* 76 (4): 757–768. https://doi.org/10.4269/ajtmh.2007.76.757.

Hygnstrom, S., and S. Craven. 1988. "Electric Fences and Commercial Repellents for Reducing Deer Damage in Cornfields." *Wildlife Society Bulletin (1973–2006)* 16 (3): 291–296. https://www.jstor.org/stable/3782102.

Hygnstrom, S., G. Larson, and R. Timm, eds. 1994. *Prevention and Control of Wildlife Damage*. Lincoln: University of Nebraska.

Hygnstrom, S., and K. VerCauteren. 2000. "Cost-Effectiveness of Five Burrow Fumigants for Managing Black-Tailed Prairie Dogs." *International Biodeterioration and Biodegradation* 45 (3–4): 159–168. https://doi.org/10.1016/S0964-8305(00)00037-8.

Ingold, D. 1998. "The Influence of Starlings on Flicker Reproduction When Both Naturally Excavated Cavities and Artificial Nest Boxes Are Available." *Wilson Bulletin* 110 (2): 218–225. https://www.jstor.org/stable/4163931.

Islam, Z., and M. Hossain. 2003. "Response of Rice Plants to Rat Damage at the Reproductive Phase." *International Rice Research Notes* 28 (1): 1. https://ejournals.ph/article.php?id=8568.

Ivaşcu, C., and A. Biro. 2020. "Coexistence through the Ages: The Role of Native Livestock Guardian Dogs and Traditional Ecological Knowledge as Key Resources in Conflict Mitigation between Pastoralists and Large Carnivores in the Romanian Carpathians." *Journal of Ethnobiology* 40 (4): 465–482. https://doi.org/10.2993/0278-0771-40.4.465.

Jackson, W., and A. D. Ashton. 1986. "Case Histories of Anticoagulant Resistance." In *Pesticide Resistance: Strategies and Tactics for Management*, edited by National Research Council, 355–369. Washington, DC: National Academies Press. https://www.nap.edu/read/619/chapter/28.

Jacobson, E., P. Ginn, J. Troutman, et al. 2005. "West Nile Virus Infection in Farmed American Alligators (*Alligator mississippiensis*) in Florida." *Journal of Wildlife Diseases* 41 (1): 96–106. https://doi.org/10.7589/0090-3558-41.1.96.

Jakes, A., P. Jones, L. C. Paige, et al. 2018. "A Fence Runs through It: A Call for Greater Attention to the Influence of Fences on Wildlife and Ecosystems." *Biological Conservation* 227: 310–318. https://doi.org/10.1016/j.biocon.2018.09.026.

Jenkins, A., J. Smallie, and M. Diamond. 2010. "Avian Collisions with Power Lines: A Global Review of Causes and Mitigation with a South African Perspective." *Bird Conservation International* 20 (3): 263–278. https://doi.org/10.1017/S0959270910000122.

Jenkins, M., S. Jose, and P. White. 2007. "Impacts of an Exotic Disease and Vegetation Change on Foliar Calcium Cycling in Appalachian Forests." *Ecological Applications* 17 (3): 869–881. https://doi.org/10.1890/06-1027.

Johnson, B., A. Mader, R. Dasgupta, et al. 2020. "Citizen Science and Invasive Alien Species: An Analysis of Citizen Science Initiatives Using Information and Communications Technology (ICT) to Collect Invasive Alien Species Observations." *Global Ecology and Conservation* 21: e00812. https://doi.org/10.1016/j.gecco.2019.e00812.

Johnson, C., J. Isaac, and D. Fisher. 2007. "Rarity of a Top Predator Triggers Continent-Wide Collapse of Mammal Prey: Dingoes and Marsupials in Australia." *Proceedings of the Royal Society B: Biological Sciences* 274 (1608): 341–346. https://doi.org/10.1098/rspb.2006.3711.

Johnson, C., and S. Wroe. 2003. "Causes of Extinction of Vertebrates during the Holocene of Mainland Australia: Arrival of the Dingo, or Human Impact?" *The Holocene* 13 (6): 941–948. https://doi.org/10.1191/0959683603hl682fa.

Johnson, D., and P. Stiling. 1998. "Distribution and Dispersal of *Cactoblastis cactorum* (*Lepidoptera: pyralidae*), an Exotic Opuntia-Feeding Moth, in Florida." *Florida Entomologist* 81 (1): 12–22. https://doi.org/10.2307/3495992.

Jolly, C., E. Kelly, G. Gillespie, et al. 2018. "Out of the Frying Pan: Reintroduction of Toad-Smart Northern Quolls to Southern Kakadu National Park." *Austral Ecology* 43 (2): 139–149. https://doi.org/10.1111/aec.12551.

Jones, H., B. Tershy, E. Zavaleta, et al. 2008. "Severity of the Effects of Invasive Rats on Seabirds: A Global Review." *Conservation Biology* 22 (1): 16–26. https://doi.org/10.1111/j.1523-1739.2007.00859.x.

Jones, J. L., and J. P. Dubey. 2010. "Waterborne Toxoplasmosis—Recent Developments." *Experimental Parasitology* 124 (1): 10–25. https://doi.org/10.1016/j.exppara.2009.03.013.

Josselyn, J. (1674) 1986. *New-Englands Rarities Discovered.* Bedford, MA: Applewood Books.

Kaeslin, E., I. Redmond, N. Dudley, et al., eds. 2012. *Wildlife in a Changing Climate.* FAO Forestry Paper 167. Rome: Food and Agriculture Organization of the United Nations.

Karanth, K. U., and J. Nichols. 1998. "Estimation of Tiger Densities in India Using Photographic Captures and Recaptures." *Ecology* 79 (8): 2852–2862. https://doi.org/10.1890/0012-9658(1998)079[2852:EOTDII]2.0.CO;2.

Kateregga, E., and T. Sterner. 2007. "Indicators for an Invasive Species: Water Hyacinths in Lake Victoria." *Ecological Indicators* 7 (2): 362–370. https://doi.org/10.1016/j.ecolind.2006.02.008.

Kats, L., and R. Ferrer. 2003. "Alien Predators and Amphibian Declines: Review of Two Decades of Science and the Transition to Conservation." *Diversity and Distributions* 9 (2): 99–110. https://doi.org/10.1046/j.1472-4642.2003.00013.x.

Kauffman, M., J. Brodie, and E. Jules. 2010. "Are Wolves Saving Yellowstone's Aspen? A Landscape-Level Test of a Behaviorally Mediated Trophic Cascade." *Ecology* 91 (9): 2742–2755. https://doi.org/10.1890/09-1949.1.

Kaufman, L. 1992. "Catastrophic Change in Species-Rich Freshwater Ecosystems." *BioScience* 42 (11): 846–858. https://doi.org/10.2307/1312084.

Kay, Charles E. 2007. "Were Native People Keystone Predators? A Continuous-Time Analysis of Wildlife Observations Made by Lewis and Clark in 1804–1806." *Canadian Field-Naturalist* 121 (1): 1–16. https://doi.org/10.22621/cfn.v121i1.386.

Keitt, B., N. Holmes, E. Hagen, et al. 2019. "Going to Scale: Reviewing Where We've Been and Where We Need to Go in Invasive Vertebrate Eradications." In *Island Invasives: Scaling Up to Meet the Challenge,* edited by C. R. Veitch, M. N. Clout, A. R. Martin, et al., 633–636. Gland, Switzerland: IUCN.

Kellert, S. R. 1993. "The Biological Basis for Human Values of Nature." In *The Biophilia Hypothesis,* edited by S. R. Kellert and E. O. Wilson, 42–69. Washington, DC: Island Press.

Kellert, S. R. 2007. "Biophilia, Children and Restoring Connections to Nature in the Modern Built Environment." PowerPoint pamphlet. https://web.archive.org/web/20100602042303/http://www.childrenandnature.org/reports/9_2006/PPTs/kellert.pdf.

Kellert, S. R., and J. K. Berry. 1982a. *Activities of the American Public Relating to Animals. Phase II.* Washington, DC: US Department of the Interior, Fish and Wildlife Service. https://hdl.handle.net/2027/umn.31951002892056m.

Kellert, S. R., and J. K. Berry. 1982b. *Public Attitudes toward Critical Wildlife and Natural Habitat Issues: Phase I.* Washington, DC: US Department of the Interior, Fish and Wildlife Service. https://hdl.handle.net/2027/umn.31951002892054q.

Kharas, H. 2017. "The Unprecedented Expansion of the Global Middle Class." *Brookings* (blog), February 28, 2017. https://www.brookings.edu/research/the-unprecedented-expansion-of-the-global-middle-class-2.

Khorozyan, I., and M. Waltert. 2019. "How Long Do Anti-Predator Interventions Remain Effective? Patterns, Thresholds and Uncertainty." *Royal Society Open Science* 6 (9): 190826. https://doi.org/10.1098/rsos.190826.

Kikillus, K., K. M. Hare, and S. Hartley. 2010. "Minimizing False-Negatives When Predicting the Potential Distribution of an Invasive Species: A Bioclimatic Envelope for the Red-Eared Slider at Global and Regional Scales." *Animal Conservation* 13: 5–15. https://doi.org/10.1111/j.1469-1795.2008.00299.x.

Kilpatrick, H., S. Spohr, and K. Lima. 2001. "Effects of Population Reduction on Home Ranges of Female White-Tailed Deer at High Densities." *Canadian Journal of Zoology* 79 (6): 949–954. https://doi.org/10.1139/z01-057.

Kim, H., E. McCloy, G. Williamson, et al. 2019. "Low Cost Autonomous Amphibious Bird Chasing Robot." In *2019 IEEE International Symposium on Measurement and Control in Robotics (ISMCR),* A2-3-1-A2-3-7, https://doi.org/10.1109/ISMCR47492.2019.8955705.

Kim, R., and M. Faisal. 2011. "Emergence and Resurgence of the Viral Hemorrhagic Septicemia Virus (*Novirhabdovirus, Rhabdoviridae, Mononegavirales*)." *Journal of Advanced Research* 2 (1): 9–23. https://doi.org/10.1016/j.jare.2010.05.007.

King, D. R., L. E. Twigg, and J. L. Gardner. 1989. "Tolerance to Sodium Monofluoroacetate in Dasyurids in Western Australia." *Australian Wildlife Research* 16 (2): 131–140. https://doi.org/10.1071/WR9890131.

King, D. T. 2005. "Interactions between the American White Pelican and Aquaculture in the Southeastern United States: An Overview." *Waterbirds* 28 (S1): 83–86. https://digitalcommons.unl.edu/icwdm_usdanwrc/39/.

King, D. T., and D. Anderson. 2005. "Recent Population Status of the American White Pelican: A Continental Perspective." *Waterbirds* 28 (SP1):48–54. https://doi.org/10.1675/1524-4695(2005)28[48:RPSOTA]2.0.CO;2.

King, L. 2019. "How Bees Can Keep the Peace between Elephants and Humans." *TEDWomen,* December. https://www.ted.com/talks/lucy_king_how_bees_can_keep_the_peace_between_elephants_and_humans/transcript?language=en.

King, L., F. Lala, H. Nzumu, et al. 2017. "Beehive Fences as a Multidimensional Conflict-Mitigation Tool for Farmers Coexisting with Elephants." *Conservation Biology* 31 (4): 743–752. https://doi.org/10.1111/cobi.12898.

Kinka, D., and J. Young. 2018. "A Livestock Guardian Dog by Any Other Name: Similar Response to Wolves across Livestock Guardian Dog Breeds." *Rangeland Ecology and Management* 71 (4): 509–517. https://doi.org/10.1016/j.rama.2018.03.004.

Kirkpatrick, J., and A. Turner. 2008. "Achieving Population Goals in a Long-Lived Wildlife Species (*Equus caballus*) with Contraception," https://doi.org/10.1071/WR07106.

Kluckhohn, C. 1951. "Values and Value Orientations in the Theory of Action: An Exploration in Definition and Classification." In *Toward a General Theory of Action*, edited by T. Parsons and E. A. Shils, 388–433. Cambridge, MA: Harvard University Press.

Koehn, J. 2004. "Carp (*Cyprinus carpio*) as a Powerful Invader in Australian Waterways." *Freshwater Biology* 49 (7): 882–894. https://doi.org/10.1111/j.1365-2427.2004.01232.x.

Koehn, J., A. Brumley, and P. Gehrke. 2000. *Managing the Impacts of Carp*. Canberra, Australia: Bureau of Rural Sciences. https://www.yumpu.com/en/document/view/3981510/managing-the-impacts-of-carp-feralorgau.

Kowalczyk, R., A. Zalewski, B. Jędrzejewska, et al. 2009. "Reproduction and Mortality of Invasive Raccoon Dogs (*Nyctereutes Procyonoides*) in the Białowieża Primeval Forest (Eastern Poland)." *Annales Zoologici Fennici* 46: 291–301. http://www.sekj.org/PDF/anzf46/anzf46-291.pdf.

Krajcarz, M., M. T. Krajcarz, M. Baca, et al. 2020. "Ancestors of Domestic Cats in Neolithic Central Europe: Isotopic Evidence of a Synanthropic Diet." *Proceedings of the National Academy of Sciences* 117 (30): 17710–17719. https://doi.org/10.1073/pnas.1918884117.

Kramer, B. 2011. "Budget Rider Gives States Wolf Control." *Spokesman-Review* [Spokane, WA], April 15. https://www.spokesman.com/stories/2011/apr/15/budget-rider-gives-states-wolf-control/.

Kraus, F., R. Stahl, and W. Pitt. 2015. "Chemical Repellents Appear Non-Useful for Eliciting Exit of Brown Tree Snakes from Cargo." *International Journal of Pest Management* 61 (2): 144–152. https://doi.org/10.1111/j.1365-2427.2004.01232.x.

Krijger, I., S. Belmain, G. Singleton, et al. 2017. "The Need to Implement the Landscape of Fear within Rodent Pest Management Strategies." *Pest Management Science* 73 (12): 2397–2402. https://doi.org/10.1002/ps.4626.

Kuykendall, K. 2016. "Local Mountain Lion Population Faces Precipitous Decline in Genetic Diversity Within 50 Years, Possible Extinction." National Park Service: Santa Monica Mountains National Recreation Area. Press release, August 30. https://www.nps.gov/samo/learn/news/local-mountain-lion-population-faces-precipitous-decline-in-genetic-diversity-within-50-years-possible-extinction.htm.

Lack, D. 1954. *The Natural Regulation of Animal Numbers*. Oxford: Oxford University Press.

Lafferty, K., and E. Hofmann. 2016. "Marine Disease Impacts, Diagnosis, Forecasting, Management and Policy." *Philosophical Transactions of the Royal Society B: Biological Sciences* 371: 20150200. https://doi.org/10.1098/rstb.2015.0200.

Lai, L., X. Yu, M. He, et al. 2020. "Impact of Michaelis–Menten Type Harvesting in a Lotka–Volterra Predator–Prey System Incorporating Fear Effect." *Advances in Difference Equations* 2020 (1): 320. https://doi.org/10.1186/s13662-020-02724-8.

Lance, N. J., S. W. Breck, C. Sime, et al. 2011. "Biological, Technical, and Social Aspects of Applying Electrified Fladry for Livestock Protection from Wolves (*Canis lupus*)." *Wildlife Research* 37 (8): 708–714. https://doi.org/10.1071/WR10022.

Langbein, J., R. Putman, and B. Pokorny. 2010. "Traffic Collisions Involving Deer and Other Ungulates in Europe and Available Measures for Mitigation." In *Ungulate Management in Europe: Problems and Practices*, edited by R. Putman, M. Apollonio, and R. Andersen, 215–259. Cambridge: Cambridge University Press. https://doi.org/10.1017/CBO9780511974137.009.

Langley, R. 2010. "Adverse Encounters with Alligators in the United States: An Update." *Wilderness and Environmental Medicine* 21 (2): 156–163. https://doi.org/10.1016/j.wem.2010.02.002.

Lapidge, S., D. Dall, J. Dawes, et al. 2005. "Starlicide®—the Benefits, Risks and Industry Need for DRC-1339 in Australia." In *Proceedings of the 13th Australasian Vertebrate Pest Conference*, Wellington, New Zealand, May 2–6, edited by J. Parkes, M. Stratham, and G. Edwards, 235–238. Lincoln: Manaaki Whenua/Landcare Research. https://avpc.net.au/wp-content/uploads/13th-AVPCProceedings2005.pdf.

Larson, L., A. Conway, S. Hernandez, et al. 2016. "Human-Wildlife Conflict, Conservation Attitudes, and a Potential Role for Citizen Science in Sierra Leone, Africa." *Conservation and Society* 14 (3): 205–217. https://doi.org/10.4103/0972-4923.191159.

LaRue, M., and C. Nielsen. 2016. "Population Viability of Recolonizing Cougars in Midwestern North America." *Ecological Modelling* 321: 121–129. https://doi.org/10.1016/j.ecolmodel.2015.09.026.

Laughlin, J., N. Greenwald, G. Dyson, et al. 2010. "Lawsuit Filed to Stop Federal, State-Sanctioned Killing of Endangered Wolves." Press release, July 1. Cascadia Wildlands, Center for Biological Diversity, Hells Canyon Preservation Council, and Oregon Wild. https://www.biologicaldiversity.org/news/press_releases/2010/wolves-07-01-2010.html.

Lawana, V., and J. Cannon. 2020. "Rotenone Neurotoxicity: Relevance to Parkinson's Disease." In *Neurotoxicity of Pesticides*, edited by M. Aschner and L. G. Costa, 209–254. Cambridge, MA: Academic Press. https://doi.org/10.1016/bs.ant.2019.11.004.

Lee, F., K. Simon, and G. W. Perry. 2018. "Prey Selectivity and Ontogenetic Diet Shift of the Globally Invasive Western Mosquitofish (*Gambusia affinis*) in Agriculturally Impacted Streams." *Ecology of Freshwater Fish* 27 (3): 822–833. https://doi.org/10.1111/eff.12395.

Lehrman, D. 1964. "The Reproductive Behavior of Ring Doves." *Scientific American* 211 (5): 48–55. https://doi.org/10.1038/scientificamerican1164-48.

Leigh, D., A. Hendry, E. Vázquez-Domínguez, et al. 2019. "Estimated Six Per Cent Loss of Genetic Variation in Wild Populations since the Industrial Revolution." *Evolutionary Applications* 12 (8): 1505–1512. https://doi.org/10.1111/eva.12810.

Leprieur, F., O. Beauchard, S. Blanchet, et al. 2008. "Fish Invasions in the World's River Systems: When Natural Processes Are Blurred by Human Activities." *PLOS Biology* 6 (2): e28. https://doi.org/10.1371/journal.pbio.0060028.

Lerner, H., and C. Berg. 2017. "A Comparison of Three Holistic Approaches to Health: One Health, EcoHealth, and Planetary Health." *Frontiers in Veterinary Science* 4: 163. https://doi.org/10.3389/fvets.2017.00163.

Lever, C. 2003. *Naturalized Reptiles and Amphibians of the World*. Oxford: Oxford University Press.

Levins, R. 1969. "Some Demographic and Genetic Consequences of Environmental Heterogeneity for Biological Control." *American Entomologist* 15 (3): 237–240. https://doi.org/10.1093/besa/15.3.237.

Lewis, S., and M. Maslin. 2015. "Defining the Anthropocene." *Nature* 519 (7542): 171–180. https://doi.org/10.1038/nature14258.

Li, K., and O. DeMasi. 2009. "It's a Coyote Eat Deer Feed Tick World: A Deterministic Model of Predator-Prey Interaction in the Northeast." *Catalyst: Rice Undergraduate Science and Engineering Review* 2 (Spring): 29–37. https://cpb-us-e1.wpmucdn.com/blogs.rice.edu/dist/e/505/files/deterministicmodelpreypredator.pdf.

Ligtvoet, W., F. Witte, T. Goldschmidt, et al. 1991. "Species Extinction and Concomitant Ecological Changes in Lake Victoria." *Netherlands Journal of Zoology* 42 (2–3): 214–232. https://doi.org/10.1163/156854291X00298.

Lindgren, P.M.F., T. P. Sullivan, and D. R. Crump. 1995. "Review of Synthetic Predator Odor Semiochemicals as Repellents for Wildlife Management in the Pacific Northwest." In *Repellents in Wildlife Management Symposium: Proceedings of the 2nd DWRC Special Symposium*, August 8–10, Denver, CO, edited by J. R. Mason, 217–230. Fort Collins, CO: National Wildlife Research Center. https://digitalcommons.unl.edu/nwrcrepellants/24/.

Linhart, S., G. Dasch, R. Johnson, et al. 1992. "Electronic Frightening Devices for Reducing Coyote Predation on Domestic Sheep: Efficacy under Range Conditions and Operational Use." In *Proceedings of the 15th Vertebrate Pest Conference*, March 3–5, Newport Beach, CA, edited by J. E. Borrecco and R. E. Marsh, 386–392. Davis: University of California. https://digitalcommons.unl.edu/vpc15/47/.

Linhart, S., and F. Knowlton. 1975. "Determining the Relative Abundance of Coyotes by Scent Station Lines." *Wildlife Society Bulletin* 3 (3): 119–124. https://www.jstor.org/stable/3781822.

Linnell, J., E. Nilsen, U. Lande, et al. 2005. "Zoning as a Means of Mitigating Conflicts with Large Carnivores: Principles and Reality." In *People and Wildlife: Conflict or Co-existence?* edited by R. Woodroffe, S. Thirgood, and A. Rabinowitz, 162–175. Cambridge: Cambridge University Press. https://doi.org/10.1017/CBO9780511614774.011.

Linz, G., and H. Homan. 2011. "Use of Glyphosate for Managing Invasive Cattail (*Typha* spp.) to Disperse Blackbird (Icteridae) Roosts." *Crop Protection* 30 (2): 98–104. https://doi.org/10.1016/j.cropro.2010.10.003.

Linz, G. M., R. Johnson, and J. Thiele. 2018. "European Starlings." In *Ecology and Management of Terrestrial Vertebrate Invasive Species in the United States*, edited by W. C. Pitt, J. C. Beasley, and G. W. Witmer, 311–332. Boca Raton, FL: CRC Press. https://digitalcommons.unl.edu/icwdm_usdanwrc/2027/.

Linz, G. M., P.E. Klug, and R.A. Dolbeer. 2017. "Ecology and Management of Red-Winged Blackbirds." In *Ecology and Management of Blackbirds (Icteridae) in North America*, edited by G. M. Linz et al., 17–41. Boca Raton, FL: CRC Press. https://digitalcommons.unl.edu/icwdm_usdanwrc/1983/.

Linz, G. M., D. Schaaf, P. Mastrangelo, et al. 2004. "Wildlife Conservation Sunflower Plots as a Dual-Purpose Wildlife Management Strategy." In *Proceedings of the 21st Vertebrate Pest Conference*, March 1–4, Visalia, CA, edited by R. M. Timm and W. P. Gorenzel, 291–294. Davis: University of California. https://digitalcommons.unl.edu/icwdm_usdanwrc/358/.

Lischka, S., T. Teel, H. Johnson, et al. 2018. "A Conceptual Model for the Integration of Social and Ecological Information to Understand Human-Wildlife Interactions." *Biological Conservation* 225: 80–87. https://doi.org/10.1016/j.biocon.2018.06.020.

Liu, X. 2019. "Rodent Biology and Management: Current Status, Opinion and Challenges in China." *Journal of Integrative Agriculture* 18 (4): 830–839. https://doi.org/10.1016/S2095-3119(18)61943-4.

Lockwood, J., P. Cassey, and T. Blackburn. 2005. "The Role of Propagule Pressure in Explaining Species Invasions." *Trends in Ecology and Evolution* 20 (5): 223–228. https://doi.org/10.1016/j.tree.2005.02.004.

Lockwood, J., M. Hoopes, M. Marchetti, et al. 2013. *Invasion Ecology*. Hoboken, NJ: John Wiley & Sons.

Long, K., and A. Robley. 2004. *Cost Effective Feral Animal Exclusion Fencing for Areas of High Conservation Value in Australia: A Report*. Melbourne: Victoria Department of Sustainability and Environment. https://www.awe.gov.au/biosecurity-trade/invasive-species/publications/cost-effective-feral-animal-exclusion-fencing.

Longcore, T., C. Rich, and L. Sullivan. 2009. "Critical Assessment of Claims Regarding Management of Feral Cats by Trap–Neuter–Return." *Conservation Biology* 23 (4): 887–894. https://doi.org/10.1111/j.1523-1739.2009.01174.x.

Loope, L., O. Hamann, and C. Stone. 1988. "Comparative Conservation Biology of Oceanic Archipelagoes: Hawaii and the Galapagos." *BioScience* 272–282. https://doi.org/10.2307/1310851.

López-Perea, J., and R. Mateo. 2018. "Secondary Exposure to Anticoagulant Rodenticides and Effects on Predators." In *Anticoagulant Rodenticides and Wildlife*, edited by N. W. van den Brink, J. E. Elliott, R. F. Shore, et al., 159–193. Cham, Switzerland: Springer International. https://doi.org/10.1007/978-3-319-64377-9_7.

Lotka, A. 1925. *Elements of Physical Biology*. Baltimore: Williams & Wilkins.

Louda, S., and P. Stiling. 2004. "The Double-Edged Sword of Biological Control in Conservation and Restoration." *Conservation Biology* 18 (1): 50–53. https://doi.org/10.1111/j.1523-1739.2004.00070.x.

Lowe, S., M. Browne, S. Boudjelas, et al. 2000. *100 of the World's Worst Invasive Alien Species: A Selection from the Global Invasive Species Database*. Vol. 12. Auckland: Invasive Species Specialist Group.

Löyttyniemi, K., and L. Mikkola. 1980. "Elephant as a Pest of Pines in Zambia." *International Journal of Pest Management*

26 (2): 167–169. https://doi.org/10.1080/09670878009414389.

Luby, S., M. Rahman, M. J. Hossain, et al. 2006. "Foodborne Transmission of Nipah Virus, Bangladesh." *Emerging Infectious Diseases* 12 (12): 1888. https://doi.org/10.3201/eid1212.060732.

Lucas, G., and H. Synge, eds. 1978. *The IUCN Plant Red Data Book*. Morges, Switzerland: IUCN.

Luque, G., C. Bellard, C. Bertelsmeier, et al. 2014. "The 100th of the World's Worst Invasive Alien Species." *Biological Invasions* 16 (5): 981–985. https://doi.org/10.1007/s10530-013-0561-5.

MacArthur, R., and E. Wilson. 1967. *The Theory of Island Biogeography*. Princeton, NJ: Princeton University Press.

MacInnes, C., S. Smith, R. Tinline, et al. 2001. "Elimination of Rabies from Red Foxes in Eastern Ontario." *Journal of Wildlife Diseases* 37 (1): 119–132. https://doi.org/10.7589/0090-3558-37.1.119.

Mack, R., D. Simberloff, W. M. Lonsdale, et al. 2000. "Biotic Invasions: Causes, Epidemiology, Global Consequences, and Control." *Ecological Applications* 10 (3): 689–710. https://doi.org/10.1890/1051-0761(2000)010[0689:BICEGC]2.0.CO;2.

MacKinnon, M., and P. Erdkamp. 2013. "Pack Animals, Pets, Pests, and Other Non-human Beings." *The Cambridge Companion to Ancient Rome*, edited by P. Erdkamp, 110–128. Cambridge: Cambridge University Press. https://doi.org/10.1017/CCO9781139025973.009.

Madden, F. 2004. "Creating Coexistence between Humans and Wildlife: Global Perspectives on Local Efforts to Address Human–Wildlife Conflict." *Human Dimensions of Wildlife* 9 (4): 247–257. https://doi.org/10.1080/10871200490505675.

Madsen, J., J. Williams, F. Johnson, et al. 2017. "Implementation of the First Adaptive Management Plan for a European Migratory Waterbird Population: The Case of the Svalbard Pink-Footed Goose Anser Brachyrhynchus." *Ambio* 46 (S2): 275–289. https://doi.org/10.1007/s13280-016-0888-0.

Mahy, B. W., and C. C. Brown. 2000. "Emerging Zoonoses: Crossing the Species Barrier." *Revue Scientifique et Technique (International Office of Epizootics)* 19: 33–40. https://doi.org/10.20506/rst.19.1.1212.

Major, R., M. Ashcroft, A. Davis, et al. 2015. "Nest Caging as a Conservation Tool for Threatened Songbirds." *Wildlife Research* 41 (7): 598–605. https://doi.org/10.1071/WR14136.

Malthus, T. (1798) 1993. *An Essay on the Principle of Population*. Oxford: Oxford University Press.

Mamboleo, A. A., C. Doscher, and A. Paterson. 2017. "Are Elephants the Most Disastrous Agricultural Pests or the Agents of Ecological Restorations?" *Journal of Biodiversity and Endangered Species* 5: 1. https://doi.org/10.4172/2332-2543.1000185.

Manfredo. M. 2008. *Who Cares about Wildlife? Social Science Concepts for Exploring Human-Wildlife Relationships and Conservation Issues*. New York: Springer.

Manfredo, M., and A. Bright. 2008. "Attitudes and the Study of Human Dimensions of Wildlife." In *Who Cares about Wildlife? Social Science Concepts for Exploring Human-Wildlife Relationships and Conservation Issues* by M. Manfredo, 75–109. New York: Springer.

Manfredo, M., and T. Teel. 2008. "Integrating Concepts: Demonstration of a Multilevel Model for Exploring the Rise of Mutualism Value Orientations in Post-industrial Society." In *Who Cares about Wildlife? Social Science Concepts for Exploring Human-Wildlife Relationships and Conservation Issues* by M. Manfredo, 191–217. New York: Springer.

Manfredo, M., T. L. Teel, and H. C. Zinn. 2009. "Understanding Global Values toward Wildlife." In *Wildlife and Society: The Science of Human Dimensions*, edited by M. J. Manfredo et al., 31–43. Washington, DC: Island Press.

Marada, P., J. Cukor, R. Linda, et al. 2019. "Extensive Orchards in the Agricultural Landscape: Effective Protection against Fraying Damage Caused by Roe Deer." *Sustainability* 11 (13): 3738. https://doi.org/10.3390/su11133738.

Marais, H. J., D. A. Hendrickson, M. Stetter, et al. 2013. "Laparoscopic Vasectomy in African Savannah Elephant (*Loxodonta africana*); Surgical Technique and Results." *Journal of Zoo and Wildlife Medicine* 44 (4S). https://doi.org/10.1638/1042-7260-44.4S.S18.

Marcus, J., J. Dinan, R. Johnson, et al. 2007. "Directing Nest Site Selection of Least Terns and Piping Plovers." *Waterbirds* 30 (2): 251–258. https://doi.org/10.1675/1524-4695(2007)30[251:DNSSOL]2.0.CO;2.

Marinthe, G., G. Brown, S. Delouvée, et al. 2020. "Looking Out for Myself: Exploring the Relationship between Conspiracy Mentality, Perceived Personal Risk, and COVID-19 Prevention Measures." *British Journal of Health Psychology* 25 (4): 957–980. https://doi.org/10.1111/bjhp.12449.

Marker, L., and S. Shivamani. 2009. "Policy for Human-Leopard Conflict Management in India." *CAT News* 50 (Spring): 23–26. https://cheetah.org/cheetah-2019/wp-content/uploads/2019/05/policy-for-human-leopard-conflict-management-in-india.pdf.

Marsden, J. E., and M. Siefkes. 2019. "Control of Invasive Sea Lamprey in the Great Lakes, Lake Champlain, and Finger Lakes of New York." In *Lampreys: Biology, Conservation and Control*, Vol. 2, edited by Margaret F. Docker, 411–479. Dordrecht, Netherlands: Springer. https://doi.org/10.1007/978-94-024-1684-8_5.

Marsh, R., A. Koehler, and T. Salmon. 1990. "Exclusionary Methods and Materials to Protect Plants from Pest Mammals—A Review." In *Proceedings of the 14th Vertebrate Pest Conference*, March 6–8, Sacramento, CA, edited by L. R. Davis and R. E. Marsh, 174–180. Davis: University of California. https://digitalcommons.unl.edu/vpc14/59/.

Martin, P. 1966. "Africa and Pleistocene Overkill." *Nature* 212 (5060): 339–342. https://doi.org/10.1038/212339a0.

Martin, P. S. 2005. *Twilight of the Mammoths: Ice Age Extinctions and the Rewilding of America*. Berkeley: University of California Press,

Martín-Torrijos, L., T. Kawai, J. Makkonen, et al. 2018. "Crayfish Plague in Japan: A Real Threat to the Endemic Cambaroides Japonicus." *PLoS One* 13 (4): e0195353. https://doi.org/10.1371/journal.pone.0195353.

Martorello, D., T. Eason, and M. Pelton. 2001. "A Sighting Technique Using Cameras to Estimate Population Size of Black Bears." *Wildlife Society Bulletin* 29 (2): 560–567. https://www.jstor.org/stable/3784181.

Mason, J. R., L. Clark, and P. S. Shah. 1992. "Taxonomic Differences between Birds and Mammals in Their Responses to Chemical Irritants." In *Chemical Signals in Vertebrates 6*, edited by R. L. Doty and D. Müller-Schwarze, 311–317. New York: Springer Science. https://doi.org/10.1007/978-1-4757-9655-1_50.

Matthews, J., A. Schipper, A. J. Hendriks, et al. 2015. "A Dominance Shift from the Zebra Mussel to the Invasive Quagga Mussel May Alter the Trophic Transfer of Metals." *Environmental Pollution* 203: 183–190. https://doi.org/10.1016/j.envpol.2015.03.032.

Mauldin, R., and P. Savarie. 2010. "Acetaminophen as an Oral Toxicant for Nile Monitor Lizards (*Varanus niloticus*) and Burmese Pythons (Python *Molurus bivittatus*)." *Wildlife Research* 37 (3): 215–222. https://digitalcommons.unl.edu/icwdm_usdanwrc/943/.

McBeath, D. 1941. "Whitetail Traps and Tags." *Michigan Conservation* 10 (11): 6–7.

McCann, S., M. R. Crossland, and R. Shine. 2019. "Pheromones Can Cull an Invasive Amphibian without Releasing Survivors from Intraspecific Competition." *Ecosphere* 10 (12): e02969. https://doi.org/10.1002/ecs2.2969.

McCoy, N. H. 2002. "Economic Tools for Managing Impacts of Urban Canada Geese." In *Human Conflicts with Wildlife: Economic Considerations: Proceedings of the Third NWRC Special Symposium*, August 1–3, 2000, Fort Collins, CO, edited by L. Clark, J. Hone, J. A. Shivik, et al., 117–122. Fort Collins, CO: National Wildlife Research Center. https://digitalcommons.unl.edu/nwrchumanconflicts/12/.

McCullough, D. 1997. "Irruptive Behavior in Ungulates." In *The Science of Overabundance: Deer Ecology and Population Management*, edited by W. McShea, H. Underwood, and J. Rappole, 69–98. Washington, DC: Smithsonian Books.

McDonald, C., and G. McPherson. 2013. "Creating Hotter Fires in the Sonoran Desert: Buffelgrass Produces Copious Fuels and High Fire Temperatures." *Fire Ecology* 9 (2): 26–39. https://doi.org/10.4996/fireecology.0902026.

McDowall, R. M. 1990. *New Zealand Freshwater Fishes: A Natural History and Guide*. Rev. ed. Auckland: Heinemann Reed.

McFadyen, R. E. C. 2000. "Successes in Biological Control of Weeds." In *Proceedings of the X International Symposium on Biological Control of Weeds*, July 4–14, 1999, Bozeman, MT, edited by N. R. Spencer, 3–14. Center for Invasive Species and Ecosystem Health. https://www.invasive.org/publications/xsymposium/proceed/01apg03.pdf.

McGlynn, T. 1999. "The Worldwide Transfer of Ants: Geographical Distribution and Ecological Invasions." *Journal of Biogeography* 26 (3): 535–548. https://doi.org/10.1046/j.1365-2699.1999.00310.x.

Mcgovern, E., and H. Kretser. 2015. "Predicting Support for Recolonization of Mountain Lions (*Puma concolor*) in the Adirondack Park." *Wildlife Society Bulletin* 39 (3): 503–511. https://doi.org/10.1002/wsb.557.

McKee, S., S. Shwiff, and A. Anderson. 2021. "Estimation of Wildlife Damage from Federal Crop Insurance Data." *Pest Management Science* 77 (1): 406–416. https://doi.org/10.1002/ps.6031.

McKinney, M. 2002. "Urbanization, Biodiversity, and Conservation: The Impacts of Urbanization on Native Species Are Poorly Studied, but Educating a Highly Urbanized Human Population about These Impacts Can Greatly Improve Species Conservation in All Ecosystems." *BioScience* 52 (10): 883–890. https://doi.org/10.1641/0006-3568(2002)052[0883:UBAC]2.0.CO;2.

McMahon, B. J., S. Morand, and J. S. Gray. 2018. "Ecosystem Change and Zoonoses in the Anthropocene." *Zoonoses and Public Health* 65: 755–765. https://doi.org/10.1111/zph.12489.

McMichael, A. 2004. "Environmental and Social Influences on Emerging Infectious Diseases: Past, Present and Future." *Philosophical Transactions of the Royal Society of London. Series B: Biological Sciences* 359 (1447): 1049–1058. https://doi.org/10.1098/rstb.2004.1480.

McNay, M., and M. Hicks. 2002. "A Case History of Wolf-Human Encounters in Alaska and Canada." Alaska Department of Fish and Game Wildlife Technical Bulletin 13. Juneau, AK: ADF&G, Wildlife Conservation. https://digitalcommons.unl.edu/wolfrecovery/26/.

Mech, L. D. 2017. "Where Can Wolves Live and How Can We Live with Them?" *Biological Conservation* 210: 310–317. https://doi.org/10.1016/j.biocon.2017.04.029.

Meehan, T., N. Michel, and H. Rue. 2019. "Spatial Modeling of Audubon Christmas Bird Counts Reveals Fine-Scale Patterns and Drivers of Relative Abundance Trends." *Ecosphere* 10 (4): e02707. https://doi.org/10.1002/ecs2.2707.

Mendia, S, M. Johnson, and J. M. Higley. 2019. "Ecosystem Services and Disservices of Bear Foraging on Managed Timberlands." *Ecosphere* 10 (7): e02816. https://doi.org/10.1002/ecs2.2816.

Mendoza-Roldan, J., D. Modry, and D. Otranto. 2020. "Zoonotic Parasites of Reptiles: A Crawling Threat." *Trends in Parasitology* 36 (8): 677–687. https://doi.org/10.1016/j.pt.2020.04.014.

Menon, M., and S. Kataria. 2018. "Monkeys Run Amok in India's Corridors of Power." Reuters News Service, December 10. https://www.reuters.com/article/us-india-monkeys/monkeys-run-amok-in-indias-corridors-of-power-idUSKBN1OA01R.

Messmer, Terry. 2019. "The Growing Business of Human-Wildlife Conflict Management." *Human–Wildlife Interactions* 13 (1): 3. https://doi.org/10.26076/a969-w636.

Meyer, J., and J. Florence. 1996. "Tahiti's Native Flora Endangered by the Invasion of *Miconia calvescens* DC (Melastomataceae)." *Journal of Biogeography* 775–781. https://doi.org/10.1111/j.1365-2699.1996.tb00038.x.

Meyer, L., L. Du Preez, E. Bonneau, et al. 2015. "Parasite Host-Switching from the Invasive American Red-Eared Slider, *Trachemys scripta elegans*, to the Native Mediterranean Pond Turtle, *Mauremys leprosa*, in Natural Environments," *Aquatic Invasions* 10 (1): 79–91 https://doi.org/10.3391/ai.2015.10.1.08.

Mhuriro-Mashapa, P., E. Mwakiwa, and C. Mashapa. 2017. "Determinants of Communal Farmers' Willingness to Pay for Human-Wildlife Conflict Management in the Periphery of Save Valley Conservancy, South Eastern Zimbabwe." *Journal of Animal and Plant Sciences* 27 (5): 1678–1688. http://www.thejaps.org.pk/docs/v-27-05/36.pdf.

Midgley, J., B. Coetzee, D. Tye, et al. 2020. "Mass Sterilization of a Common Palm Species by Elephants in Kruger

National Park, South Africa." *Scientific Reports* 10 (1): 11719. https://doi.org/10.1038/s41598-020-68679-8.

Miller, J. E. 2018. "Muskrats." Wildlife Damage Management Technical Series. Fort Collins, CO: USDA, APHIS, WS National Wildlife Research Center. https://digitalcommons.unl.edu/nwrcwdmts/14/.

Miller, J. E., and G. K. Yarrow. 1994. "Beavers." In *Prevention and Control of Wildlife Damage*, edited by S. E. Hygnstrom, R. M. Timm, and G. M. Larson, B1–B11. Lincoln: University of Nebraska.

Miller, M. A., W. Miller, P. Conrad, et al. 2008. "Type X *Toxoplasma gondii* in a Wild Mussel and Terrestrial Carnivores from Coastal California: New Linkages between Terrestrial Mammals, Runoff and Toxoplasmosis of Sea Otters." *International Journal for Parasitology* 38 (11): 1319–1328. https://doi.org/10.1016/j.ijpara.2008.02.005.

Mohajan, H. 2019. "The First Industrial Revolution: Creation of a New Global Human Era." *Journal of Social Sciences and Humanities* 5 (4): 377–387. https://mpra.ub.uni-muenchen.de/96644/.

Molenaar, F., J. Jaffe, I. Carter, et al. 2017. "Poisoning of Reintroduced Red Kites (*Milvus milvus*) in England." *European Journal of Wildlife Research* 63 (6): 1–8. https://doi.org/10.1007/s10344-017-1152-z.

Moloney, P. D., and J. W. Hearne. 2009. "The Population Dynamics of Converting Properties from Cattle to Kangaroo Production." In *Proceedings of the 18th World IMACS Congress and MODSIM09 International Congress on Modelling and Simulation*, July 13–17, Cairns, Australia, edited by R. S. Anderssen, R. D. Braddock, and L. T. H. Newham, 561–566. Modelling and Simulation Society of Australia and New Zealand and International Association for Mathematics and Computers in Simulation. https://mssanz.org.au/modsim09/B1/moloney.pdf.

Montag, J. 2003. "Compensation and Predator Conservation: Limitations of Compensation." *Carnivore Damage Prevention News* 6: 2–6.

Morey, D. 1994. "The Early Evolution of the Domestic Dog." *American Scientist* 82 (4): 336–347. https://doi.org/10.2307/29775234.

Morey, P., E. Gese, and S. Gehrt. 2007. "Spatial and Temporal Variation in the Diet of Coyotes in the Chicago Metropolitan Area." *American Midland Naturalist* 158 (1): 147–161.

Morrison, P., and R. I. Allcorn. 2006. "The Effectiveness of Different Methods to Deter Large Gulls Larus Spp from Competing with Nesting Terns Sterna Spp on Coquet Island RSPB Reserve, Northumberland, England." *Conservation Evidence* 3: 84–87. https://www.conservationevidence.com/individual-study/2232.

Moseby, K. E., H. McGregor, and J. L. Read. 2020. "Effectiveness of the Felixer Grooming Trap for the Control of Feral Cats: A Field Trial in Arid South Australia." *Wildlife Research* 47 (8): 599–609. https://doi.org/10.1071/WR19132.

Moser, B., and G. Witmer. 2000. "The Effects of Elk and Cattle Foraging on the Vegetation, Birds, and Small Mammals of the Bridge Creek Wildlife Area, Oregon." *International Biodeterioration and Biodegradation* 45 (3–4): 151–157. https://doi.org/10.1016/S0964-8305(00)00036-6.

Murie, A. 1944. *The Wolves of Mount McKinley*. Fauna Series No. 5. US Department of the Interior, National Park Service. Washington, DC: US Government Printing Office. https://www.nps.gov/parkhistory/online_books/fauna5/fauna.htm.

Mushtaq, M., I. Hussain, A. Mian, et al. 2013. "Field Evaluation of Some Bait Additives against Indian Crested Porcupine (*Hystrix indica*) (Rodentia: Hystricidae)." *Integrative Zoology* 8 (3): 285–292. https://doi.org/10.1111/1749-4877.12014.

Musiani, M., C. Mamo, L. Boitani, et al. 2003. "Wolf Depredation Trends and the Use of Fladry Barriers to Protect Livestock in Western North America." *Conservation Biology* 17 (6): 1538–1547. https://doi.org/10.1111/j.1523-1739.2003.00063.x.

Mutze, G. 2016. "Barking up the Wrong Tree? Are Livestock or Rabbits the Greater Threat to Rangeland Biodiversity in Southern Australia?" *Rangeland Journal* 38 (6): 523–531. https://doi.org/10.1071/RJ16047.

Nadin-Davis, S., T. Buchanan, L. Nituch, et al. 2020. "A Long-Distance Translocation Initiated an Outbreak of Raccoon Rabies in Hamilton, Ontario, Canada." *PLOS Neglected Tropical Diseases* 14 (3): e0008113. https://doi.org/10.1371/journal.pntd.0008113.

Nagano, N., S. Oana, Y. Nagano, et al. 2006. "A Severe *Salmonella enterica* Serotype Paratyphi B Infection in a Child Related to a Pet Turtle, *Trachemys scripta elegans*." *Japanese Journal of Infectious Diseases* 59 (2): 132. https://www.niid.go.jp/niid/images/JJID/59/132.pdf.

Nagashima, K., T. Shimomura, and K. Tanaka. 2019. "Early-Stage Vegetation Recovery in Forests Damaged by Oak Wilt Disease and Deer Browsing: Effects of Deer-Proof Fencing and Clear-Cutting." *Landscape and Ecological Engineering* 15 (2): 155–166. https://doi.org/10.1007/s11355-019-00372-z.

Nation, T. H. 2007. "The Influence of Flowering Dogwood (*Cornus florida*) on Land Snail Diversity in a Southern Mixed Hardwood Forest." *American Midland Naturalist* 157 (1): 137–148. https://doi.org/10.1674/0003-0031(2007)157[137:TIOFDC]2.0.CO;2.

Naz, R., and A. Saver. 2016. "Immunocontraception for Animals: Current Status and Future Perspective." *American Journal of Reproductive Immunology* 75 (4): 426–439. https://doi.org/10.1111/aji.12431.

Nelson, L. H. 2009. "The Great Famine (1315–1317) and the Black Death (1346–1351)." *Lectures in Medieval History*. http://www.vlib.us/medieval/lectures/black_death.html.

Nicholson, L., J. Mahar, T. Strive, et al. 2017. "Benign Rabbit Calicivirus in New Zealand." *Applied and Environmental Microbiology* 83 (11): e00090-17. https://doi.org/10.1128/AEM.00090-17.

Nico, L. G., P. Fuller, and J. Li. 2019. "*Hypophthalmichthys molitrix* (Valenciennes in Cuvier and Valenciennes, 1844)." In *Nonindigenous Aquatic Species Database*. Gainesville, FL: US Geological Survey. https://nas.er.usgs.gov/queries/factsheet.aspx?SpeciesID=549.

Nico, L. G., P. Fuller, M. Neilson, et al. 2021. "*Cyprinella lutrensis* (Baird and Girard, 1853)." In *Nonindigenous Aquatic Species*

Database. Gainesville, FL: US Geological Survey. https://nas
.er.usgs.gov/queries/FactSheet.aspx?speciesID=518.

Nicolaus, L., J. F. Cassel, R. Carlson, et al. 1983. "Taste-
Aversion Conditioning of Crows to Control Predation on
Eggs." *Science* 220 (4593): 212–214. https://doi.org/10.1126
/science.220.4593.212.

Nikolopoulos, K., S. Punia, A. Schäfers, et al. 2021. "Forecasting and Planning during a Pandemic: COVID-19 Growth
Rates, Supply Chain Disruptions, and Governmental
Decisions." *European Journal of Operational Research* 290 (1):
99–115. https://doi.org/10.1016/j.ejor.2020.08.001.

Nogales, M., A. Martín, B. Tershy, et al. 2004. "A Review of
Feral Cat Eradication on Islands." *Conservation Biology* 18 (2):
310–319. https://doi.org/10.1111/j.1523-1739.2004.00442.x.

Nogueira-Filho, S., S. Nogueira, and J. Fragoso. 2009.
"Ecological Impacts of Feral Pigs in the Hawaiian Islands."
Biodiversity and Conservation 18 (14): 3677. https://doi.org/10
.1007/s10531-009-9680-9.

Nolte, D., and M. Dykzeul. 2002. "Wildlife Impacts on Forest
Resources." In *Human Conflicts with Wildlife: Economic
Considerations: Proceedings of the Third NWRC Special
Symposium,* August 1–3, 2000, Fort Collins, CO, edited by L.
Clark, J. Hone, J. A. Shivik, et al., 163–168. Fort Collins, CO:
National Wildlife Research Center. https://digitalcommons
.unl.edu/nwrchumanconflicts/20/.

Nugent, B., K. Gagne, and M. Dillingham. 2008. "Managing
Gulls to Reduce Fecal Coliform Bacteria in a Municipal
Drinking Water Source." In *Proceedings of the 23rd Verte-
brate Pest Conference,* edited by R. M. Timm and M. B.
Madon, 26–30. Davis: University of California. https://doi
.org/10.5070/V423110534.

Nussey, D., L. Kruuk, A. Morris, et al. 2007. "Environmental
Conditions in Early Life Influence Ageing Rates in a Wild
Population of Red Deer." *Current Biology* 17 (23): R1000–
R1001. https://doi.org/10.1016/j.cub.2007.10.005.

Nyahongo, J., and E. Røskaft. 2011. "Perception of People
Towards Lions and Other Wildlife Killing Humans,
around Selous Game Reserve, Tanzania." *International
Journal of Biodiversity and Conservation* 3 (4): 110–115.
https://doi.org/10.5897/IJBC.9000104.

Nyhus, P. 2016. "Human–Wildlife Conflict and Coexistence."
Annual Review of Environment and Resources 41 (1): 143–171.
https://doi.org/10.1146/annurev-environ-110615-085634.

O'Dowd, D., P. Green, and P. S. Lake. 2003. "Invasional
'Meltdown' on an Oceanic Island." *Ecology Letters* 6 (9):
812–817. https://doi.org/10.1046/j.1461-0248.2003.00512.x.

O'Gara, B., and D. Getz. 1986. "Capturing Golden Eagles
Using a Helicopter and Net Gun." *Wildlife Society Bulletin*
14 (4): 400–402. https://www.jstor.org/stable/3782276.

Ogden, D. M. 1971. "How National Policy Is Made." In
*Increasing Understanding of Public Problems and Policies,
1971,* 510. Chicago: Farm Foundation. https://ageconsearch
.umn.edu/bitstream/17268/1/ar710005.pdf.

Ogutu-Ohwayo, R. 2004. "Management of the Nile Perch,
Lates niloticus Fishery in Lake Victoria in Light of the
Changes in Its Life History Characteristics." *African
Journal of Ecology* 42 (4): 306–314. https://doi.org/10.1111/j
.1365-2028.2004.00527.x.

O'Hare, J., J. Eisemann, K. Fagerstone, et al. 2007. "Use of
Alpha-Chloralose by USDA Wildlife Services to Immobilize
Birds." In *Proceedings of the 12th Wildlife Damage Management
Conference,* April 9–12, Corpus Christi, TX, edited by D. L.
Nolte, W. M. Arjo, and D. H. Stalman, 103–113. Fort
Collins, CO: National Wildlife Research Center. https://
digitalcommons.unl.edu/icwdm_usdanwrc/769/.

Öhman, A. 1986. "Face the Beast and Fear the Face: Animal
and Social Fears as Prototypes for Evolutionary Analyses
of Emotion." *Psychophysiology* 23 (2): 123–145. https://doi
.org/10.1111/j.1469-8986.1986.tb00608.x.

Okarma, H., and W. Jędrzejewski. 1997. "Live-trapping
Wolves with Nets." *Wildlife Society Bulletin* 25 (1): 78–82.

Olsen, B., V. Munster, A. Wallensten, et al. 2006. "Global
Patterns of Influenza A Virus in Wild Birds." *Science* 312
(5772): 384–388. https://doi.org/10.1126/science.1122438.

Oppel, S., B. Beaven, M Bolton, et al. 2011. "Eradication of
Invasive Mammals on Islands Inhabited by Humans and
Domestic Animals." *Conservation Biology* 25 (2): 232–240.
https://doi.org/10.1111/j.1523-1739.2010.01601.x.

Osgood, D., and J. Zieman. 1998. "The Influence of
Subsurface Hydrology on Nutrient Supply and Smooth
Cordgrass (*Spartina alterniflora*) Production in a Developing Barrier Island Marsh." *Estuaries* 21 (4): 767–783.
https://doi.org/10.2307/1353280.

Oswalt, W. H. 1999. *Eskimos and Explorers.* 2nd ed. Lincoln:
University of Nebraska Press.

Pachauri, R., M. Allen, V. Barros, et al. 2014. *Climate Change
2014: Synthesis Report. Contribution of Working Groups I, II
and III to the Fifth Assessment Report of the Intergovernmental
Panel on Climate Change.* Geneva: IPCC. https://www.ipcc
.ch/site/assets/uploads/2018/05/SYR_AR5_FINAL_full
_wcover.pdf.

Paine, R. 1966. "Food Web Complexity and Species
Diversity." *American Naturalist* 100 (910): 65–75. https://doi
.org/10.1086/282400.

Palmateer, S. 1987. "Current and Future Status of Rodenti-
cides and Predacides." In *8th Great Plains Wildlife Damage
Control Workshop,* April 26–30, Rapid City, SD, edited by
F. R. Henderson and R. Timm, 16–17. Fort Collin, CO: US
Department of Agriculture, Forest Service, Rocky
Mountain Forest and Range Experiment Station.
https://digitalcommons.unl.edu/gpwdcwp/87/.

Palminteri, S. 2015. "Taking Technology out in the Cold:
Working to Conserve Snow Leopards." *Mongabay
Environmental News,* June 30. https://news.mongabay.com
/2015/06/taking-technology-out-in-the-cold-working-to
-conserve-snow-leopards/.

Panagiotakopulu, E. 2004. "Pharaonic Egypt and the Origins of
Plague." *Journal of Biogeography* 31 (2): 269–275. https://doi
.org/10.1046/j.0305-0270.2003.01009.x.

Patel, R., and T. Burke. 2009. "Urbanization—An Emerging
Humanitarian Disaster." *New England Journal of Medicine*
361 (8): 741–743. https://doi.org/10.1056/NEJMp0810878.

Pauchard, A., and K. Shea. 2006. "Integrating the Study of
Non-native Plant Invasions across Spatial Scales."
Biological Invasions 8 (3): 399–413. https://doi.org/10.1007
/s10530-005-6419-8.

Pedler, R., R. Brandle, J. Read, et al. 2016. "Rabbit Biocontrol and Landscape-Scale Recovery of Threatened Desert Mammals." *Conservation Biology* 30 (4): 774–782. https://doi.org/10.1111/cobi.12684.

Peh, K., and N. Sodhi. 2002. "Characteristics of Nocturnal Roosts of House Crows in Singapore." *Journal of Wildlife Management* 66 (4): 1128–1133. https://doi.org/10.2307/3802944.

Penteriani, V., M. Delgado, F. Pinchera, et al. 2016. "Human Behaviour Can Trigger Large Carnivore Attacks in Developed Countries." *Scientific Reports* 6 (1): 20552. https://doi.org/10.1038/srep20552.

Pereira, H., and L. Navarro, eds. 2015. *Rewilding European Landscapes*. Cham, Switzerland: Springer Nature. https://doi.org/10.1007/978-3-319-12039-3.

Peterson, S., and M. Colwell. 2014. "Experimental Evidence That Scare Tactics and Effigies Reduce Corvid Occurrence." *Northwestern Naturalist* 95 (2): 103–112. https://doi.org/10.1898/NWN13-18.1.

Pfeiffer, M., J. Kougher, and T. DeVault. 2018. "Civil Airports from a Landscape Perspective: A Multi-Scale Approach with Implications for Reducing Bird Strikes." *Landscape and Urban Planning* 179: 38–45. https://doi.org/10.1016/j.landurbplan.2018.07.004.

Pimentel, D., S. McNair, J. Janecka, et al. 2001. "Economic and Environmental Threats of Alien Plant, Animal, and Microbe Invasions." *Agriculture, Ecosystems and Environment* 84 (1): 1–20. https://doi.org/10.1016/S0167-8809(00)00178-X.

Pimentel, D., M. Whitecraft, Z. Scott, et al. 2010. "Will Limited Land, Water, and Energy Control Human Population Numbers in the Future?" *Human Ecology* 38 (5): 599–611. https://www.jstor.org/stable/40928150.

Pitt, W., P. Box, and F. Knowlton. 2003. "An Individual-Based Model of Canid Populations: Modelling Territoriality and Social Structure." *Ecological Modelling* 166 (1–2): 109–121. https://digitalcommons.unl.edu/icwdm_usdanwrc/267/.

Pitt, W., D Vice, and M. Pitzler. 2005. "Challenges of Invasive Reptiles and Amphibians." In *Proceedings of the 11th Wildlife Damage Management Conference*, May 16–19, Traverse City, MI, edited by D. L. Nolte and K. A. Fagerstone, 113–119. Fort Collins, CO: National Wildlife Research Center. https://digitalcommons.unl.edu/icwdm_wdmconfproc/84/.

Pochop, P., J. Cummings, and R. Engeman. 2001. "Field Evaluation of a Visual Barrier to Discourage Gull Nesting." *Pacific Conservation Biology* 7 (2): 143–145.

Porta, J. (1658) 1959. *15th Booke of Natural Magick. Of Fishing, Fowling and Hunting*. New York: Basic Books.

Portney, P. R. 2020. "Benefit-Cost Analysis." *Econlib*, July 16. https://www.econlib.org/library/Enc/BenefitCostAnalysis.html.

Pourrut, X., B. Kumulungui, T. Wittmann, et al. 2005. "The Natural History of Ebola Virus in Africa." *Microbes and Infection* 7 (7–8): 1005–1014. https://doi.org/10.1016/j.micinf.2005.04.006.

Power, A., R. Walker, K. Payne, et al. 2004. "First Occurrence of the Nonindigenous Green Mussel, *Perna viridis* (Linnaeus, 1758) in Coastal Georgia, United States." *Journal of Shellfish Research* 23 (3): 741–745.

Prugh, L., C. Stoner, C. Epps, et al. 2009. "The Rise of the Mesopredator." *BioScience* 59 (9): 779–791. https://doi.org/10.1525/bio.2009.59.9.9.

Pullins, C., T. Guerrant, S. Beckerman, et al. 2018. "Mitigation Translocation of Red-Tailed Hawks to Reduce Raptor–Aircraft Collisions." *Journal of Wildlife Management* 82 (1): 123–129. https://doi.org/10.1002/jwmg.21332.

Pummerer, L., R. Böhm, L. Lilleholt, et al. 2021. "Conspiracy Theories and Their Societal Effects during the COVID-19 Pandemic." *Social Psychological and Personality Science* 13 (1): 49–59. https://doi.org/10.1177/19485506211000217.

Pyšek, P., and D. Richardson. 2006. "The Biogeography of Naturalization in Alien Plants." *Journal of Biogeography* 33 (12): 2040–2050. https://doi.org/10.1111/j.1365-2699.2006.01578.x.

Raffaelli, D., and H. Moller. 1999. "Manipulative Field Experiments in Animal Ecology: Do They Promise More than They Can Deliver?" *Advances in Ecological Research* 30: 299–338. https://doi.org/10.1016/S0065-2504(08)60020-3.

Rattner, B., S. Volker, J. Lankton, et al. 2020. "Brodifacoum Toxicity in American Kestrels (Falco Sparverius) with Evidence of Increased Hazard on Subsequent Anticoagulant Rodenticide Exposure." *Environmental Toxicology and Chemistry* 39 (2): 468–481. https://doi.org/10.1002/etc.4629.

Ravallion, M. 2010. "The Developing World's Bulging (but Vulnerable) Middle Class." *World Development* 38 (4): 445–454. https://doi.org/10.1016/j.worlddev.2009.11.007.

Raven, P. H., and G. B. Johnson. 1992. *Biology*. 3rd ed. St. Louis: Mosby Year Book.

Rayner, T., and R. Creese. 2006. "A Review of Rotenone Use for the Control of Non-Indigenous Fish in Australian Fresh Waters, and an Attempted Eradication of the Noxious Fish, *Phalloceros Caudimaculatus*." *New Zealand Journal of Marine and Freshwater Research* 40 (3): 477–486. https://doi.org/10.1080/00288330.2006.9517437.

Reddy, P., D. Gorelick, C. Brasher, et al. 1970. "Progressive Disseminated Histoplasmosis as Seen in Adults." *American Journal of Medicine* 48 (5): 629–636. https://doi.org/10.1016/0002-9343(70)90014-8.

Redman, M., A. King, C. Watson, et al. 2016. "What Is CRISPR/Cas9?" *Archives of Disease in Childhood: Education and Practice* 101 (4): 213–215. https://doi.org/10.1136/archdischild-2016-310459.

Redpath, S., and S. Thirgood. 2009. "Hen Harriers and Red Grouse: Moving Towards Consensus?" *Journal of Applied Ecology* 46 (5): 961–963. https://doi.org/10.1111/j.1365-2664.2009.01702.x.

Reeves, C. 1992. *Egyptian Medicine*. London: Osprey.

Reidinger, R. F. 1995. "Recent Studies on Flavor Aversion Learning in Wildlife Damage Management." In *National Wildlife Research Center Repellents Conference 1995*, 31.

Reidinger, R. F., Jr., G. Beauchamp, and M. Barth. 1982. "Conditioned Aversion to a Taste Perceived while Grooming." *Physiology and Behavior* 28 (4): 715–723. https://doi.org/10.1016/0031-9384(82)90057-9.

Reidinger, R. F., Jr., J. L. Jibay, and A. L. Kolz. 1985. "Field Trial of an Electric Barrier for Protecting Rice Fields from Rat Damage." *Philippine Agriculturist* 68: 168–179.

Reidinger, R. F., Jr., and J. R. Mason. 1983. "Exploitable Characteristics of Neophobia and Food Aversions for Improvements in Rodent and Bird Control." In *Vertebrate Pest Control and Management Materials: Fourth Symposium.* ASTM International,

Regnery, J., A. Friesen, A. Geduhn, et al. 2019. "Rating the Risks of Anticoagulant Rodenticides in the Aquatic Environment: A Review." *Environmental Chemistry Letters* 17 (1): 215–240. https://doi.org.10.1007/s10311-018-0788-6.

Richardson, W., and T. West. 2000. "Serious Bird Strike Accidents to Military Aircraft: Updated List and Summary." In *Proceedings of the 25th International Bird Strike Committee,* April 17–21, Amsterdam, 67–97 (WP SA1). https://lgl.com/images/pdf/Richardson_West _2000_IBSC25-Amsterdam-as-publ.pdf.

Richardson, S., A. Mill, D. Davis, et al. 2020. "A Systematic Review of Adaptive Wildlife Management for the Control of Invasive, Non-Native Mammals, and Other Human–Wildlife Conflicts." *Mammal Review* 50 (2): 147–156. https://doi.org/10.1111/mam.12182.

Rigg, R. 2001. "Livestock Guarding Dogs: Their Current Use World Wide." IUCN/SSC Canid Specialist Group Occasional Paper No. 1. http://www.slovakwildlife.org /pdf/Rigg_LGDs_worldwide.pdf.

Riley, S., and D. Decker. 2000. "Wildlife Stakeholder Acceptance Capacity for Cougars in Montana." *Wildlife Society Bulletin* 28 (4): 931–939. https://www.jstor.org /stable/3783850.

Rivera, N., A. Brandt, J. Novakofski, et al. 2019. "Chronic Wasting Disease in Cervids: Prevalence, Impact and Management Strategies." *Veterinary Medicine: Research and Reports* 10: 123–139. https://doi.org/10.2147/VMRR.S197404.

Robardet, E., D. Bosnjak, L. Englund, et al. 2019. "Zero Endemic Cases of Wildlife Rabies (Classical Rabies Virus, RABV) in the European Union by 2020: An Achievable Goal." *Tropical Medicine and Infectious Disease* 4 (4): 124. https://doi.org/10.3390/tropicalmed4040124.

Rodenburg, J., M. Demont, A. Sow, et al. 2014. "Bird, Weed and Interaction Effects on Yield of Irrigated Lowland Rice." *Crop Protection* 66: 46–52. https://doi.org/10.1016/j .cropro.2014.08.015.

Rodger, H. 2016. "Fish Disease Causing Economic Impact in Global Aquaculture." In *Fish Vaccines,* edited by A. Adams, 1–34. Basel: Springer. https://doi.org/10.1007 /978-3-0348-0980-1_1.

Rodgers, L., C. Mason, M. Bodle, et al. 2017. "Chapter 7: Status of Nonindigenous Species," *2017 South Florida Environment Report.* Vol. 1: *The South Florida Environment.* 1–58. West Palm Beach: South Florida Water Management District. https://apps.sfwmd.gov/sfwmd/SFER /2017_sfer_final/v1/chapters/v1_ch7.pdf.

Roemer, Gary W., Paul S. Martin, and C. J. Donlan. 2007. "Lessons from Land Present and Past Signs of Ecological Decay and the Overture to Earth's Sixth Mass Extinction." In *Whales, Whaling, and Ocean Ecosystems,* edited by J. A. Estes, D. P. DeMaster, D. F. Doak, et al., 14–26. Berkeley: University of California Press.

Rogers, L. 2011. "Does Diversionary Feeding Create Nuisance Bears and Jeopardize Public Safety?" *Human–Wildlife Interactions* 5 (2): 287–295. https://doi.org/10.26077/sg4g-k319.

Romney, A. K., S. Weller, and W. Batchelder. 1986. "Culture as Consensus: A Theory of Culture and Informant Accuracy." *American Anthropologist,* N.S., 88 (2): 313–338. https://www.jstor.org/stable/677564.

Ronconi, R. A., Z. T. Swaim, H. A. Lane, et al. 2010. "Modified Hoop-Net Techniques for Capturing Birds at Sea and Comparison with Other Capture Methods." *Marine Ornithology* 38: 23–29. https://marineornithology .org/~marineor/PDF/38_1/38_1_23-29.pdf.

Root, J. J., R. Puskas, J. Fischer, et al. 2009. "Landscape Genetics of Raccoons (*Procyon lotor*) Associated with Ridges and Valleys of Pennsylvania: Implications for Oral Rabies Vaccination Programs." *Vector-Borne and Zoonotic Diseases* 9 (6): 583–588. https://doi.org/10.1089/vbz.2008.0110.

Roy, S., G. Smith, and J. Russell. 2009. "The Eradication of Invasive Mammal Species: Can Adaptive Resource Management Fill the Gaps in Our Knowledge?" *Human–Wildlife Conflicts* 3 (1): 8. https://doi.org/10.26077/5dac-aw65.

Rudd, J., S. McMillin, M. Kenyon, et al. 2018. "Prevalence of First and Second-Generation Anticoagulant Rodenticide Exposure in California Mountain Lions (*Puma concolor*)." In *Proceedings of the 28th Vertebrate Pest Conference,* edited by D. M. Woods, 254–257. Davis: University of California. https://doi.org/10.5070/V42811046.

Ruell, E., C. Niebuhr, R. Sugihara, et al. 2019. "An Evaluation of the Registration and Use Prospects for Four Candidate Toxicants for Controlling Invasive Mongooses (*Herpestes javanicus* Auropunctatus)." *Management of Biological Invasions* 10 (3): 573–596. https://doi.org/10.3391/mbi.2019.10.3.11.

Ruha, A., K. Kleinschmidt, S. Greene, et al. 2017. "The Epidemiology, Clinical Course, and Management of Snakebites in the North American Snakebite Registry." *Journal of Medical Toxicology* 13 (4): 309–320. https://doi.org /10.1007/s13181-017-0633-5.

Russell, D. J., P. A. Thuesen, and F. E. Thomson. 2012. "A Review of the Biology, Ecology, Distribution and Control of Mozambique Tilapia, *Oreochromis mossambicus* (Peters 1852) (Pisces: Cichlidae) with Particular Emphasis on Invasive Australian Populations." *Reviews in Fish Biology and Fisheries* 22 (3): 533–554. https://doi.org/10.1007 /s11160-011-9249-z.

Rutberg, A. T. 2005. "Deer Contraception: What We Know and What We Don't." *Humane Wildlife Solutions: The Role of Immunocontraception,* edited by A. T. Rutberg, 23–42. Washington, DC: Humane Society Press.

Sáenz-de-Santa-María, and J. Tellería. 2015. "Wildlife-Vehicle Collisions in Spain." *European Journal of Wildlife Research* 61 (3): 399–406. https://doi.org/10.1007/s10344-015-0907-7.

Sallam, M., D. Dababseh, A. Yaseen, et al. 2020. "COVID-19 Misinformation: Mere Harmless Delusions or Much More? A Knowledge and Attitude Cross-Sectional Study among the General Public Residing in Jordan." *PLoS One* 15 (12): e0243264. https://doi.org/10.1371/journal.pone.0243264.

Samuel, M., W. Liao, C. Atkinson, et al. 2020. "Facilitated Adaptation for Conservation—Can Gene Editing Save Hawaii's Endangered Birds from Climate Driven Avian Malaria?" *Biological Conservation* 241: 108390. https://doi.org/10.1016/j.biocon.2019.108390.

Sandom, C., S. Faurby, B. Sandel, et al. 2014. "Global Late Quaternary Megafauna Extinctions Linked to Humans, Not Climate Change." *Proceedings of the Royal Society B: Biological Sciences* 281 (1787): 20133254. https://doi.org/10.1098/rspb.2013.3254.

Sargeant, A., R. Greenwood, M. Sovada, et al. 1993. "Distribution and Abundance of Predators That Affect Duck Production—Prairie Pothole Region." Jamestown, ND: Fish and Wildlife Service, Northern Prairie Wildlife Research.

Sauer, J., K. Pardieck, D. Ziolkowski Jr., et al. 2017. "The First 50 Years of the North American Breeding Bird Survey." *The Condor* 119 (3): 576–593. https://doi.org/10.1650/Condor-17-83.1.

Saul, E. 1967. "Birds and Aircraft: A Problem at Auckland's New International Airport." *Journal of the Royal Aeronautical Society* 71 (677): 366–376. https://doi.org/10.1017/S0001924000067737.

Savage, A., B. Gratwicke, K. Hope, et al. 2020. "Sustained Immune Activation Is Associated with Susceptibility to the Amphibian Chytrid Fungus." *Molecular Ecology* 29 (15): 2889–2903. https://doi.org/10.1111/mec.15533.

Savarie, P., D. Vice, L. Bangerter, et al. 2004. "Operational Field Evaluation of a Plastic Bulb Reservoir as a Tranquilizer Trap Device for Delivering Propiopromazine Hydrochloride to Feral Dogs, Coyotes, and Gray Wolves." In *Proceedings of the 21st Vertebrate Pest Conference*, March 1–4, Visalia, CA, edited by R. M. Timm and W. P. Gorenzel, 64–69. Davis: University of California. https://digitalcommons.unl.edu/icwdm_usdanwrc/383/.

Schaller, M. 2002. "Evaluation of Wildlife Damage to Forests in Germany." In *Human Conflicts with Wildlife: Economic Considerations: Proceedings of the Third NWRC Special Symposium*, August 1–3, 2000, Fort Collins, CO, edited by L. Clark, J. Hone, J. A. Shivik, et al., 123–126. Fort Collins, CO: National Wildlife Research Center. https://digitalcommons.unl.edu/nwrchumanconflicts/14/.

Schaller, M. 2007. "Forests and Wildlife Management in Germany: A Mini-Review." *Eurasian Journal of Forest Research* 10 (1): 59–70. https://eprints.lib.hokudai.ac.jp/dspace/bitstream/2115/24487/1/10(1)_P59-70.pdf.

Schartel, T., and C. Brooks. 2018. "Biotic Constraints on *Cactoblastis cactorum* (Berg) Host Use in the Southern US and Their Implications for Future Spread." *Food Webs* 15: e00083. https://doi.org/10.1016/j.fooweb.2018.e00083.

Scheele, B., F. Pasmans, L. Skerratt, et al. 2019. "Amphibian Fungal Panzootic Causes Catastrophic and Ongoing Loss of Biodiversity." *Science* 363 (6434): 1459–1463. https://doi.org/10.1126/science.aav0379.

Schneider, D., C. Ellis, and K. Cummings. 1998. "A Transportation Model Assessment of the Risk to Native Mussel Communities from Zebra Mussel Spread." *Conservation Biology* 12 (4): 788–800. https://www.jstor.org/stable/2387539.

Schuhmann, P., and K. Schwabe. 2002. "Fundamentals of Economic Principles and Wildlife Management." In *Human Conflicts with Wildlife: Economic Considerations: Proceedings of the Third NWRC Special Symposium*, August 1–3, 2000, Fort Collins, CO, edited by L. Clark, J. Hone, J. A. Shivik, et al., 1–16. Fort Collins: National Wildlife Research Center. https://digitalcommons.unl.edu/nwrchumanconflicts/.

Schulte, B. 2016. "Learning and Applications of Chemical Signals in Vertebrates for Human–Wildlife Conflict Mitigation." In *Chemical Signals in Vertebrates 13*, edited by B. A. Schulte, T. E. Goodwin, and M. H. Ferkin, 499–510. Cham, Switzerland: Springer International. https://doi.org/10.1007/978-3-319-22026-0_32.

Schutgens, M., J. Hanson, N. Baral, et al. 2019. "Visitors' Willingness to Pay for Snow Leopard Panthera Uncia Conservation in the Annapurna Conservation Area, Nepal." *Oryx* 53 (4): 633–642. https://doi.org/10.1017/S0030605317001636.

Schwarzländer, M., H. Hinz, R. L. Winston, et al. 2018. "Biological Control of Weeds: An Analysis of Introductions, Rates of Establishment and Estimates of Success, Worldwide." *BioControl* 63 (3): 319–331. https://doi.org/10.1007/s10526-018-9890-8.

Scyphers, S., S. Powers, J. L. Akins, et al. 2015. "The Role of Citizens in Detecting and Responding to a Rapid Marine Invasion." *Conservation Letters* 8 (4): 242–250. https://doi.org/10.1111/conl.12127.

Seamans, T., S. Barras, and G. Bernhardt. 2007. "Evaluation of Two Perch Deterrents for Starlings, Blackbirds and Pigeons." *International Journal of Pest Management* 53 (1): 45–51. https://doi.org/10.1080/09670870601058890.

Seamans, T., and A. L. Gosser. 2016. *Bird Dispersal Techniques*. Wildlife Damage Management Technical Series. Fort Collins, CO: US Department of Agriculture, Animal & Plant Health Inspection Service, Wildlife Services. https://digitalcommons.unl.edu/nwrcwdmts/2/.

Seamans, T., and K. VerCauteren. 2006. "Evaluation of ElectroBraid Fencing as a White-Tailed Deer Barrier." *Wildlife Society Bulletin* 34 (1): 8–15. https://doi.org/10.2193/0091-7648(2006)34[8:EOEFAA]2.0.CO;2.

Selye, H. 1936. "A Syndrome Produced by Diverse Nocuous Agents." *Nature* 138 (3479): 32. https://doi.org/10.1038/138032a0.

Sengl, J., J. Spencer Jr., and Z. Bowers. 2008. "Developing Standard Operating Procedures for Wildlife Damage Management Activities in Urban and Suburban Areas in Southern Nevada." In *Proceedings of the 23rd Vertebrate Pest Conference*, March 17–20, San Diego, CA, edited by R. M. Timm and M. B. Madon, 201–205. Davis: University of California. https://doi.org/10.5070/V423110416.

Séquin, E., M. Jaeger, P. Brussard, et al. 2003. "Wariness of Coyotes to Camera Traps Relative to Social Status and Territory Boundaries." *Canadian Journal of Zoology* 81 (12): 2015–2025. https://doi.org/10.1139/z03-204.

Serieys, L., A. Lea, M. Epeldegui, et al. 2018. "Urbanization and Anticoagulant Poisons Promote Immune Dysfunction in Bobcats." *Proceedings of the Royal Society B: Biological*

Sciences 285 (1871): 20172533. https://doi.org/10.1098/rspb
.2017.2533.

Shaffer, L. J., K. Khadka, J. Van Den Hoek, et al. 2019. "Human-
Elephant Conflict: A Review of Current Management
Strategies and Future Directions." *Frontiers in Ecology and
Evolution* 6: 235. https://doi.org/10.3389/fevo.2018.00235.

Shanmuganathan, T., J. Pallister, S. Doody, et al. 2010.
"Biological Control of the Cane Toad in Australia: A
Review." *Animal Conservation* 13: 16–23. https://doi.org
/10.1111/j.1469-1795.2009.00319.x.

Sharma, V., S. Kaushik, R. Kumar, et al. 2019. "Emerging
Trends of Nipah Virus: A Review." *Reviews in Medical
Virology* 29 (1): e2010. https://doi.org/10.1002/rmv.2010.

Shave, M., S. Shwiff, J. Elser, et al. 2018. "Falcons Using
Orchard Nest Boxes Reduce Fruit-Eating Bird Abun-
dances and Provide Economic Benefits for a Fruit-
Growing Region." *Journal of Applied Ecology* 55 (5):
2451–2460. https://doi.org/10.1111/1365-2664.13172.

Sherley, G. 2000. *Invasive Species in the Pacific*. Samoa: South
Pacific Regional Environment Programme.

Shirk, J., H. Ballard, C. Wilderman, et al. 2012. "Public
Participation in Scientific Research: A Framework for
Deliberate Design." *Ecology and Society* 17 (2): 29.
https://doi.org/10.5751/ES-04705-170229.

Shivik, J., D. Martin, M. Pipas, et al. 2005. "Initial Comparison: Jaws, Cables, and Cage-Traps to Capture Coyotes."
Wildlife Society Bulletin 33 (4): 1375–1383. https://
digitalcommons.unl.edu/icwdm_usdanwrc/519/.

Shivik, J., P. Savarie, and L. Clark. 2002. "Aerial Delivery of
Baits to Brown Treesnakes." *Wildlife Society Bulletin* 30 (4):
1062–1067. https://www.jstor.org/stable/3784274.

Shope, R. 1992. "Impacts of Global Climate Change on Human
Health: Spread of Infectious Disease." *Global Climate
Change: Implications, Challenges, and Mitigation Measures*,
edited by S. K. Majumdar, L. S. Kalkstein, B. Yarnal, et al.,
363–370. Easton, PA: Pennsylvania Academy of Science.

Shwiff, S. 2004. "Economics in Wildlife Damage Management Studies: Common Problems and Solutions." In
Proceedings of the 21st Vertebrate Pest Conference, March 1–4,
Visalia, CA, edited by R. M. Timm and W. P. Gorenzel,
346–349. Davis: University of California. https://
digitalcommons.unl.edu/icwdm_usdanwrc/387/.

Shwiff, S., and R. J. Merrell. 2004. "Coyote Predation
Management: An Economic Analysis of Increased
Antelope Recruitment and Cattle Production in South
Central Wyoming." *Sheep and Goat Research Journal* 15.
https://digitalcommons.unl.edu/icwdmsheepgoat/15/.

Shwiff, S., A. Pelham, S. Shwiff, et al. 2020. "Framework for
Assessing Vertebrate Invasive Species Damage: The Case
of Feral Swine in the United States." *Biological Invasions* 22
(10): 3101–3117. https://doi.org/10.1007/s10530-020-02311-8.

Shwiff, S, and R. Sterner. 2002. "An Economic Framework for
Benefit-Cost Analysis in Wildlife Damage Studies." In
Proceedings of the 20th Vertebrate Pest Conference, March 4–7,
Reno, NV, edited by R. M. Timm and R. H. Schmidt,
340–344. Davis: University of California. https://doi.org
/10.5070/V420110089.

Sicard, B., W. Diarra, and H. M. Cooper. 1999. "Ecophysiology and Chronobiology Applied to Rodent Pest Manage-

ment in Semi-Arid Agricultural Areas in Sub-Saharan
West Africa." In *Ecologically-based Rodent Management*,
edited by G. Singleton, H. Leirs, L. Hinds, et al., 409–440.
Canberra: Australian Centre for International Agricultural Research.

Sillitoe, P., J. Pottier, and A. Bicker, eds. 2019. *Investigating
Local Knowledge: New Directions, New Approaches*. New
York: Routledge.

Simberloff, D. 2005. "The Politics of Assessing Risk for
Biological Invasions: The USA as a Case Study." *Trends in
Ecology and Evolution* 20 (5): 216–222. https://doi.org/10
.1016/j.tree.2005.02.008.

Sims, B. 1995. "Predator Politics in Texas." In *Symposium
Proceedings—Coyotes in the Southwest: A Compendium of Our
Knowledge*, December 13–14, San Angelo, TX, edited by
D. Rollins, C. Richardson, T. Blankenship, et al., 141–142.
Austin: Texas Parks and Wildlife Department. https://
digitalcommons.unl.edu/coyotesw/8/.

Singleton, G. R. 2014. "Ecologically-Based Rodent Manage-
ment 15 Years On: A Pathway to Sustainable Agricultural
Production." *Proceedings of the 26th Vertebrate Pest Conference*,
March 3–6, Waikoloa, HI, edited by R. M. Timm and J. M.
O'Brien, 176–179. https://doi.org/10.5070/V426110594.

Singleton, G. R., P. Brown, and J. Jacob. 2004. "Ecologically-
Based Rodent Management: Its Effectiveness in Cropping
Systems in South-East Asia." *Njas-Wageningen Journal of
Life Sciences* 52 (2): 163–171. https://doi.org/10.1016/S1573
-5214(04)80011-3.

Singleton, G. R., H. Leirs, L. Hinds, et al. 1999. *Ecologically-
Based Management of Rodent Pests-Re-evaluating Our Approach
to an Old Problem*. Canberra: Australian Centre for International Agricultural Research.

Singleton, G. R., Sudarmaji, and S. Suriapermana. 1998. "An
Experimental Field Study to Evaluate a Trap-Barrier
System and Fumigation for Controlling the Rice Field
Rat, *Rattus argentiventer*, in Rice Crops in West Java." *Crop
Protection* 17 (1): 55–64. https://doi.org/10.1016/S0261
-2194(98)80013-6.

Slate, D., R. Chipman, C. Rupprecht, et al. 2002. "Oral Rabies
Vaccination: A National Perspective on Program Development and Implementation." USDA National Wildlife
Research Center: Staff Publications, 476. https://digital
commons.unl.edu/icwdm_usdanwrc/476/.

Slate, D., C. Rupprecht, J. Rooney, et al. 2005. "Status of Oral
Rabies Vaccination in Wild Carnivores in the United
States." *Virus Research* 111 (1): 68–76. https://doi.org/10
.1016/j.virusres.2005.03.012.

Smith, B., M. Cherkiss, K. Hart, et al. 2016. "Betrayal:
Radio-Tagged Burmese Pythons Reveal Locations of
Conspecifics in Everglades National Park." *Biological
Invasions* 18 (11): 3239–3250. https://doi.org/10.1007
/s10530-016-1211-5.

Smith, B., N. Jaques, R. Appleby, et al. 2020. "Automated
Shepherds: Responses of Captive Dingoes to Sound and
an Inflatable, Moving Effigy." *Pacific Conservation Biology*
27 (2): 195–201. https://doi.org/10.1071/PC20022.

Smith, K., A. Dobson, F. E. McKenzie, et al. 2005. "Ecological
Theory to Enhance Infectious Disease Control and Public
Health Policy." *Frontiers in Ecology and the Environment* 3

(1): 29–37. https://doi.org/10.1890/1540-9295(2005)003
[0029:ETTEID]2.0.CO;2.

Smith, M. 2016. "The Substance and Symbolism of Long-
Distance Exchange: Textiles as Desired Trade Goods in the
Bronze Age Middle Asian Interaction Sphere." In *Connec-
tions and Complexity: New Approaches to the Archaeology of
South Asia*, edited by S. A. Abraham, P. Gullapalli, T. P.
Raczek, et al., 143–160. New York: Routledge.

Smith, T., S. Herrero, T. Debruyn, et al. 2008. "Efficacy of Bear
Deterrent Spray in Alaska." *Journal of Wildlife Management*
72 (3): 640–645. https://doi.org/10.2193/2006-452.

Snow, N., and G. Witmer. 2010. "American Bullfrogs as
Invasive Species: A Review of the Introduction, Subse-
quent Problems, Management Options, and Future
Directions." *Proceedings of the 24th Vertebrate Pest Conference*,
February 22–25, Sacramento, CA, edited by R. M. Timm
and K. A. Fagerstone, 86–89. Davis: University of
California. https://doi.org/10.5070/V424110490.

Soltani, N., J. A. Dille, I. Burke, et al. 2017. "Perspectives on
Potential Soybean Yield Losses from Weeds in North
America." *Weed Technology* 31 (1): 148–154. https://doi.org
/10.1017/wet.2016.2.

Soltis, J., L King, F. Vollrath, et al. 2016. "Accelerometers and
Simple Algorithms Identify Activity Budgets and Body
Orientation in African Elephants *Loxodonta africana*."
Endangered Species Research 31: 1–12. https://doi.org/10
.3354/esr00746.

Sowka, P. 2013. "Practical Electric Fencing Resource Guide:
Controlling Predators." Living with Predators Resource
Guide Series. Arlee, MT: Living with Wildlife Foundation. https://drive.google.com/file/d/1RghqUL2UfbEHR
_logT-_wDxk1n7uC-Di/view.

Spinage, C. 2003. *Cattle Plague: A History*. New York: Kluwer
Academic/Plenum.

Spurr, E. B., and J. D. Coleman. 2005a. "Cost-Effectiveness of
Bird Repellents for Crop Protection." 2009. In *Proceedings
of the 13th Australasian Vertebrate Pest Conference*, Welling-
ton, New Zealand, May 2–6, edited by J. Parkes,
M. Stratham, and G. Edwards, 227–233. Lincoln, New
Zealand: Manaaki Whenua Press/Landcare Research.
https://avpc.net.au/wp-content/uploads/13th-AVPC
Proceedings2005.pdf.

Spurr, E. B., and J. D. Coleman. 2005b. *Review of Canada Goose
Population Trends, Damage, and Control in New Zealand*.
Landcare Research Science Series No. 30. Lincoln, New
Zealand: Manaaki Whenua Press. http://www.mwpress.co
.nz/__data/assets/pdf_file/0013/70501/LRSS_30_-Review
_Canada_Goose.pdf.

Steiner, K., J. Westbrook, F. Hebard, et al. 2017. "Rescue of
American Chestnut with Extraspecific Genes Following
Its Destruction by a Naturalized Pathogen." *New Forests*
48 (2): 317–336. https://doi.org/10.1007/s11056-016-9561-5.

Stengel, C., R. Chipman, K. Nelson, et al. 2019. "The Path to
Eliminating Raccoon Rabies in the Eastern US-Obstacles
and Opportunities in Urban-Suburban Landscapes." In
Proceedings of the 18th Wildlife Damage Management Conference,
April 6–9, Mount Berry, GA, edited by G. R. Gallagher and
J. B. Armstrong, 46. https://digitalcommons.usu.edu
/wdmconference/2019/all2019/31/.

Sterner, R. 1994. "Zinc Phosphide: Implications of Optimal
Foraging Theory and Particle-Dose Analyses Efficacy,
Acceptance, Bait Shyness, and Non-Target Hazards." In
Proceedings of the 16th Vertebrate Pest Conference, March 1–3,
Santa Clara, CA, edited by W. S. Halverson and A. C.
Crabb, 152–159. Davis: University of California. https://
escholarship.org/uc/item/35j5f892.

Sterner, R. 2002. "Spreadsheets, Response Surfaces, and
Intervention Decisions in Wildlife Damage Management."
In *Human Conflicts with Wildlife: Economic Considerations:
Proceedings of the Third NWRC Special Symposium*, Au-
gust 1–3, 2000, Fort Collins, CO, edited by L. Clark,
J. Hone, J. A. Shivik, et al., 42–47. Fort Collins, CO:
National Wildlife Research Center. https://digitalcom
mons.unl.edu/nwrchumanconflicts/5/.

Sterner, R. 2008. "The IPM Paradigm: Vertebrates, Economics,
and Uncertainty." In *Proceedings of the 23rd Vertebrate Pest
Conference*, March 17–20, San Diego, CA, edited by R. M.
Timm and M. B. Madon, 194–200. Davis: University of
California. https://doi.org/10.5070/V423110517.

Sterner, R. 2009. "The Economics of Threatened Species
Conservation: A Review and Analysis." In *Handbook of
Nature Conservation: Global, Environmental, and Economic
Issues*, edited by J. B. Aronoff, 213–235. Hauppauge, NY:
Nova Science. https://digitalcommons.unl.edu/icwdm
_usdanwrc/978/.

Sterner, R., and K. Crane. 2000. "Sheep-Predation Behaviors of
Wild-Caught, Confined Coyotes: Some Historical Data." In
Proceedings of the 19th Vertebrate Pest Conference, March 6–9,
San Diego, CA, edited by T. P. Salmon and A. C. Crabb,
325–330. Davis: University of California Agriculture and
Natural Resources. https://digitalcommons.unl.edu/icwdm
_usdanwrc/816/.

Sterner, R., and K. Tope. 2002. "Repellents: Projections of
Direct Benefit-Cost Surfaces." In *Proceedings of the 20th Verte-
brate Pest Conference*, March 4–7, Reno, NV, edited by R. M.
Timm and R. H. Schmidt, 319–325. Davis: University of
California. https://doi.org/10.5070/V420110206.

Steuber, J. E., M. E. Pitzler, and J. Oldenburg. 1995. "Protecting Juvenile Salmonids from Gull Predation Using Wire
Exclusion below Hydroelectric Dam." In *Proceedings of
the 12th Great Plains Wildlife Damage Control Workshop*,
April 10–13, Tulsa, OK, edited by R. E. Masters and J. G.
Huggins, 38–41. Ardmore, OK: Noble Foundation.
https://digitalcommons.unl.edu/gpwdcwp/452/.

Stevens, G., J. Rogue, R. Weber, et al. 2000. "Evaluation of a
Radar-Activated, Demand-Performance Bird Hazing
System." *International Biodeterioration and Biodegradation*
45 (3–4): 129–137. https://doi.org/10.1016/S0964
-8305(00)00065-2.

Stickley, A. R, Jr., and J. O. King. 1993. "Long-Term Trial of
an Inflatable Effigy Scare Device for Repelling Cormorants from Catfish Ponds." In *Proceedings of the 6th Eastern
Wildlife Damage Control Conference*, October 3–6, Asheville,
NC, edited by M. M. King, 89–92. Raleigh: North Carolina
Cooperative Extension Service. https://digitalcommons
.unl.edu/ewdcc6/33/.

Stoddart, D. M., and P. A. Smith. 1986. "Recognition of
Odour-Induced Bias in the Live-Trapping of *Apodemus*

sylvaticus." *Oikos* 46 (2): 194–199. https://doi.org/10.2307/3565467.

Stohlgren, T. J., C. Jarnevich, G. W. Chong, et al. 2006. "Scale and Plant Invasions: A Theory of Biotic Acceptance." *Preslia* 78 (4): 405–426. http://www.preslia.cz/P064CSto.pdf.

Stolze, A., A. Wanke, N. van Deenen, et al. 2017. "Development of Rubber-Enriched Dandelion Varieties by Metabolic Engineering of the Inulin Pathway." *Plant Biotechnology Journal* 15 (6): 740–753. https://doi.org/10.1111/pbi.12672.

Stone, C., and S. Anderson. 1988. "Introduced Animals in Hawaii's Natural Areas." In *Proceedings of the 13th Vertebrate Pest Conference*, Monterey, CA, March 1–3, edited by A. C. Crabb and R. E. Marsh, 134–140. Davis: University of California. https://digitalcommons.unl.edu/vpcthirteen/28/.

Storm, D., C. Nielsen, E. Schauber, et al. 2007. "Deer–Human Conflict and Hunter Access in an Exurban Landscape." *Human–Wildlife Conflicts* 1 (1): 53–59. https://doi.org/10.26077/zs0n-s822.

Strayer, D. 2009. "Twenty Years of Zebra Mussels: Lessons from the Mollusk That Made Headlines." *Frontiers in Ecology and the Environment* 7 (3): 135–141. https://doi.org/10.1890/080020.

Strayer, D., N. Caraco, J. Cole, et al. 1999. "Transformation of Freshwater Ecosystems by Bivalves: A Case Study of Zebra Mussels in the Hudson River." *BioScience* 49 (1): 19–27. https://doi.org/10.1525/bisi.1999.49.1.19.

Stuart, I., and A. Conallin. 2018. "Control of Globally Invasive Common Carp: An 11-Year Commercial Trial of the Williams' Cage." *North American Journal of Fisheries Management* 38 (5): 1160–1169. https://doi.org/10.1002/nafm.10221.

Stuart, J. 2000. "Additional Notes on Native and Non-native Turtles of the Rio Grande Drainage Basin, New Mexico." *Bulletin of the Chicago Herpetological Society* 35: 229–235.

Sullivan, T., and W. Klenner. 1993. "Influence of Diversionary Food on Red Squirrel Populations and Damage to Crop Trees in Young Lodgepole Pine Forest." *Ecological Applications* 3 (4): 708–718. https://doi.org/10.2307/1942102.

Sullivan, T., D. Sullivan, and E. Hogue. 2001. "Influence of Diversionary Foods on Vole (*Microtus montanus* and *Microtus longicaudus*) Populations and Feeding Damage to Coniferous Tree Seedlings." *Crop Protection* 20 (2): 103–112. https://doi.org/10.1016/S0261-2194(00)00062-4.

Sutton, W., D. Larson, and L. Jarvis. 2004. "A New Approach for Assessing the Costs of Living with Wildlife in Developing Countries." University of California–Davis Agricultural and Resource Economics Working Paper 04-001. https://escholarship.org/uc/item/2rg95396.

Svenning, J.-C., P. Pedersen, C. J. Donlan, et al. 2016. "Science for a Wilder Anthropocene: Synthesis and Future Directions for Trophic Rewilding Research." *Proceedings of the National Academy of Sciences* 113 (4): 898–906. https://doi.org/10.1073/pnas.1502556112.

Svoboda, J., A. Mrugała, E. Kozubíková-Balcarová, et al. 2017. "Hosts and Transmission of the Crayfish Plague Pathogen *Aphanomyces astaci*: A Review." *Journal of Fish Diseases* 40 (1): 127–140. https://doi.org/10.1111/jfd.12472.

Swanepoel, L., C. Swanepoel, P. Brown, et al. 2017. "A Systematic Review of Rodent Pest Research in Afro-Malagasy Small-Holder Farming Systems: Are We Asking the Right Questions?" *PLoS One* 12 (3): e0174554. https://doi.org/10.1371/journal.pone.0174554.

Swelum, A., M. Shafi, N. Albaqami, et al. 2020. "COVID-19 in Human, Animal, and Environment: A Review." *Frontiers in Veterinary Science* 7. https://doi.org/10.3389/fvets.2020.00578.

Takami, T., T. Yoshihara, Y. Miyakoshi, et al. 2002. "Replacement of White-Spotted Charr *Salvelinus leucomaenis* by Brown Trout *Salmo trutta* in a Branch of the Chitose River, Hokkaido." *Nihon-suisan-gakkai-shi* [*Bulletin of the Japanese Society of Scientific Fisheries*] 68 (1): 23–28. https://doi.org/10.2331/suisan.68.24.

Talbot, P. 1912. *In the Shadow of the Bush*. New York: George H. Doran.

Tan, S. 1990. "The Watchtower Casts No Shadow: Nonliability of Federal and State Governments for Property Damage Inflicted by Wildlife." *University of Colorado Law Review* 61 (2): 427–454.

Taylor, B., and R. Irwin. 2004. "Linking Economic Activities to the Distribution of Exotic Plants." *Proceedings of the National Academy of Sciences of the United States of America* 101 (51): 17725–17730. https://doi.org/10.1073/pnas.0405176101.

Taylor, J., W. Courtenay, and J. McCann. 1984. "Known Impact of Exotic Fishes in the Continental United States." In *Distribution, Biology, and Management of Exotic Fish*, edited by W. Courtenay and J. Stauffer, 322–373. Baltimore: Johns Hopkins University Press.

Taylor, J., and B. Dorr. 2003. "Double-Crested Cormorant Impacts to Commercial and Natural Resources." In *Proceedings of the 10th Wildlife Damage Management Conference*, April 6–9, Hot Springs, AR, edited by K. A. Fagerstone and G. W. Witmer, 43–51. Bethesda, MD: The Wildlife Society. https://digitalcommons.unl.edu/icwdm_usdanwrc/278/.

Taylor, J., G. Yarrow, and J. Miller. 2017. *Beavers*. Wildlife Damage Management Technical Series. Ft. Collins, CO: USDA, APHIS, WS National Wildlife Research Center. https://www.aphis.usda.gov/wildlife_damage/reports/Wildlife%20Damage%20Management%20Technical%20Series/Beaver-WDM-Technical-Series.pdf.

Teel, T., R. Krannich, and R. Schmidt. 2002. "Utah Stakeholders' Attitudes toward Selected Cougar and Black Bear Management Practices." *Wildlife Society Bulletin* 30 (1): 2–15. https://www.jstor.org/stable/3784630.

Teem, J., L. Alphey, S. Descamps, et al. 2020. "Genetic Biocontrol for Invasive Species." *Frontiers in Bioengineering and Biotechnology* 8: 452. https://doi.org/10.3389/fbioe.2020.00452.

Thamis. 2012. "Herodotus on the Egyptians." *World History Encyclopedia*, January 18. https://www.worldhistory.org/article/86/herodotus-on-the-egyptians/.

Thibault, M., E. Vidal, M. Potter, et al. 2018. "The Red-Vented Bulbul (*Pycnonotus cafer*): Serious Pest or Understudied Invader?" *Biological Invasions* 20 (1): 121–136. https://doi.org/10.1007/s10530-017-1521-2.

Thiele, J., G. Linz, H. J. Homan, et al. 2012. "Developing an Effective Management Plan for Starlings Roosting in

Downtown Omaha, Nebraska." USDA Wildlife Services: Staff Publications, 1196. https://digitalcommons.unl.edu/icwdm_usdanwrc/1196/.

Thirgood, S., and S. Redpath. 2005. "Hen Harriers and Red Grouse: The Ecology of a Conflict." In *People and Wildlife: Conflict or Co-existence?* edited by R. Woodroffe, S. Thirgood, and A. Rabinowitz, 192–208. Cambridge: Cambridge University Press. https://doi.org/10.1017/CBO9780511614774.013.

Thirgood, S., and S. Redpath. 2008. "Hen Harriers and Red Grouse: Science, Politics and Human–Wildlife Conflict." *Journal of Applied Ecology* 45 (5): 1550–1554. https://doi.org/10.1111/j.1365-2664.2008.01519.x.

Thomas, C., A. Cameron, R. Green, et al. 2004. "Extinction Risk from Climate Change." *Nature* 427 (6970): 145–148. https://doi.org/10.1038/nature02121.

Thompson, D., R. Stuckey, and E. Thompson. 1987. "Spread, Impact, and Control of Purple Loosestrife (*Lythrum salicaria*) in North American Wetlands." US Fish and Wildlife Service. Jamestown, ND: Northern Prairie Wildlife Research Center Online. http://stoppinginvasives.com/dotAsset/670d2f92-cd0c-41ab-9955-7204f1a9a192.pdf.

Thompson, R. C., S. Kutz, and A. Smith. 2009. "Parasite Zoonoses and Wildlife: Emerging Issues." *International Journal of Environmental Research and Public Health* 6 (2): 678–693. https://doi.org/10.3390/ijerph6020678.

Thornton, C., and M. Quinn. 2009. "Coexisting with Cougars: Public Perceptions, Attitudes, and Awareness of Cougars on the Urban-Rural Fringe of Calgary, Alberta, Canada." *Human-Wildlife Conflicts* 3 (2): 282–295. https://doi.org/10.26077/xvx2-ba39.

Thorpe, J. 2003. "Fatalities and Destroyed Aircraft due to Bird Strikes, 1912–2002." In *Proceedings of the 26th International Bird Strike Committee Meeting*, vol. 1, May 5–9, Warsaw, Poland, 85–113.

Tierkel, E. 1975. "Canine Rabies." In *The Natural History of Rabies*, Vol. 1, edited by G. Baer, 123–137. New York: Academic Press.

Tillman, E., J. Humphrey, and M. Avery. 2002. "Use of Vulture Carcasses and Effigies to Reduce Vulture Damage to Property and Agriculture." In *Proceedings of the 20th Vertebrate Pest Conference*, March 4–7, Reno, NV, edited by R. M. Timm and R. H. Schmidt, 123–128. Davis: University of California. https://doi.org/10.5070/V420110216.

Tindall, S. D., C. J. Ralph, and M. N. Clout. 2007. "Changes in Bird Abundance Following Common Myna Control on a New Zealand Island." *Pacific Conservation Biology* 13 (3): 202–212. https://doi.org/10.1071/PC070202.

Tinker, M. T., V. Gill, G. Esslinger, et al. 2019. "Trends and Carrying Capacity of Sea Otters in Southeast Alaska." *Journal of Wildlife Management* 83 (5): 1073–1089. https://doi.org/10.1002/jwmg.21685.

Tipping, P., L. Gettys, C. Minteer, et al. 2017. "Herbivory by Biological Control Agents Improves Herbicidal Control of Water Hyacinth (*Eichhornia crassipes*)." *Invasive Plant Science and Management* 10 (3): 271–276. https://doi.org/10.1017/inp.2017.30.

Tisdell, C., and X. Zhu. 1998. "Protected Areas, Agricultural Pests and Economic Damage: Conflicts with Elephants and Pests in Yunnan, China." *Environmentalist* 18 (2): 109–118. https://doi.org/10.1023/A:1006674425017.

Tobajas, J., P. Gómez-Ramírez, P. María-Mojica, et al. 2019a. "Conditioned Food Aversion Mediated by Odour Cue and Microencapsulated Levamisole to Avoid Predation by Canids." *European Journal of Wildlife Research* 65 (3): 32. https://doi.org/10.1007/s10344-019-1271-9.

Tobajas, J., P. Gómez-Ramírez, P. María-Mojica, et al. 2019b. "Selection of New Chemicals to Be Used in Conditioned Aversion for Non-Lethal Predation Control." *Behavioural Processes* 166: 103905. https://doi.org/10.1016/j.beproc.2019.103905.

Tobin, M. E., P. P. Woronecki, R. A. Dolbeer, et al. 1988. "Reflecting Tape Fails to Protect Ripening Blueberries from Bird Damage." *Wildlife Society Bulletin* 16 (3): 300–203. https://www.jstor.org/stable/3782104.

Tom-Aba, D., P. Nguku, C. Arinze, et al. 2018. "Assessing the Concepts and Designs of 58 Mobile Apps for the Management of the 2014–2015 West Africa Ebola Outbreak: Systematic Review." *JMIR Public Health and Surveillance* 4 (4): e9015. https://doi.org/10.2196/publichealth.9015.

Toomey, A., and M. Domroese. 2013. "Can Citizen Science Lead to Positive Conservation Attitudes and Behaviors?" *Human Ecology Review* 20 (1): 50–62. https://www.jstor.org/stable/24707571.

Tourenq, C., S. Aulagnier, L. Durieux, et al. 2001. "Identifying Rice Fields at Risk from Damage by the Greater Flamingo." *Journal of Applied Ecology* 38 (1): 170–179. https://www.jstor.org/stable/2655742.

Treves, A. 2009. "The Human Dimensions of Conflicts with Wildlife around Protected Areas." In *Wildlife and Society: The Science of Human Dimensions*, edited by M. J. Manfredo et al., 214–228. Washington, DC: Island Press.

Treves, A., R. Jurewicz, L. Naughton-Treves, et al. 2009. "The Price of Tolerance: Wolf Damage Payments after Recovery." *Biodiversity and Conservation* 18 (14): 4003–4021. https://doi.org/10.1007/s10531-009-9695-2.

Tribe, A., J. Hanger, I. McDonald, et al. 2014. "A Reproductive Management Program for an Urban Population of Eastern Grey Kangaroos (*Macropus giganteus*)." *Animals* 4 (3): 562–582. https://doi.org/10.3390/ani4030562.

Troll, C. 1939. "Luftbildplan und ökologische Bodenforschung. Ihr zweckmäßiger Einsatz für die wissenschaftliche Erforschung und praktische Erschließung wenig bekannter Länder." *Zeitschrift der gesellschaft für erdkunde zu Berlin* 1939 (7, 8): 241–298.

Trouwborst, A. 2018. "Wolves Not Welcome? Zoning for Large Carnivore Conservation and Management under the Bern Convention and EU Habitats Directive." *RECIEL: Review of European, Comparative and International Environmental Law* 27: 306–319. https://doi.org/10.1111/reel.12249.

Tsunoda, H., and H. Enari. 2020. "A Strategy for Wildlife Management in Depopulating Rural Areas of Japan." *Conservation Biology* 34 (4): 819–828. https://doi.org/10.1111/cobi.13470.

Turner, C. E., T. D. Center, D. W. Burrows, et al. 1997. "Ecology and Management of *Melaleuca quinquenervia*, an Invader of Wetlands in Florida, USA." *Wetlands Ecology and Management* 5 (3): 165–178. https://doi.org/10.1023/A:1008205122757.

Turner, M. G., R. H. Gardner, and R. V. O'Neill. 2001. *Landscape Ecology in Theory and Practice: Pattern and Process*. New York: Springer.

Unestam, T. 1975. "Defence Reactions in and Susceptibility of Australian and New Guinean Freshwater Crayfish to European-Crayfish-Plague Fungus." *Australian Journal of Experimental Biology and Medical Science* 53 (5): 349–359. https://doi.org/10.1038/icb.1975.40.

Unestam, T., and D. W. Weiss. 1970. "The Host-Parasite Relationship between Freshwater Crayfish and the Crayfish Disease Fungus *Aphanomyces astaci*: Responses to Infection by a Susceptible and a Resistant Species." *Journal of General Microbiology* 60 (1): 77–90. https://doi.org/10.1099/00221287-60-1-77.

United Nations, Department of Economic and Social Affairs, and Population Division. 2019. *World Population Prospects 2019: Highlights*. ST/ESA/SER.A/423. New York: United Nations. https://population.un.org/wpp/Publications/Files/WPP2019_Highlights.pdf.

United Nations, Food and Agriculture Organization (FAO). 2017. "Guidelines for the Export, Shipment, Import and Release of Biological Control Agents and Other Beneficial Organisms." International Standard for Phytosanitary Measures 3 (2005), International Plant Protection Convention (IPPC). https://www.fao.org/3/j5365e/J5365E.pdf.

US Centers for Disease Control and Prevention (CDC). 2019. "2009 H1N1 Pandemic (H1N1pdm09 Virus)." National Center for Immunization and Respiratory Diseases. https://www.cdc.gov/flu/pandemic-resources/2009-h1n1-pandemic.html.

US Centers for Disease Control and Prevention (CDC). 2020. "West Nile Virus Disease Cases and Deaths Reported to CDC by Year and Clinical Presentation, 1999–2019." https://www.cdc.gov/westnile/resources/pdfs/data/WNV-Disease-Cases-by-Year_1999-2019-P.pdf.

US Department of Agriculture (USDA). 2015. "Sheep and Lamb Predator and Nonpredator Death Loss in the United States." Report #721.0915. Fort Collins, CO: Animal and Plant Health Inspection Service (APHIS), Veterinary Service (VS)–CEAH–National Animal Health Monitoring System (NAHMS). https://www.aphis.usda.gov/animal_health/nahms/sheep/downloads/sheepdeath/SheepDeathLoss2015.pdf.

US Department of Agriculture (USDA). 2020a. "Biological Control Program." Animal and Plant Health Inspection Service (APHIS). https://www.aphis.usda.gov/aphis/ourfocus/planthealth/plant-pest-and-disease-programs/biological-control-program.

US Department of Agriculture (USDA). 2020b. "USDA Begins 2020 Oral Rabies Vaccine Efforts in Eastern United States." Animal and Plant Health Inspection Service (APHIS) press release, August 3. https://www.aphis.usda.gov/aphis/newsroom/news/sa_by_date/sa-2020/orv-efforts.

US Department of Agriculture (USDA). 2021. "Sodium Lauryl Sulfate: European Starling and Blackbird Wetting Agent." Wildlife Services Tech Note WS-21-001A. Animal and Plant Health Inspection Service. Fort Collins, CO:

National Wildlife Research Center. https://www.aphis.usda.gov/wildlife_damage/nwrc/publications/Tech_Notes/TN_SodiumLaurylSulfate.pdf.

US Department of Agriculture (USDA). Undated. "Sterile Fly Release Programs." Animal and Plant Health Inspection Service. (APHIS). https://www.aphis.usda.gov/aphis/ourfocus/internationalservices/sterile-fly-release-programs.

US Environmental Protection Agency (EPA). 2013. "Data Requirements for Pesticide Registration." Modified March 30, 2021. https://www.epa.gov/pesticide-registration/data-requirements-pesticide-registration.

US Environmental Protection Agency (EPA). 2018. "Acetaminophen for Brown Treesnake Control, Pesticide Product Label, 06/07/2018." Office of Chemical Safety and Pollution Prevention, June 7. EPA Registration No. 56228-34, Decision No. 536641. https://www3.epa.gov/pesticides/chem_search/ppls/056228-00034-20180607.pdf.

Van Aarde, R. J., and T. Jackson. 2007. "Megaparks for Metapopulations: Addressing the Causes of Locally High Elephant Numbers in Southern Africa." *Biological Conservation* 134 (3): 289–297. https://doi.org/10.1016/j.biocon.2006.08.027.

Van Aarde, R. J., T. Jackson, and S. M. Ferreira. 2006. "Conservation Science and Elephant Management in Southern Africa." *South African Journal of Science*, 102 (9): 385–388.

Van der Lee, and L. M. Boot. 1955. "Spontaneous Pseudopregnancy in Mice." *Acta Physiologica et Pharmacologica Neerlandica* 4 (3): 442–444.

Van Eeden, L., M. Crowther, C. Dickman, et al. 2018. "Managing Conflict between Large Carnivores and Livestock." *Conservation Biology* 32 (1): 26–34. https://doi.org/10.1111/cobi.12959.

Van Eeden, L, A. Eklund, J. Miller, et al. 2018. "Carnivore Conservation Needs Evidence-Based Livestock Protection." *PLoS Biology* 16 (9): e2005577. https://doi.org/10.1371/journal.pbio.2005577.

Van Sant, F., S. Hassan, D. Reavill, et al. 2019. "Evidence of Bromethalin Toxicosis in Feral San Francisco 'Telegraph Hill' Conures." *PloS One* 14 (3): e0213248. https://doi.org/10.1371/journal.pone.0213248.

Veitch, C. R. 2001. "The Eradication of Feral Cats (*Felis catus*) from Little Barrier Island, New Zealand." *New Zealand Journal of Zoology* 28 (1): 1–12. https://doi.org/10.1080/03014223.2001.9518252.

Veldhuis, D., and S. Underdown. 2017. "Human Biology of Migration." *Annals of Human Biology* 44 (5): 393–396. https://doi.org/10.1080/03014460.2017.1352186.

VerCauteren, K. C., S. Hygnstrom, R. Timm, et al. 2002. "Development of a Model to Assess Rodent Control in Swine Facilities." In *Human Conflicts with Wildlife: Economic Considerations: Proceedings of the Third NWRC Special Symposium*, August 1–3, 2000, Fort Collins, CO, edited by L. Clark, J. Hone, J. A. Shivik, et al., 59–62. Fort Collins, CO: National Wildlife Research Center. https://digitalcommons.unl.edu/nwrchumanconflicts/7/.

VerCauteren, K. C., M. J. Lavelle, and S. Hygnstrom. 2006a. "From the Field: Fences and Deer-Damage Management: A Review of Designs and Efficacy." *Wildlife Society Bulletin*

34 (1): 191–200. https://doi.org/10.2193/0091-7648(2006)34 [191:FADMAR]2.0.CO;2.

VerCauteren, K. C., M. J. Lavelle, and S. Hygnstrom. 2006b. "A Simulation Model for Determining Cost-effectiveness of Fences for Reducing Deer Damage." *Wildlife Society Bulletin* 34 (1): 16–22. https://doi.org/10.2193/0091 -7648(2006)34[16:ASMFDC]2.0.CO;2.

VerCauteren, K. C., M. J. Lavelle, and G. Phillips. 2008. "Livestock Protection Dogs for Deterring Deer from Cattle and Feed." *Journal of Wildlife Management* 72 (6): 1443–1448. https://doi.org/10.2193/2007-372.

VerCauteren, K. C., M. Pipas, P. Peterson, et al. 2003. "Stored-Crop Loss due to Deer Consumption." *Wildlife Society Bulletin* 31 (2): 578–582. https://www.jstor.org /stable/3784342.

VerCauteren, K. C., N. Seward, D. Hirchert, et al. 2005. "Dogs for Reducing Wildlife Damage to Organic Crops: A Case Study." In *Proceedings of the 11th Wildlife Damage Management Conference*, May 16–19, Traverse City, MI, edited by D. L. Nolte and K. A. Fagerstone, 286–293. Fort Collins, CO: National Wildlife Research Center. https://digitalcommons .unl.edu/icwdm_wdmconfproc/130/.

Verhulst, P. 1838. "Notice sur la loi que la population suit dans son accroissement." *Correspondence Mathematique et Physique de l'Observatoire de Bruxelles, Brussels* 10: 113–126.

Vermont Fish and Wildlife Department. 2017. "Best Management Practices for Resolving Human-Beaver Conflicts in Vermont." Montpelier, VT. https://vtfishandwildlife.com /sites/fishandwildlife/files/documents/Learn%20More /Library/REPORTS%20AND%20DOCUMENTS /FURBEARER%20AND%20TRAPPING/BMP-FOR -BEAVER-HUMAN-CONFLICTS-2017.pdf.

Vilizzi, L., A. S. Tarkan, and G. H. Copp. 2015. "Experimental Evidence from Causal Criteria Analysis for the Effects of Common Carp Cyprinus Carpio on Freshwater Ecosystems: A Global Perspective." *Reviews in Fisheries Science and Aquaculture* 23 (3): 253–290. https://doi.org/10.1080 /23308249.2015.1051214.

Vitasek, J. 2004. "A Review of Rabies Elimination in Europe." *Veterinarni Medicina* 49 (5): 171–185. https://doi.org/10 .17221/5692-VETMED.

Vollset, S., E. Goren, C-W. Yuan, et al. 2020. "Fertility, Mortality, Migration, and Population Scenarios for 195 Countries and Territories from 2017 to 2100: A Forecasting Analysis for the Global Burden of Disease Study." *The Lancet* 396 (10258): 1285–1306. https://doi.org/10.1016 /S0140-6736(20)30677-2.

Volterra, V. 1926. "Variazioni e fluttuazioni del numero d'individui in specie animali conviventi." *Memoria della Reale Accademia Nazionale dei Lincei*, series 6, 2: 31–113.

Von Hagen, R. L., P. Norris, and B. A. Schulte. 2020. "Quantifying Capsaicinoids from Chili Pepper and Motor Oil Mixtures Used in Elephant Deterrent Fences." *Chromatographia* 83 (9): 1153–1157. https://doi.org/10.1007/s10337-020 -03934-8.

Walton, M., and C. A. Feild. 1989. "Use of Donkeys to Guard Sheep and Goats in Texas." In *Proceedings of the 4th Eastern Wildlife Damage Control Conference*, September 25–28, Madison, WI, edited by S. R. Craven, 87–94. Madison:

Wisconsin Department of Natural Resources. https:// digitalcommons.unl.edu/ewdcc4/43/.

Wang, Z., D. Fahey, A. Lucas, et al. 2020. "Bird Damage Management in Vineyards: Comparing Efficacy of a Bird Psychology-Incorporated Unmanned Aerial Vehicle System with Netting and Visual Scaring." *Crop Protection* 137: 105260. https://doi.org/10.1016/j.cropro.2020.105260.

Wang, Z., A. Griffin, A. Lucas, et al. 2019. "Psychological Warfare in Vineyard: Using Drones and Bird Psychology to Control Bird Damage to Wine Grapes." *Crop Protection* 120: 163–170. https://doi.org/10.1016/j.cropro.2019.02.025.

Ward, A., K. VerCauteren, W. D. Walter, et al. 2009. "Options for the Control of Disease 3: Targeting the Environment." In *Management of Disease in Wild Mammals*, edited by R. J. Delahay, G. C. Smith, and M. R. Hutchings, 147–168. Tokyo: Springer Science.

Ward, M. G. 2016. "The Regulatory Landscape for Biological Control Agents." *EPPO Bulletin* 46 (2): 249–253. https://doi .org/10.1111/epp.12307.

Warren, R., M. Candeias, A. Labatore, et al. 2019. "Multiple Mechanisms in Woodland Plant Species Invasion." *Journal of Plant Ecology* 12 (2): 201–209. https://doi.org/10.1093/jpe /rty010.

Washburn, B. 2016. *Hawks and Owls*. Ft. Collins, CO: USDA, APHIS, WS National Wildlife Research Center. https:// digitalcommons.unl.edu/nwrcwdmts/6/.

Water, A., L. King, R. Arkajak, et al. 2020. "Beehive Fences as a Sustainable Local Solution to Human-Elephant Conflict in Thailand." *Conservation Science and Practice* 2 (10): e260. https://doi.org/10.1111/csp2.260.

Welsh, R. G., and D. Muller-Schwarze. 1989. "Experimental Habitat Scenting Inhibits Colonization by Beaver, *Castor canadensis*." *Journal of Chemical Ecology* 15 (3): 887–893. https://doi.org/10.1007/bf01015184.

Werner, S., and M. Avery. 2017. "Chemical Repellents." *Ecology and Management of Blackbirds (Icteridae) in North America*, edited by G. M. Linz, M. L. Avery, and R. A. Dolbeer, 135–158. Boca Raton, FL: CRC Press. https:// digitalcommons.unl.edu/icwdm_usdanwrc/1979.

Werner, S., J. Carlson, S. Tupper, et al. 2009. "Threshold Concentrations of an Anthraquinone-Based Repellent for Canada Geese, Red-Winged Blackbirds, and Ring-Necked Pheasants." *Applied Animal Behaviour Science* 121 (3–4): 190–196. https://doi.org/10.1016/j.applanim.2009.09.016.

Werner, S., and F. Provenza. 2011. "Reconciling Sensory Cues and Varied Consequences of Avian Repellents." *Physiology and Behavior* 102 (2): 158–163. https://doi.org/10.1016/j .physbeh.2010.10.012.

West, B. C., A. L. Cooper, and J. B. Armstrong. 2009. *Managing Wild Pigs: A Technical Guide*. Human-Wildlife Interactions Monograph 1. Starkville, MS: Berryman Institute.

West, P. 2018. *Guide to Introduced Pest Animals of Australia*. Clayton, Australia: CSIRO Publishing.

Westbrook, J., J. Holliday, A. Newhouse, et al. 2020. "A Plan to Diversify a Transgenic Blight-Tolerant American Chestnut Population Using Citizen Science." *Plants, People, Planet* 2 (1): 84–95. https://doi.org/10.1002/ppp3.10061.

Westman, K., and R. Savolainen. 2001. "Long Term Study of Competition between Two Co-occurring Crayfish

Species, the Native *Astacus astacus* L. and the Introduced *Pacifastacus leniusculus* Dana, in a Finnish Lake." *Bulletin Français de la Pêche et de la Pisciculture* 361: 613–627. https://doi.org/10.1051/kmae:2001008.

Wheat, L., T. Slama, H. Eitzen, et al. 1981. "A Large Urban Outbreak of Histoplasmosis: Clinical Features." *Annals of internal Medicine* 94 (3): 331–337. https://doi.org/10.7326/0003-4819-94-3-331.

White, H. B., G. Batcheller, E. Boggess, et al. 2021. "Best Management Practices for Trapping Furbearers in the United States." *Wildlife Monographs* 207 (1): 3–59. https://doi.org/10.1002/wmon.1057.

Whitten, W. K. 1956. "Modification of the Oestrous Cycle of the Mouse by External Stimuli Associated with the Male." *Journal of Endocrinology* 13 (4): 399–404. https://doi.org/10.1677/joe.0.0130399.

Whyte, I. 2004. "Ecological Basis of the New Elephant Management Policy for Kruger National Park and Expected Outcomes." *Pachyderm* 36: 99–108.

Wildlife Society, The. 2020. "TWS Issue Statement: Baiting and Supplemental Feeding of Game Wildlife Species." March 24. https://wildlife.org/tws-issue-statement-baiting-and-supplemental-feeding-of-game-wildlife-species/.

Wilkinson, C. E., A. McInturff, J. R. B. Miller, et al. 2020. "An Ecological Framework for Contextualizing Carnivore–Livestock Conflict." *Conservation Biology* 34 (4): 854–867. https://doi.org/10.1111/cobi.13469.

Williams, A. F., and J. K. Wells. 2005. "Characteristics of Vehicle-Animal Crashes in Which Vehicle Occupants Are Killed." *Traffic Injury Prevention* 6 (1): 56–59. https://doi.org/10.1080/15389580590903186.

Williams, J. M. 1974. "The Effect of Artificial Rat Damage on Coconut Yields in Fiji." *PANS Pest Articles and News Summaries* 20 (3): 275–282. https://doi.org/10.1080/09670877409411851.

Williams, T. 1999. "The Terrible Turtle Trade." *Audubon-New York* 101: 44–51. http://nytts.org/asia/twilliams.htm.

Williams, T. 2009. "Feline Fatales." *Audubon* (September–October): 30–38. https://web.archive.org/web/20111004140310/http://archive.audubonmagazine.org/incite/incite0909.html.

Wilson, D. 2005. "Recent Advances in the Control of Oak Wilt in the United States." *Plant Pathology Journal* 4 (2): 177–191. https://www.fs.usda.gov/treesearch/pubs/21644.

Wilson, D., R. Cole, J. Nichols, et al. 1996. *Measuring and Monitoring Biological Diversity: Standard Methods for Mammals.* Washington, DC: Smithsonian Books.

Wilson, E. O. 1984. *Biophillia: The Human Bond with Other Species.* Cambridge, MA: Harvard University Press.

Wilson, E. O. 1993. "Biophilia and the Conservation Ethic." In *The Biophilia Hypothesis,* edited by S. R. Kellert and E. O. Wilson, 31–40. Washington, DC: Island Press.

Wilson, S., E. Bradley, and G. Neudecker. 2017. "Learning to Live with Wolves: Community-Based Conservation in the Blackfoot Valley of Montana." *Human–Wildlife Interactions* 11 (3): 4. https://doi.org/10.26077/bf8e-6f56.

Winston, R. L., M. Schwarzländer, H. L. Hinz, et al. 2014. "Biological Control of Weeds: A World Catalogue of Agents and Their Target Weeds." In *Biological Control of Weeds: A World Catalogue of Agents and Their Target Weeds,* 5th ed. CABI Direct. https://www.cabdirect.org/cabdirect/abstract/20153151446.

Winton, J., G. Kurath, and W. Batts. 2008. "Molecular Epidemiology of Viral Hemorrhagic Septicemia Virus in the Great Lakes Region." USGS Numbered Series. *Molecular Epidemiology of Viral Hemorrhagic Septicemia Virus in the Great Lakes Region.* Vol. 2008–3003. Fact Sheet. Reston, VA: US Geological Survey. https://doi.org/10.3133/fs20083003.

Witmer, G. 2007. "The Ecology of Vertebrate Pests and Integrated Pest Management (IPM)." In *Perspectives in Ecological Theory and Integrated Pest Management,* edited by M. Kogan and P. Jepson, 393–410. Cambridge: Cambridge University Press. https://digitalcommons.unl.edu/icwdm_usdanwrc/730/.

Witmer, G., F. Boyd, and Z. Hillis-Starr. 2007. "The Successful Eradication of Introduced Roof Rats (*Rattus rattus*) from Buck Island Using Diphacinone, Followed by an Irruption of House Mice (*Mus musculus*)." *Wildlife Research* 34 (2): 108–115. https://digitalcommons.unl.edu/icwdm_usdanwrc/674/.

Witmer, G., and S. Jojola. 2006. "What's Up with House Mice? A Review." *Proceedings of the 22nd Vertebrate Pest Conference,* March 6–9, 2006, Berkeley, CA. Davis: University of California–Davis. https://doi.org/10.5070/V422110126.

Witmer, G., R. Moulton, and C. Samura. 2017. "Cage Efficacy Study of an Experimental Rodenticide Using Wild-Caught House Mice." In *Proceedings of the 17th Wildlife Damage Management Conference,* February 26–March 1, Orange Beach, AL, edited by D. J. Morin and M. J. Cherry, 44–53. Bethesda, MD: The Wildlife Society. https://digitalcommons.usu.edu/wdmconference/2017/session5/3.

Witmer, G., and K. VerCauteren. 2001. "Understanding Vole Problems in Direct Seeding—Strategies for Management." In *Proceedings of the Northwest Direct Seed Cropping Systems Conference,* January 17–19, Spokane, WA, edited by R. Veseth, 104–110. Pasco, WA: Northwest Direct Seed Conference. https://digitalcommons.unl.edu/icwdm_usdanwrc/582/.

Wittenberg, R., and M. Cock. 2001. *Invasive Alien Species: A Toolkit of Best Prevention and Management Practices.* Wallingford, UK: CAB International. https://www.cbd.int/doc/pa/tools/Invasive%20Alien%20Species%20Toolkit.pdf.

Wood, W. (1634) 1865. *Wood's New-England's Prospect.* Vol. 1. Boston: John Wilson & Son. https://www.gutenberg.org/ebooks/47082.

Woodroffe, R., S. Thirgood, and A. Rabinowitz. 2005a. "The Future of Coexistence: Resolving Human-Wildlife Conflicts in a Changing World." In *People and Wildlife: Conflict or Co-existence?* edited by R. Woodroffe, S. Thirgood, and A. Rabinowitz, 388–405. Cambridge: Cambridge University Press. https://doi.org/10.1017/CBO9780511614774.025.

Woodroffe, R., S. Thirgood, and A. Rabinowitz. 2005b. "The Impact of Human-Wildlife Conflict on Natural Systems." In *People and Wildlife: Conflict or Co-existence?* edited by

R. Woodroffe, S. Thirgood, and A. Rabinowitz, 1–12. Cambridge: Cambridge University Press. https://doi.org /10.1017/CBO9780511614774.002.

Woolhouse, M.E.J., D. T. Haydon, and R. Antia. 2005. "Emerging Pathogens: The Epidemiology and Evolution of Species Jumps." *Trends in Ecology and Evolution* 20 (5): 238–244. https://doi.org/10.1016/j.tree.2005.02.009.

World Health Organization. 1980. "A Revision of the System of Nomenclature for Influenza Viruses: A WHO Memorandum." *Bulletin of the World Health Organization* 58 (4): 585–591. https://apps.who.int/iris/handle/10665/262025.

World Health Organization. 2020. "Managing the COVID-19 Infodemic: Promoting Healthy Behaviours and Mitigating the Harm from Misinformation and Disinformation." Accessed May 1, 2021. https://www.who.int/news/item /23-09-2020-managing-the-covid-19-infodemic-promoting -healthy-behaviours-and-mitigating-the-harm-from -misinformation-and-disinformation.

World Health Organization. 2021. "Rabies." Online fact sheet. https://www.who.int/news-room/fact-sheets/detail/rabies.

Wywialowski, A. 1991. "Implications of the Animal Rights Movement for Wildlife Damage Management." In *Proceedings of the 10th Great Plains Wildlife Damage Control Workshop*, April 15–18, Lincoln, NE, edited by S. E. Hygnstrom, R. M. Case, and R. J. Johnson, 28–32. Lincoln: University of Nebraska. https://digitalcommons.unl.edu/gpwdcwp/7/.

Xiong, W., H. Wang, Q. Wang, et al. 2018. "Non-Native Species in the Three Gorges Dam Reservoir: Status and Risks." *BioInvasions Records* 7 (2): 153–158. https://doi.org /10.3391/bir.2018.7.2.06.

Xu, J., J. Wei, and W. Liu. 2019. "Escalating Human–Wildlife Conflict in the Wolong Nature Reserve, China: A Dynamic and Paradoxical Process." *Ecology and Evolution* 9 (12): 7273–7283. https://doi.org/10.1002/ece3.5299.

Yang, H., F. Lupi, J. Zhang, et al. 2020. "Hidden Cost of Conservation: A Demonstration Using Losses from Human-Wildlife Conflicts under a Payments for Ecosystem Services Program." *Ecological Economics* 169: 106462. https://doi.org/10.1016/j.ecolecon.2019 .106462.

Yoder, J. 2002. "Damage Abatement and Compensation Programs as Incentives for Wildlife Management on Private Land." In *Human Conflicts with Wildlife: Economic Considerations: Proceedings of the Third NWRC Special Symposium*, August 1–3, 2000, Fort Collins, CO, edited by L. Clark, J. Hone, J. A. Shivik, et al., 17–28. Fort Collins, CO: National Wildlife Research Center. https:// digitalcommons.unl.edu/nwrchumanconflicts/2/.

Young, J., J. Draper, and S. Breck. 2019. "Mind the Gap: Experimental Tests to Improve Efficacy of Fladry for Nonlethal Management of Coyotes." *Wildlife Society Bulletin* 43 (2): 265–271. https://doi.org/10.1002/wsb.970.

Zavaleta, E., R. Hobbs, and H. Mooney. 2001. "Viewing Invasive Species Removal in a Whole-Ecosystem Context." *Trends in Ecology and Evolution* 16 (8): 454–459. https://doi.org/10.1016/S0169-5347(01)02194-2.

Zemanova, M. 2020. "Towards More Compassionate Wildlife Research through the 3Rs Principles: Moving from Invasive to Non-invasive Methods." *Wildlife Biology* 2020 (1). https://doi.org/10.2981/wlb.00607.

Zhang, Y., T. Song, Q. Jin, et al. 2020. "Status of an Alien Turtle in City Park Waters and Its Potential Threats to Local Biodiversity: The Red-Eared Slider in Beijing." *Urban Ecosystems* 23 (1): 147–157. https://doi.org/10.1007 /s11252-019-00897-z.

Zhu, G., J. Illan, C. Looney, et al. 2020. "Assessing the Ecological Niche and Invasion Potential of the Asian Giant Hornet." *Proceedings of the National Academy of Sciences of the United States of America* 117 (40): 24646– 24648. https://doi.org/10.1073/pnas.2011441117.

Zieglrum, G. 2004. "Efficacy of Black Bear Supplemental Feeding to Reduce Conifer Damage in Western Washington." *Journal of Wildlife Management* 68 (3): 470–474. https://www.jstor.org/stable/3803379.

Zieglrum, G. 2008. "Impacts of the Black Bear Supplemental Feeding Program on Ecology in Western Washington." *Human–Wildlife Conflicts* 2 (2): 153–159. https://digitalcom mons.unl.edu/hwi/60/.

Zieglrum, G., and D. Nolte. 1995. "Black Bear Damage Management in Washington State." In *Proceedings of the 7th Eastern Wildlife Damage Management Conference*, November 5–8, Jackson, MI, edited by J. B. Armstrong, 104–107. Charlotte: North Carolina Cooperative Extension Service. https://digitalcommons.unl.edu/ewdcc7/28/.

Zimmermann, H., P. Brandt, J. Fischer, et al. 2014. "The Human Release Hypothesis for Biological Invasions: Human Activity as a Determinant of the Abundance of Invasive Plant Species." *F1000Research* 3: 109. https://doi .org/10.12688/f1000research.3740.2.

Zimmermann, H. G., V. C. Moran, and J. H. Hoffmann. 2000. "The Renowned Cactus Moth, *Cactoblastis cactorum*: Its Natural History and Threat to Native Opuntia Floras in Mexico and the United States of America." *Diversity and Distributions* 6 (5): 259–269. https://doi.org/10.1046/j.1472-4642.2000.00088.x.

Zinn, H., and M. J. Manfredo. 2000. "An Experimental Test of Rational and Emotional Appeals about a Recreation Issue." *Leisure Sciences* 22 (3): 183–194. https://doi.org/10 .1080/01490409950121852.

Zinn, H. and C. Pierce. 2002. "Values, Gender, and Concern about Potentially Dangerous Wildlife." *Environment and Behavior* 34 (2): 239–256. https://doi.org/10.1177 /0013916502034002005.

Zinsstag, J., E. Schelling, F. Roth, et al. 2007. "Human Benefits of Animal Interventions for Zoonosis Control." *Emerging Infectious Diseases* 13 (4): 527–531. https://doi.org /10.3201/eid1304.060381.

Zuckerman, P. 2003. "Activists Sue U.S. Wildlife Service over the Killing of Bears." Associated Press, May 7. https:// theworldlink.com/news/local/activists-sue-u-s-wildlife -service-over-the-killing-of-bears/article_5d059685-9194 -59ef-a46e-0d337a22cf7e.html.

Index

WILDLIFE SCIENCE BOOKS FROM HOPKINS PRESS

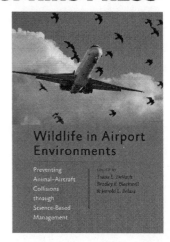